MAGNETIC RESONANCE IN BIOLOGY

VOLUME ONE

Magnetic Resonance in Biology

VOLUME ONE

edited by

JACK S. COHEN
National Institutes of Health
Bethesda, Maryland

A WILEY-INTERSCIENCE PUBLICATION

JOHN WILEY & SONS
New York • Chichester • Brisbane • Toronto

Library of Congress Cataloging in Publication Data:
Magnetic resonance in biology.

  "A Wiley-Interscience publication."
  Includes index.
  1. Nuclear magnetic resonance. 2. Electron
paramagnetic resonance. 3. Biology—Technique.
I. Cohen, Jack S. [DNLM: 1. Biology—
Periodicals. 2. Electron spin resonance—Methods—
Periodicals. 3. Molecular biology—Periodicals.
4. Nuclear magnetic resonance—Methods—Periodicals.
W1 MA34I]

QH324.9.N8M33   574.19′285   80-13070
ISBN 0-471-05176-4

Printed in the United States of America

10 9 8 7 6 5 4 3 2 1

Contributors

WILLIAM EGAN
Division of Bacterial Products, Bureau of Biologics, Federal Drug Administration, Bethesda, MD. 20205.

DAVID I. HOULT
Biomedical Engineering and Instrumentation Branch, Division of Research Services, National Institutes of Health, Bethesda, MD. 20205.

ROBERT E. LONDON
Los Alamos Scientific Laboratory, University of California, Los Alamos, NM. 87545.

MARTIN P. SCHWEIZER
Medicinal Chemistry, College of Pharmacy, University of Utah, Salt Lake City, UT. 84112.

BRIAN D. SYKES
Department of Biochemistry and NMR Group on Protein Structure and Function, University of Alberta, Edmonton, Alberta T6G 2H7, Canada.

JOEL H. WEINER
Department of Biochemistry and NMR Group on Protein Structure and Function, University of Alberta, Edmonton, Alberta T6G 2H7, Canada.

GERALD ZON
Department of Chemistry, The Catholic University of America, Washington, DC. 20064.

Magnetic Resonance in Biology

The aim of this series is to provide a forum where experts in magnetic resonance methods can describe the results of applications to biological systems in a format that is comprehensible to the nonexpert.

The need for such a forum is two-fold: first, there are an ever-increasing number and range of magnetic resonance applications to biology appearing in the scientific literature; second, there is an apparent reluctance by biologists and biochemists to assimilate these results into the general corpus of their specific fields. Such results are often not referenced except by other spectroscopists. One reason for this attitude may be the relative ease with which some spectra can be run and the superficial nature of many of the earlier analyses in the area. Also, the mathematical-physical basis of the method and the specialized technology involved are unfamiliar to most biologists. It is my intention that researchers in a given field will be able to find an accessible account of what information magnetic resonance methods have contributed to their subject in the articles published in this series, *Magnetic Resonance in Biology*. The articles will be written by those actively pursuing research and publishing articles in the scientific literature.

The methods in question, nuclear magnetic resonance (NMR) and electron spin resonance (ESR), have provided valuable, and often unique, information on a wide range of systems, from small molecules (drugs and hormones), through macromolecules (proteins and nucleic acids), to membranes, cells, and whole organisms. The nature of the information obtained is dynamic as well as structural and spatial. It is intended to cover the whole range of biological applications currently to be found in the literature. This includes studies of cell metabolism (with NMR observation of ^{31}P, ^{13}C, etc.), whole organism imaging by NMR (so-called zeugmatography), combinations of NMR and ESR approaches (including use of spin labels and paramagnetic metal ions), as well as the more familiar molecular studies. That is why the title of the series was chosen to include the term "biology," since the methods in question are indeed adding to our knowledge of biology in general.

January 1980   JACK S. COHEN

Preface

Volume 1 of this series contains discussions of two topics which, to my knowledge, have not been reviewed before in the context of magnetic resonance methods, namely the articles on "Polysaccharide Antigens" by William Egan and "Drug Metabolism" by Gerald Zon. These are both subjects of intense biological research activity, to which nuclear magnetic resonance has made significant contributions in terms of the correlation of chemical structure with biological mechanism.

Two other chapters are the first reviews by authors who have made major contributions to their fields of interest; these are "Intramolecular Dynamics of Proteins and Peptides" by Robert London, and "Medical Imaging" by David Hoult. These may well come to be regarded as seminal reviews of the respective topics.

The other two chapters that make up this volume are devoted to more established subjects, yet both also cover new material. The article by Brian Sykes and Joel Weiner concentrates on the biosynthetic route to ^{19}F-labeled proteins. By contrast, the article by Martin Schweizer is a review of a much larger field, "Nucleic Acid Structure, Interactions, and Functions," in which the author has done an excellent job of selecting among the most recent work and communicating the excitement in this most important area.

These six articles represent a collection of first class reviews of important topics by active researchers, which will be valuable to any student or specialist of the respective biological field, as well as being invaluable to the expert of magnetic resonance in biology.

JACK S. COHEN

Bethesda, Maryland
April 1980

Contents

MAGNETIC RESONANCE IN BIOLOGY

VOLUME ONE

One

Intramolecular Dynamics of Proteins and Peptides as Monitored by Nuclear Magnetic Relaxation Measurements

Robert E. London

Los Alamos Scientific Laboratory
University of California
Los Alamos, NM 87545

By acceptance of this article for publication, the publisher recognizes the Government's (license) rights in any copyright and the Government and its authorized representatives have unrestricted right to reproduce in whole or in part said article under any copyright secured by the publisher. This work was performed under the auspices of the USERDA.

1. INTRODUCTION

Nuclear magnetic resonance (NMR) shares with X-ray crystallography the capability of detecting individual nuclei in a complex biomolecule. Although the structure derived from NMR measurements is in general much less detailed, it contains dynamic information which may be manifested in several ways. For example, hydrogen–deuterium exchange experiments monitored by NMR require that the folded structure of many proteins deduced from a crystallographic analysis be in equilibrium with an "open" structure in which amide protons of the peptide backbone can exchange with the deuterons of the D_2O solvent (Englander *et al.*, 1972). As another illustration, rapid 180° flips of the phenylalanine and tyrosine side chains observed in a number of proteins are inconsistent with the rotational barriers deduced from the rigid crystal structure. However, rotational barriers calculated on the basis of a flexible protein geometry are in many cases more than an order of magnitude lower, and consistent with the NMR observations (Gelin and Karplus, 1975). Viewed from a thermodynamic standpoint, significant fluctuations of protein energy and volume are quite reasonable (Cooper, 1976). Differences between the crystallographic and solution structures of peptides are even more significant. Relaxation measurements indicate that most of the peptides which have been studied undergo rapid conformational fluctuations.

Dynamic processes which can be monitored using NMR cover an exceedingly great range of rates. For the slowest processes, successive NMR spectra can be obtained following the application of a chemical, thermal, or magnetic perturbation to the system under study. Such "time lapse spectroscopy" has provided considerable insight into the exchange rates of amide protons in peptides (Molday *et al.*, 1972) and proteins (Karplus *et al.*, 1973) as well as the slow conformational processes which occur in concanavalin A (Grimaldi and Sykes, 1975; Brown *et al.*, 1977) where spectral changes occurring over a period of many hours have been monitored. Intermediate rate processes, for example, the isomerization of peptide bonds or the internal diffusion of aromatic side chains in proteins, characterized by rates in the range 10^0–10^6 s^{-1} produce spectral perturbations that can be described as a chemical exchange phenomenon and can be treated quantitatively by application of the well-known McConnell equations (McConnell, 1958; for a recent review of dynamic NMR see Steigel, 1978). Finally, diffusive motions of whole protein molecules and internal motion of side chains with rates 10^7–10^{12} s^{-1} can be studied by analysis of the nuclear magnetic relaxation rates and nuclear Overhauser enhancements.

Having noted the value of dynamic data in obtaining a complete understanding of biological molecules, it must be emphasized that not all of the dynamic behavior to which NMR measurements are sensitive is equally significant in a biological context. For example, the rapid internal motion of protein methyl groups will reduce the observed linewidth by an order of magnitude, so that such motion has a high profile as far as the technique is concerned, but this motion is probably not important biochemically. The phenomenon is important because the enhanced resolution corresponding to the line narrowing resulting from rapid internal rotation of methyl groups makes such resonances suitable for observation in macromolecules. Thus, a proposal concerning the biochemical importance of a particular methionine residue in a protein might be evaluated by an NMR study of the methionine methyl resonance. In addition, a few cases may arise in which a methyl group expected to undergo rapid internal diffusion is, in fact, highly restricted and such data would indicate the nature of the interactions between the particular methyl goup and other groups in the molecule.

In summary, a valid picture of the structure of a protein or peptide in solution must include an answer to the question: Over what time scale? Different results will generally be obtained as the time scale is varied so that the protein or peptide molecule is allowed to sample different classes of conformationally accessible states. Not all of the dynamic processes that can be studied are of equal importance in terms of the biochemical properties of the protein, but most can make an indirect, if not a direct contribution to our understanding of the biochemical properties of proteins and peptides.

This article contains a discussion of how conclusions, dealing with dynamic aspects of proteins and peptides can be derived from relaxation measurements [T_1, T_2, NOE (see definitions in Section 2)]. The material is presented in three sections. Section 2 provides a brief account of the different relaxation mechanisms which have been found to be significant for various nuclei most commonly studied in connection with these systems. Such considerations are of the utmost importance since interpretation of the relaxation data requires either an *a priori* assumption or experimental separation of the important contributions to the relaxation. Section 3 describes a number of physical models which have proven useful for the analysis of protein and peptide relaxation data. No attempt at mathematical rigor has been made; only the spectral density function, which embodies the essence of the physical model, is provided. A number of figures have been included illustrating the dynamic models involved; several illustrative calculations for ^{13}C are also given. The latter were selected because these relaxation measurements are most directly interpretable in terms of local motions of the nuclei. Section 4 contains a number of illustrative applications which have been selected primarily from studies done by the Los Alamos group in collaboration with Professor R. L. Blakley and his students at the University of Iowa. No attempt was made to provide an exhaustive review of the many dynamic studies which have been reported.

2. RELAXATION CHARACTERISTICS OF COMMONLY STUDIED NUCLEAR SPECIES

The information content of an NMR experiment is, in general, dependent on the particular nuclear species being studied. Resolution of resonances corresponding to individual nuclei is clearly desirable if the structural and dynamic properties of a molecule are to be understood in any detail. In peptides, many of the resonances corresponding to individual nuclei can typically be resolved; in proteins, the large number of resonances and increased linewidths severely limit the resolution. A complete discussion of the many factors determining the resolution of resonances corresponding to individual nuclei in proteins (Allerhand, 1978) is well beyond the scope of the present review; however, one factor, the resonance linewidth, is relevant to the present discussion of relaxation behavior. In lower molecular weight peptides, linewidths are often dominated by the magnetic field inhomogeneity and, therefore, reflect instrumental limitations. In proteins, the linewidths are dominated by the spin–spin relaxation rate of the particular nucleus under observation. Resonances corresponding to individual nuclei are most frequently resolved if the transverse relaxation (T_2) mechanisms for the particular nucleus are relatively inefficient due to either (1) large internuclear distances between dipolar coupled nuclei present or (2) the presence of a substantial degree of internal motion leading to narrower resonances. The first consideration favors the observation of unprotonated carbons or aromatic protons such as His H-2 for which the intraresidue ^{1}H distances are relatively large. The second consideration favors the observation of methyl groups or the ϵ carbon of lysine.

The relaxation parameters which will be considered in this review are the spin lattice relaxation time (T_1), the spin–spin relaxation time (T_2) and the nuclear Overhauser enhancement (NOE). Although the first two parameters are familiar to most workers in the field of NMR (Pople *et al.*, 1959), the nuclear Overhauser enhancement is less generally familiar. This phenomenon reflects a change in the intensity of a particular resonance when another resonance is saturated. To avoid confusion, we adopt the notation of Schaefer (1973) in which the nuclear Overhauser enhancement factor (NOEF = η) is the additive change in intensity in units of the unperturbed intensity. The nuclear Overhauser enhancement (NOE $= 1 + \eta$) then represents a multiplicative factor applied to the original intensity to obtain the enhanced value. It should be noted that despite the terminology, the "enhancement" can be positive, negative, or zero. The magnitude and sign of the NOE are sensitive to three characteristics of the nuclei: (1) the magnetogyric ratios of the spins being observed and saturated, (2) the relative importance of dipolar and nondipolar relaxation mechanisms for the nucleus under observation (Kuhlmann et al., 1970), and (3) the detailed dynamics of the internuclear dipolar vector. More detailed treatment of the NOE can be found in Noggle and Schirmer (1971) and Kuhlmann et al. (1970), as well as in papers dealing with specific dynamic models, a number of which are discussed in Section 3.

Analysis of the NMR data in terms of dynamic processes is dependent on the nuclear species under observation. In particular, interpretation of the observed relaxation parameters requires (1) a determination of the particular relaxation mechanisms which are important and (2) evaluation of inter- and intramolecular (or inter- and intraresidue) contributions corresponding to the dipolar relaxation mechanism. Probably the simplest data to interpret correspond to protonated ^{13}C nuclei since the relaxation is generally, but apparently not always (Cutnell et al., 1975), dominated by a single mechanism: the intramolecular ^{13}C$-^{1}$H dipolar interaction with directly bonded protons. In this case, the $(r_{CH})^{-6}$ dependence of the interaction (Appendix A) leads to negligible contributions from nonbonded protons (however, see Stilbs and Moseley, 1979). Further, since the distance between the carbon and proton(s) is fixed, the dynamic analysis is greatly simplified. The application of detailed dynamic models to the interpretation of NMR relaxation data in recent years has therefore largely involved either ^{13}C or quadrupolar nuclei for which only orientational perturbations lead to relaxation (Huntress, 1970). Although ^{13}C$-^{1}$H dipolar interaction is also important for nonprotonated carbons as indicated, for example, by the fact that at lower magnetic field strengths the relaxation rates are proportional to the number of geminal protons (Oldfield and Allerhand, 1975), other relaxation mechanisms can also make significant contributions. The most important of these is the modulation of the local field produced by tumbling of a molecule with chemical shift anisotropy (CSA) (Appendix A). The relaxation rate resulting from this mechanism depends on the square of the magnetic field so that field-dependent studies can be useful for separating out this contribution (Norton et al., 1977; Jeffers et al., 1978). The ^{13}C$-^{14}$N dipolar interaction is also significant for nonprotonated carbons directly bonded to nitrogen nuclei (Oldfield et al., 1975). Finally, we note that the NOE of ^{13}C resulting from proton noise decoupling, abbreviated ^{13}C$-\{^{1}$H$\}$, is simplest to evaluate if the ^{13}C$-^{1}$H dipolar interaction dominates the relaxation. This dominance is not necessarily proportional to the number of bonded protons since generally the existence of a single bound proton is sufficient to ensure dominance of this relaxation mechanism. Furthermore, even nonprotonated nuclei may exhibit a maximum NOE if the $^{13}C-^{1}H$ dipolar interaction is the most significant relative to the other mechanisms available. The major factor is therefore the size of the $^{13}C-^{1}H$ dipolar contribution to the relaxation rate relative to the size of the contributions of the other possible mechanisms.

Dynamic information derived from ^{1}H NMR studies has been most enlightening for slower motions which give rise to chemical exchange behavior. In addition to simulation of the spectrum of a chemically exchanging nucleus, the exchange rate can be monitored by the method of saturation transfer if the rate is slow compared to $\Delta\nu$ (Hoffman and Forsen, 1966). Of particular relevance to the field of protein and peptide dynamics are the 180° flips of phenylalanine and tyrosine residues about the C_β—C_γ bond (Wüthrich and Wagner, 1975; Campbell *et al.*, 1975c; Moore and Williams, 1975; Hull and Sykes, 1975b) and the slow *cis* ↔ *trans* peptide-bond isomerism (Maia *et al.*, 1976; Cheng and Bovey, 1977; Roques *et al.*, 1977), both of which have been studied primarily by ^{1}H NMR. However, the evaluation of ^{1}H relaxation parameters is, in general, much more complex than the ^{13}C case. In contrast with ^{13}C relaxation of protonated carbons, in which only orientational perturbations of the internuclear $^{13}C-^{1}H$ vectors occur, variations of both the orientation and the length of internuclear $^{1}H-^{1}H$ vectors which contribute significantly to the ^{1}H relaxation must generally be considered, leading to a much more difficult problem (Woessner, 1965; Bovee and Smidt, 1974; Rowan *et al.*, 1974). For typical geometries, the contributions of many $^{1}H-^{1}H$ interactions may be similar. This result contrasts with the result for protonated ^{13}C nuclei in which case the contribution of a proton two bonds removed is reduced by $\sim \frac{1}{60}$ relative to the directly bonded proton. A method of obtaining dynamic information from ^{1}H relaxation studies by combining nonselective, mono- and diselective pulse measurements has recently been discussed by Niccolai et al. (1978).

^{1}H spin lattice relaxation in larger proteins (the motion of which is in the slow tumbling range with correlation time longer than the inverse Larmor frequency), is dominated by spin diffusion, that is, diffusion of the nuclear polarization rather than of the nuclei (Kimmich and Noack, 1970, Noack, 1971; Kalk and Berendsen, 1976; Hull and Sykes, 1975a; Sykes *et al.*, 1978; Andree, 1978; Edzes and Samulski, 1977; 1978). In this case the spin lattice relaxation rate is not controlled by the local reorientation rate of the $^{1}H-^{1}H$ vectors, but by the presence of a few relaxation sinks which correspond to internally mobile groups with a more efficient coupling to the lattice. It is important to note in this context that relaxation via spin diffusion is not generally an all-or-none process, and significant differences in proton T_1 values can still be observed in many systems in which spin diffusion is expected to be important (Sykes *et al.*, 1978; Jardetzky *et al.*, 1978). Rapid internal motion of some groups or relatively large internuclear $^{1}H-^{1}H$ distances will reduce the coupling of particular spins to the protein sinks.

The same cross relaxation mechanisms which complicate the interpretation of spin-lattice relaxation in proteins can be exploited to provide interesting information about the system. Interproton NOEs have been applied frequently to obtain conformational information in peptides (Bothner-By and Johner, 1978; Krishna *et al.*, 1978; Jones *et al.*, 1978b and references therein). Gordon and Wüthrich (1978) have shown how cross relaxation pathways in proteins can be determined using transient NOE experiments; the resulting data are useful for making assignments and for evaluating the importance of various dipolar couplings. Evaluation of the relative importance of various cross relaxation interactions using selective pulse techniques has been pioneered by Hill and Freeman (Freeman *et al.*, 1974; Campbell and Freeman, 1973; Hall and Hill, 1976), and the approach has recently been applied to the analysis of ^{1}H spin-lattice relaxation in peptides (Niccolai *et al.*, 1978; Jones *et al.*, 1978a). In summary, the suitability of a particular nuclear species for the experimental analysis of proteins and peptides by NMR is strongly dependent on the type of information sought, and interactions which are a liability in one application may be an asset in another.

Despite the preeminence of ^{1}H and ^{13}C NMR in the study of proteins and peptides, valuable information has been obtained using several other nuclear species. ^{19}F is particularly attractive for protein labeling due to its high magnetogyric ratio, the absence of background resonances, and extreme sensitivity to local environmental perturbations leading to greater resolution of ^{19}F resonances corresponding to individual nuclei. Dynamic studies of ^{19}F labeled proteins have been particularly informative. Theoretical analyses by Hull and Sykes (1974, 1975a, b) indicate two important relaxation mechanisms for ^{19}F: (1) the ^{19}F–^{1}H dipolar interaction and (2) chemical shift anisotropy. As in the case of nonprotonated ^{13}C nuclei, the contribution of relaxation due to chemical shift anisotropy can be separated by studies of the magnetic-field dependence of the relaxation rates. One of the interesting aspects of ^{19}F relaxation is that the two relaxation mechanisms can exhibit different geometric dependences on the motion enabling a more complete characterization of the dynamics. For example, the dominant ^{19}F–^{1}H dipolar interaction in 5-fluorotyrosine residues is essentially independent of motion about the C_β—C_γ bond, but the chemical shift anisotropy mechanism is not. The ^{19}F–^{1}H dipolar interaction is qualitatively similar to the ^{13}C–^{1}H interaction, but exhibits several unique quantitative features. Saturation of the proton resonance leads to a ^{19}F-{^{1}H} NOEF that is dependent on the motional characteristics of the protein, and which for several cases of practical interest is -1.0, leading to nulling of the resonance. For this reason, measurements are typically carried out without decoupling the protons. Failure to decouple the protons does not suppress the ^{19}F–^{1}H scalar couplings which, in the case of alkaline phosphatase labeled with 5-fluorotyrosine studied by Hull and Sykes (1974) are not resolvable due to the broad ^{19}F linewidths. Elimination of the ^{19}F–^{1}H scalar coupling without the undesirable NOE might, in principle, be achieved by pulsed decoupling methods used in ^{13}C NMR (Freeman *et al.*, 1972). In addition, the spin-lattice relaxation of ^{19}F coupled to ^{1}H can be nonexponential due to the presence of cross relaxation interactions as discussed above for the case of protons. Hull and Sykes (1975a) find, however, that for a simulated system containing many interacting protons analogous to the situation in a protein, the relaxation is approximately exponential and has the same value as would be obtained under conditions of complete proton decoupling.

^{15}N NMR studies of a number of peptides (Sogn *et al.*, 1973; Irving and Lapidot, 1975; Kricheldorf *et al.*, 1977; Hawkes *et al.*, 1975a, b, 1977; Hull *et al.*, 1978) and proteins (Lapidot and Irving, 1975, 1977) have been reported. In systems devoid of paramagnetic metal ions, relaxation is predominantly via the ^{15}N–^{1}H dipolar mechanism. This interaction is relatively large since, as occurs in the case of ^{1}H relaxation in ^{13}C labeled molecules (London *et al.*, 1977b), the lower magnetogyric ratio of ^{15}N (or ^{13}C) is compensated for by the shorter internuclear N—H (or C—H) bond lengths. For small molecules rotating rapidly, the theoretical NOE approaches -3.93; however, for larger molecules it approaches 0.88. As a result, intermediate rates of motion corresponding to an NOEF close to -1.0 (NOE close to 0) can lead to nulling of the resonances (Hawkes *et al.*, 1975a, b). Other effects such as competing relaxation mechanisms, and the possibility of chemical exchange between species exhibiting a positive and a negative NOE could also result in nulling of the resonances. Alternatively, observation of the proton coupled ^{15}N resonances is complicated by chemical exchange of the nitrogen bound protons which will broaden the ^{15}N resonance to a degree dependent on the exchange rate (Blomberg *et al.*, 1976). The potential of ^{15}N NMR for obtaining dynamic information has not been fully exploited as yet, but the possibility of obtaining relatively inexpensive ^{15}N enriched precursors should lead to increased applications.

3. DEPENDENCE OF NUCLEAR MAGNETIC RELAXATION PARAMETERS ON THE DETAILS OF MOLECULAR MOTION

The dynamic processes of interest in proteins and peptides can be divided into two categories: those which are slow and those which are fast relative to the overall tumbling rate of the molecule. Slow processes are most often manifested as a chemical exchange process which is treated in most introductory NMR texts and in several review articles (Hoffman and Forsen, 1966, Steigel, 1978). In contrast, the relaxation behavior resulting from internal motion at a rate greater than or equal to that of overall molecular tumbling has only recently been treated in detail. In general, there are many specialized motions such as methyl rotation, puckering of the pyrrolidine ring of proline, 180° jumps of aromatic side chains about the C_β—C_γ bond, multiple internal rotations characterizing the long side chains of lysine, leucine, etc., and limited amplitude diffusion which must be considered in the interpretation of protein and peptide relaxation behavior. Recent investigations have also considered the effects of overall anisotropic rotation of the protein or peptide molecule on the observed relaxation parameters (Wilbur *et al.*, 1976; Somorjai and Deslauriers, 1976).

The approach reviewed in this section is based on the construction of a physically reasonable model for the internal or anisotropic motion and subsequent prediction of the corresponding relaxation parameters. The measurements do not uniquely define the dynamic behavior. Despite the fact that additional parameters are inevitably required to define the details of more complex internal motion leading to an underdetermined problem, many examples in the literature indicate that it is frequently difficult and occasionally impossible to fit given data sets. Even these failures are interesting since they indicate that one or more of the approximations made in constructing the motional model are not valid for the experimental system. In addition, the range of certain parameters can be restricted by other physical considerations. For example, the degree of puckering allowed for a proline residue is severely restricted by the geometry of the pyrrolidine ring. Finally, recent studies of protein and peptide relaxation at different frequencies should reduce the underdetermined nature of the problem, leading to a more precise understanding of the motions studied.

In this section, the predictions of a number of motional models in terms of the observed relaxation behavior are summarized. Illustrative calculations are given for ^{13}C at 25.2 MHz since, as discussed in Section 2, this nucleus is most amenable to this type of analysis. However, the results can be readily generalized to ^{1}H, ^{2}H, ^{19}F, ^{15}N, and ^{31}P, although the analysis of such data is, in general, more complex.

3.1. Relation of Nuclear Relaxation to Molecular Motion

Nuclear relaxation theory was originally formulated by Bloembergen *et al.* (1948) and Solomon (1955); current formulations of the theory are given by Abragam (1961) and Spiess (1978). Treatments for the nonspecialist are given in many texts and reviews, as, for example, Lyerla and Levy (1974). Nuclear magnetic relaxation may be viewed as arising from fluctuating local magnetic or electric fields. The possible sources of these fields include other nuclear dipoles, dipolar fields produced by unpaired electrons, electric charges which interact with the nuclear quadrupole moment for nuclei with spin greater than $\frac{1}{2}$, anisotropy of the chemical shielding tensor, fluctuating scalar coupling interactions, and molecular (spin) rotation. In most cases, the fluctuation arises as a result of the molecular motion. Thus, measurements of relaxation rates can be related to molecular dynamics. This motion can be characterized statistically by an autocorrelation function, $G(t)$, which is a measure of how similar a local field measured at time t is to the local field measured at $t + \Delta t$:

$$G(\Delta t) = \langle \mathbf{H}_{\mathrm{loc}}(t) \cdot \mathbf{H}_{\mathrm{loc}}(t + \Delta t) \rangle = \langle \mathbf{H}_{\mathrm{loc}}^{2}(t) \rangle \exp(-\Delta t/\tau) \qquad (1)$$

where the angular brackets denote an ensemble average. For a stationary system, the above expression will depend on Δt, but not on t, the absolute time of the measurement. Several models for molecular motion, including a model for a rigid, isotropically diffusing sphere, lead to an exponential time dependence as indicated in equation (1), where τ is a correlation time describing how rapidly $\mathbf{H}_{loc}$ is changing. Thus, for times $\Delta t \ll \tau$, $\mathbf{H}_{loc}$ will, on the average, not vary appreciably, while for $\Delta t \gg \tau$, the final local field is uncorrelated with the initial value.

The observed relaxation times as well as the NOE, are calculated from the spectral density $J(\omega)$, which is the Fourier transform of the autocorrelation function. An exponential autocorrelation function as given above leads to a spectral density of the form

$$J(\omega) \propto \frac{\tau}{1 + \omega^2 \tau^2} \qquad (2)$$

A few of the more commonly encountered formulas relating the observed relaxation rates to the spectral densities are summarized in Appendix A. An important characteristic of the spectral density of the form illustrated in equation (2) is the extremum corresponding to $\tau = 1/\omega$, leading to the well-known T_1 minimum behavior of the spin-lattice relaxation time.

In many cases of interest, local fields arising from more than one interaction may contribute to the relaxation process. For example, this situation can arise if several dipolar interactions are significant, as in the case of ^{13}C relaxation of a methyl carbon, or if several different relaxation mechanisms contribute similarly to the relaxation (Matson, 1977). Denoting the corresponding local fields by $\mathbf{H}_{loc}$ and $\mathbf{H}'_{loc}$, it is then possible to define a cross-correlation function(s), $\langle \mathbf{H}_{loc}(t) \cdot \mathbf{H}'_{loc}(t + \Delta t) \rangle$ and to obtain a cross-correlation spectral density which is the Fourier transform of the cross-correlation function. In a few cases, measurements sensitive to cross-correlation functions have been made on peptide (Cutnell and Glasel, 1976a, b) and protein (Uiterkamp *et al.*, 1978) systems. The sensitivity of various relaxation measurements to these terms varies, and in general such effects have been neglected in the majority of peptide and protein studies. Except for a brief discussion in Section 3.7, this neglect will be perpetuated in the present article; however, it is important to emphasize that experiments sensitive to cross-correlation functions promise a great extension of the dynamic information derivable from relaxation measurements. In particular, proton-coupled ^{13}C NMR studies (Prestegard and Grant, 1978), and measurements of zero and multiple quantum transitions by means of two-dimensional spectroscopy (Wokaun and Ernst, 1978) have been shown to be particularly informative about dynamic processes.

The explicit dependence of the autocorrelation function on Δt is in fact determined by the physical model used to describe the molecular motion. Relaxation due to intermolecular interactions is dependent on the translational and rotational motion of the molecules, the distance of closest approach, and other parameters which make a very detailed analysis difficult. In general, however, relaxation in peptides and proteins is dominated by intramolecular interactions. In this case, the local fields are modulated by molecular rotation. Equation (1) then corresponds to an *orientational* autocorrelation function for dipolar, chemical shift anisotropy, or quadrupolar mechanisms, or to an *angular velocity* autocorrelation function for the case of a spin rotation interaction. In the discussion below, the correlation times for molecular orientation and angular velocity are written as τ and τ_J, respectively.

Recent interpretations of relaxation data have utilized an extended diffusion model (McClung, 1969; McClung and Versmold, 1972), the essential features of which are summarized by Spiess (1978). Changes in the angular velocity of the molecule, Ω, are assumed to result from collisions which randomize the magnitude and/or orientation of Ω. In general, the autocorrelation function derived from this model is not a single exponential as assumed in equation (1), even for isotropic motion. Two limiting cases of this model are of particular interest. In the dilute-gas limit, the dynamic behavior approaches that of a perturbed free rotor in which the molecule undergoes many complete rotations between collisions. In this limit, the orientational correlation time τ in equation (1) is proportional to the angular velocity autocorrelation time τ_J, and hence to $\sqrt{I/kT}$, where I is the moment of inertia of the molecule (Somorjai and Deslauriers, 1976). Alternatively, in the limit of small angular steps between collisions (small-step Brownian diffusion), a Stokes–Einstein relation for the orientational correlation time is predicted, with $\tau \sim \eta/kT$, where η is the viscosity of the medium (equation 26). Recent treatments of Brownian motion diffusion have modified the simple theory in two respects. (1) The rotational (translational) angular-velocity autocorrelation function does not decay exponentially, but has an asymptotic power-law dependence on time, $G(t) \sim t^{-5/2}$ ($t^{-3/2}$) (Berne, 1972). This effect is important for relaxation due to spin rotation (a mechanism which is not of general importance for proteins and peptides). However, the orientational changes occur over a much slower time scale, so that the exponential behavior of the orientational autocorrelation function remains valid (Berne, 1972). (2) The "stick" boundary conditions used to obtain the friction coefficient for a rotating Brownian particle, and ultimately the Stokes–Einstein equation, have been replaced by "slip" boundary conditions (Hu and Zwanzig, 1974). The resulting friction coefficients reflect the displacement of solvent by a rotating molecule. For a spherical particle, the friction coefficient vanishes, and the motion is described as a perturbed free rotor according to the inertial limit discussed above. Consequently, the rotational friction, and thus the components of the diffusion tensor, are strongly dependent on molecular shape (Steele, 1976).

The question of applicability of the Brownian motion versus inertial diffusion models has been attacked experimentally by a number of groups. One approach is based on a determination of the components of the rotational diffusion tensor using the Woessner equations summarized in Section 3.2. For the inertial model, the ratio of the principal components of this tensor, $D_x{:}D_y{:}D_z$ should be related to the ratio $(I_x)^{-1/2}{:}(I_y)^{-1/2}{:}(I_z)^{-1/2}$. Grant *et al.* (1973) have performed such an analysis on several organic molecules. For the case of *trans*-decalin in which the motion is found to be highly anisotropic, the above proportionality is not observed. Instead the motion is related to molecular shape (D_i values vary as $e_i{}^{-1}$, where e_i are the estimated ellipticities about the principal axes). Perhaps the studies most relevant to the present discussion are those of Somorjai and Deslauriers (1976), in which rotational correlation times calculated on the basis of a free rotor model are one to two orders of magnitude smaller than those deduced from relaxation data for a series of cyclic peptides. In contrast, the Stokes–Einstein equation has been applied with success to many proteins and peptides. For the systems covered in the present discussion, molecular rotation generally reflects a Brownian diffusion process, and the application of diffusion equations is justified.

It must be emphasized that the application of hydrodynamic models on a molecular level is at best a gross approximation (Steele, 1976). Specific solute–solvent interactions such as hydrogen bonding or electrostatic interactions can be important determinants of the motional anisotropy (Stilbs and Moseley, 1979). This point is illustrated by the effects of hydrogen bonding on the motional anisotropy of pyridine, discussed in Section 3.2, and by differences in the internal motion of the phenylalanyl and tyrosyl side chains in peptides (Section 4.3.1). Studies of model systems analogous to the pyridine study of Campbell *et al.* (1975b) are of obvious value in determining the effect of specific solvent–functional-group interactions on the motional characteristics of peptides and proteins.

Before concluding this discussion, it is important to mention another class of dynamic models which is particularly relevant to proteins and peptides. Relaxation due to jumps between relatively stable conformations is of interest in relation to internal motion occurring in these systems. Wittebort and Szabo (1978) have emphasized the general applicability of jump models involving relatively few conformations, since librational motions of low amplitude are relatively ineffective in perturbing the relaxation times (Section 3.4). In the remainder of Section 3, the effects of rigid anisotropic motion, as well as various models for internal motion which have been used to interpret relaxation data in a variety of systems, will be discussed. Both types of motion lead to autocorrelation functions which are a sum of exponential terms, so that the relaxation times are expressed as a weighted sum of spectral-density terms (King and Jardetzky, 1978), of the form of equation (2).

3.2. Anisotropic Motion of Ellipsoidal Molecules

The relaxation rates corresponding to anisotropic diffusion of an ellipsoidal molecule (i.e., possessing D_{2h} symmetry) have been derived by Woessner

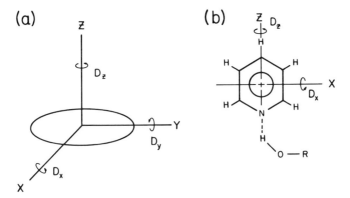

Figure 1. (*a*) Anisotropic ellipsoidal molecule undergoing diffusion characterized by $D_x \neq D_y \neq D_z$; the coordinate system shown corresponds to the principal axes system of the diffusion tensor. (*b*) Anisotropic motion of pyridine resulting from solvent interactions which reduce D_x relative to D_z.

(1962b) and Shimizu (1962). The motion is described by a set of three diffusion coefficients characterizing the diffusion rates about the principal axes of the ellipsoid (Figure 1*a*). As noted above, for motion describable by small-step Brownian diffusion, the diffusion rates about the various axes will be related to molecular shape, as well as to specific solvent interactions. For the case in which the three diffusion coefficients differ, $D_x \neq D_y \neq D_z$, the autocorrelation function is a sum of five exponentials leading to a spectral density of the form

$$J(\omega) = \sum_{n=1}^{5} \frac{C_n \tau_n}{1 + \omega^2 \tau_n^2} \tag{3}$$

where in terms of the diffusion coefficients illustrated in Figure 1*a*, the correlation times in equation (3) are given by: $(\tau_n)^{-1} = 6D_s[1 + D^*/(2D_s)]$, $6D_s[1 \pm D^*A/(2D_s)]$, $6D_s[1 - D^*(1 \pm B)/(4D_s)]$, where $D_s = \frac{1}{3}(D_x + D_y + D_z)$, $D^* = \frac{1}{3}(2D_z - D_x - D_y)$, $B = (D_y - D_x)/D^*$, and $A = (1 + B^2/3)^{1/2}$ (Spiess, 1978). The coefficients C_n contain geometric factors relating the elements of the tensor describing the relaxation interaction to the principal axis system of the diffusion tensor. For many cases of interest, it is only necessary to consider a relaxation vector, for example, the internuclear vector for dipolar relaxation. In this case, the C_n are functions of the direction cosines (Woessner, 1962b) or the polar angles (Huntress, 1970) which define the orientation of the relaxation vector in the principal axes system. These relations have been used to interpret ^{13}C relaxation data for small organic molecules in terms of anisotropic diffusion (Grant *et al.*, 1973; Berger *et al.*, 1975) as well as data for quadrupolar nuclei (Huntress, 1970; Spiess, 1978 and references therein). More recent applications to peptides by Deslauriers and coworkers are described in Section 4.1.

A substantial simplification of the spectral density equations occurs in the axially symmetric case $(D_x = D_y)$. The resulting spectral density can be expressed as a sum of three terms

$$J(\omega) = \sum_{i=0}^{2} B_{i0}(\beta) \cdot \frac{[6D_x + i^2(D_z - D_x)]^{-1}}{1 + \omega^2 [6D_x + i^2(D_z - D_x)]^{-2}} \quad (4)$$

where the elements of the B matrix (London and Avitabile, 1976, 1977a) are given in Appendix B. This equation contains only three adjustable parameters: D_z, $D_x = D_y$, and β, the angle between the relaxation vector, that is, the $^{13}C-^{1}H$ bond, and the z axis (Figure 2a). Dependence of the calculated relaxation times on the degree of motional anisotropy is most readily illustrated for the extreme narrowing case in which the second term of the denominator in equation (4) is much smaller than 1 for the ω values of interest (Appendix A). In this case, it becomes possible to write equation (4) in the form

$$J(\omega) \xrightarrow[\substack{\text{extreme} \\ \text{narrowing}}]{} \left[1 + \frac{3(1-\rho)}{5+\rho} \sin^2 \beta + \frac{9}{2} \frac{(\rho-1)^2}{(5+\rho)(1+2\rho)} \sin^4 \beta \right] \cdot \frac{1}{6D_x}$$

$$\equiv \chi(\rho,\beta)/(6D_x) \quad (5)$$

where the parameter $\rho = D_z/D_x = D_z/D_y$.

The form of equation (5) is identical to the result for a rigid, isotropically reorienting sphere with an effective rotational correlation time $\tau_{\text{eff}} = \chi(\rho,\beta)(6D_x)^{-1}$. As is evident from equation (5), $\chi = 1$ for $\beta = 0$. Thus, if the relaxation vector is aligned parallel to the symmetry axis of the motion, the calculated relaxation rates will be independent of D_z.

The use of relaxation measurements to study the dynamic behavior of pyridine provides a beautiful application of the formalism discussed above. Kintzinger and Lehn (1971) find that the motional anisotropy studied in liquid pyridine changes from solid-like, with faster motion about the y axis

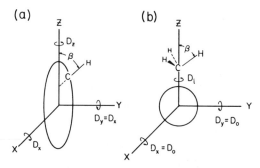

Figure 2. (a) Anisotropic motion of an axially symmetric ellipsoid $(D_z \neq D_x = D_y)$. (b) Internal motion with diffusion coefficient D_i in a molecule undergoing overall isotropic motion with diffusion coefficient D_o. The calculations based on the diffusive processes indicated in (a) and (b) are mathematically equivalent if the C–H vector fixed in the axially symmetric molecule makes an angle β with the z axis, $D_o = D_x = D_y$ and $D_i = D_z - D_x$.

(fig. 1b) to gas-like, with faster motion about the in-plane axes, as the temperature increases. Campbell *et al.* (1975b) have used the observed motional anisotropy to study transient solution complexes of pyridine with hydrogen bond donors. The relative magnitudes of the diffusion coefficients D_z and D_r are dependent on the hydrogen bonding properties of the solvent which will decrease D_r relative to D_z. Campbell *et al.* define the ratio

$$R = \frac{T_1(C-2) + T_1(C-3)}{2\,T_1(C-4)} \tag{6}$$

which serves as a measure of the motional anisotropy. Based on the above treatment, we can write:

$$R = \frac{\chi(\rho, \beta = 0°)}{\chi(\rho, \beta = 60°)} = \chi(\rho, \beta = 60°)^{-1} \tag{7}$$

Ratios in the range 1.0 for neat pyridine to ~2 in a 50% pyridine–glycerol mixture were measured. The data have been interpreted based on the hypothesis that the hydrogen bonding interactions result in a pyridine–solvent aggregate which is asymmetric. The calculated diffusion coefficient ratio D_r/D_z can be used to determine the ratio of the major to minor axes, a/b, of an axially symmetric ellipsoid (Perrin, 1936; Woessner, 1962b). Results for pyridine–water mixtures correspond to a relatively large a/b ratio of 2.9, indicating extensive cross linking of the water hydrogen bonded to the pyridine.

3.3. Free Internal Diffusion

We next consider the effect of free internal (rotational) diffusion on the calculated relaxation parameters. By far the most widely used model is the free internal diffusion model originally developed by Woessner (1962a). Assuming overall isotropic motion, this model is mathematically equivalent to the axially symmetric anisotropic diffusion model discussed above (Figure 2). In the present case, β is the angle between the internal diffusion axis, taken to be the z axis, and the relaxation vector. We therefore have $D_i = D_z - D_r = D_z - D_y$. Thus, if the internal diffusion coefficient approaches zero, we obtain $D_z = D_r = D_y$, as expected for an isotropically rotating sphere. Setting the overall isotropic diffusion coefficient equal to D_o, we then have $D_o = D_r = D_y$ giving:

$$J(\omega) = \sum_{i=0}^{2} B_{i0} \frac{(6D_o + i^2 D_i)^{-1}}{1 + \omega^2 (6D_o + i^2 D_i)^{-2}} \tag{8}$$

In the extreme narrowing limit, equation (5) also applies with $\rho = D_z/D_r = 1 + D_i/D_o$. It should be noted that while in the rigid anisotropic diffusion case ρ can be greater than 1 or less than 1, corresponding to a prolate or oblate ellipsoid, respectively, ρ is always greater than or equal to 1 in the internal diffusion case.

In the limit $\rho \gg 1$, that is, very rapid internal motion, $\chi(\rho,\beta)$ approaches the limit

$$\chi \xrightarrow[\rho\to\infty]{} \left(\frac{3\cos^2\beta - 1}{2}\right)^2 \qquad (9)$$

as can be seen most readily by noting that the B_{00} term in equation (4) will dominate. This effect has been observed for methyl groups on relatively large molecules which satisfy the condition $\rho \gg 1$. For example, a comparison of the NT_1 values (where N is the number of directly bonded protons) for methyl carbons C-18, C-19, and C-21 in cholesteryl chloride with the value for the backbone methine carbons gives a ratio ~ 9, in agreement with equation (9) using a tetrahedral angle (Allerhand et al., 1971b). In general, χ only approaches the limit indicated in equation (9) when $\rho \gtrsim 100$, the required ratio being dependent on the value of β. Thus, for example, $\chi(\rho,\beta) = \chi(10, 109.5°)^{-1} \sim 3$, rather than the asymptotic value of 9.0, so that the T_1 values will only increase by a factor of 3 if the internal diffusion rate is an order of magnitude greater than the overall diffusion rate. Numerical results for a methyl group corresponding to $\beta = 90°$ (χ_{HH}) and $\beta = 109.5°$ (χ_{CH}) are given in Figure 3.

As noted in the example of cholesteryl chloride discussed above, the conclusions regarding internal motion are generally based on a comparison of the spin-lattice relaxation times for immobile and mobile nuclei. Such a

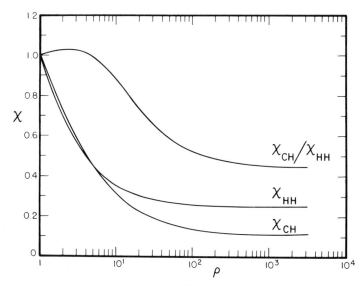

Figure 3. Anisotropic motion factor [equation (6)] χ plotted as a function of $\rho = D_z/D_x$ (Figure 2a) or $\rho = 1 + D_i/D_o$ (Figure 2b). χ_{CH} has $\beta = 109.5°$ (or 70.5°), and χ_{HH} has $\beta = 90°$ [reprinted with permission from London et al. (1977a) J. Phys. Chem., **80**, 884, copyright by the American Chemical Society].

comparison is, however, not always feasible for a variety of reasons. An alternative approach is a measurement of relaxation times for two different nuclear species on the group undergoing internal motion. This approach has been utilized to study the anisotropic motion of BCl_3 by measuring the quadrupolar relaxation rates of ^{11}B and ^{35}Cl (Allerhand, 1970) and in the case of internal motion in DMSO by studying the dipolar relaxation of ^{13}C and ^1H nuclei (London *et al.*, 1977a). By considering the ratio of relaxation times measured for the two nuclei, the overall molecular correlation time τ_o drops out of the relation, and the ratio of the χ factors, in the extreme narrowing limit, allows a solution for ρ (Figure 3). In the case of the methyl group, the ratio is given by

$$\frac{T_1 \, ^{13}C}{T_1 \, ^1H} = \left(\frac{\gamma_H}{\gamma_C}\right)^2 \left(\frac{r_{CH}}{r_{HH}}\right)^6 \frac{\chi_{HH}(\rho,\beta = 90°)}{\chi_{CH}(\rho,\beta = 109.5°)} \tag{10}$$

As can be seen from Figure 3, the T_1 ratio will be sensitive to the presence of internal motion; however, a reliable value for ρ can only be obtained in the region $10 \leq \rho \leq 100$. Application of this formalism requires that only intramolecular $^{13}C-^1H$ and $^1H-^1H$ dipole–dipole interactions are significant and that for the two nuclei considered the values of β be different. Thus, a comparison of the carbon and deuteron relaxation times in a deuterated methyl assuming quadrupolar relaxation of the deuterons will not provide information about the internal motion since the β angles are identical. In contrast, measurement of ^{13}C and ^2H relaxation times for a methyl group allows comparison of the interaction constants and hence deduction of the quadrupolar coupling constant, if the latter is not known (Saito *et al.*, 1973). More elegant methods of studying internal diffusion of methyl groups based on cross-correlation effects are also available (Cutnell *et al.*, 1976a, b; Haslinger and Lynden-Bell, 1978).

For motion not in the extreme narrowing limit, as is the case in many proteins of interest, the approximation of equation (5) is not valid and equation (8) must be applied. Extensive numerical calculations have been given by Doddrell *et al.* (1972), Led *et al.* (1975), Howarth and Lilley (1978), and Hull and Sykes (1975a) corresponding to various geometries. It is important to note that in equation (8) the $i = 0$ term contains only D_o, while the $i = 1,2$ terms will depend primarily on D_i if $D_i \gg D_o$. The relative importance of these terms is dependent on how close the various correlation times are to $1/\omega$. In general, if $\omega > D_o$ and $\omega < D_i$, all terms may be similar in magnitude and approximations involving, for example, only the $i = 0$ terms are invalid. Consequently, for fast internal diffusion, the contribution of the internal diffusion process to $J(\omega)$ will increase as D_i decreases, as long as the condition $\omega/D_i < 1$ is satisfied. Thus, for example, if the overall motion is slow ($\omega > D_o$), the calculated NOE value will be much larger for $D_i = 10^9 \, s^{-1}$ than for $D_i = 10^{11} \, s^{-1}$. An analogous effect holds for ^{19}F (Hull and Sykes, 1975b).

3.4. Restricted Amplitude Internal Diffusion

The free internal diffusion model discussed in the previous section has been extensively applied in the analysis of internal motion of peptides and proteins. In general, however, few groups other than terminal methyl groups undergo such unrestricted diffusion. Recent theoretical investigations of McCammon *et al.* (1977) emphasize the importance of restricted internal diffusion in proteins. It is therefore of considerable practical importance to determine, at least qualitatively, how such restricted diffusion will affect the observed NMR relaxation rates.

One of the simplest approaches to this problem is the use of a square-well potential imposed by the application of boundary conditions to the diffusion equation (Figure 4a). This approach has recently been followed (Wittebort and Szabo, 1978; London and Avitabile, 1978), and the resulting spectral density has the form

$$J(\omega) = \sum_{i=-2}^{+2} \sum_{n=0}^{\infty} \mid d_{i0}(\beta) \mid^2 \cdot \mid E(i,n) \mid^2 \cdot \frac{\tau_i}{1 + \omega^2 \tau_i^2} \tag{11}$$

where

$$\tau_i = \left(6D_o + \frac{n^2 \pi^2 D_i}{4\theta^2} \right)^{-1} \tag{12}$$

the $d_{i0}(\beta)$ are reduced Wigner rotation matrices (Spiess, 1978), the matrices $E(i,n)$ are functions of the angle θ defining the allowed range of motion (Figure 4a) and β is defined as in the free internal diffusion case. The first term in the series has the form: $E(i,n = 0) = [\sin(i\theta)]/(i\theta)$ so that, since the sum in equation (12) runs from $i = -2$ to $+2$, terms of the form $[\sin(2\theta)]/(2\theta)$

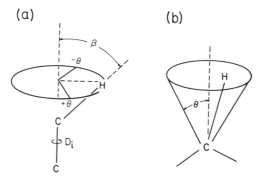

Figure 4. Models for restricted internal motion. (*a*) Internal diffusion as in Figure 2*b* but subject to boundary conditions limiting the range of motion to $(+\theta, -\theta)$. β is defined as in Figure 2*b*. (*b*) Diffusional range of the C–H vector restricted by conic boundary conditions with half-angle θ. Note that in both cases θ is a limitation on the allowed range of motion. In (*a*) the internal motion must be characterized by an additional parameter β.

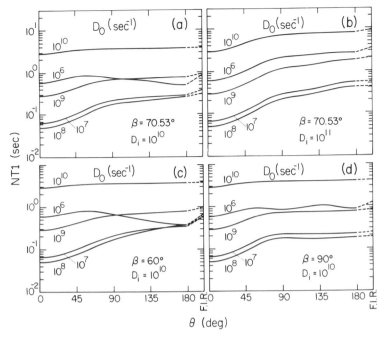

Figure 5. NT_1 values for the restricted internal diffusion model [Figure 3a, equation (11)] corresponding to the values of D_o, D_i, and β indicated, and plotted as a function of θ. Relaxation times corresponding to abscissa labeled F.I.R. are obtained using the free internal rotation model without boundary conditions. Small discrepancies between the NT_1 values using the restricted model with a full range of motion ($\theta = 180°$) and the unrestricted internal diffusion model are discussed in the text [reprinted with permission from London and Avitabile (1978), *J. Am. Chem. Soc.*, **100**, 7159, copyright by the American Chemical Society].

can lead to nonmonotonic behavior as θ increases from 0 to 180° (corresponding to a full range of motion). Calculations are straighforward due to the very rapid convergence of the sum. Numerical results for calculated ^{13}C NT_1 values (where N is the number of directly bonded protons) are summarized in Figure 5. Relaxation times vary from the values expected in the absence of internal motion to values slightly below those calculated using the free internal rotation model as the range of motion (2θ) is increased from 0 to 360°. The discrepancy in the latter comparison arises from the boundary condition preventing diffusion from $+180$ to -180. Calculated NT_2 and NOE values exhibit qualitatively similar behavior (London and Avitabile, 1978). For certain parameters the calculated values are markedly nonmonotonic functions of θ (London and Avitabile, 1978). Perhaps the most interesting feature of these calculations is that the calculated relaxation parameters change very slowly for small values of θ. There is nearly no change for a range $2\theta < 40°$. This result is of considerable practical importance since it suggests that many low-amplitude diffusive motions which can occur in a globular protein will have little effect on the measured relaxation rates which will then reflect only the overall tumbling of the molecule.

Another simple model for restricted diffusion in a biological macromolecule is illustrated in Figure 4*b*. In this case, the relaxation vector is restricted to a conical volume as illustrated. As in the above case, this problem can be solved numerically by the imposition of the appropriate boundary conditions on a diffusion equation; however, the calculation is considerably more difficult. Bull *et al.* (1978) have recently given an approximate solution of this problem in connection with the interpretation of data for halogen ions exchanging with sites on proteins. For overall isotropic diffusion, the result is

$$J(\omega) = A \cdot \frac{\tau_o}{1 + \omega^2 \tau_o^2} + (1 - A) \cdot \frac{(1/\tau_o + 1/\tau_i)^{-1}}{1 + \omega^2 (1/\tau_o + 1/\tau_i)^{-2}} \tag{13}$$

where τ_o is the overall molecular rotation correlation time, τ_i the correlation time for internal motion and A is given by:

$$A = \frac{\cos^2\theta \, \sin^4\theta}{4(1 - \cos\theta)^2} \tag{14}$$

In keeping with the consistency of the presentation, results for NT_1 values for protonated ^{13}C nuclei calculated using the above result and parameters analogous to those used for the previous model are given in Figure 6. The behavior of the spin-lattice relaxation time is qualitatively similar to that observed for the corresponding diffusion rates in the previous case. In particular, low-amplitude internal diffusion has a relatively small effect on the calculated relaxation rate. In contrast, rapid internal motion can lead to a much larger increase in T_1 in this case. This is not surprising since in the present case the limit $\theta \rightarrow 180°$ corresponds to a full solid angle for the internal motion while in the previous case $\theta \rightarrow 180°$ corresponds only to free internal diffusion about an axis fixed in the macromolecule. Equivalence of the results for $\theta = 90°$ and $\theta = 180°$ may reflect the approximations inherent in the derivation (Bull, 1978).

3.5. Jump Models

In the previous sections, we have considered anisotropic or internal motion as a diffusive process over a continuum of available orientations. As a result of either local (intraresidue) or long-range (interresidue) constraints, the internal motion is more typically described as a jump between several stable conformations. It then becomes the rate of jumping among these conformations, or, equivalently, the lifetimes of the various conformations, which determine the relaxation rates. The fact that low-amplitude diffusion is ineffective in causing relaxation suggests that jump models involving relatively few conformations are generally suitable to describe the relaxation. The first such case to be considered corresponds to a methyl group jumping

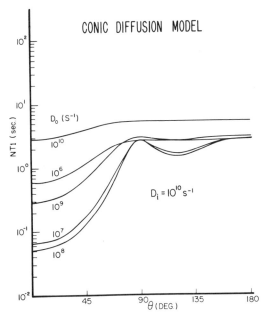

Figure 6. NT_1 values for the restricted internal motion model with conic boundary conditions [Figure 3b; equations (13), (14)]. $D_i = 10^{10}$ s^{-1} ($\tau_i = 1.67 \times 10^{-11}$ s) and D_o values in the range 10^6–10^{10} s^{-1} plotted as a function of θ. Calculations correspond to $\omega_c = 25.2$ MHz.

between three equally stable positions (Woessner, 1962a; Wallach, 1967; Marshall *et al.*, 1972; Bovee and Smidt, 1974):

$$J(\omega) = B_{00} \frac{(6D_o)^{-1}}{1 + \omega^2 (6D_o)^{-2}} + (1 - B_{00}) \frac{(6D_o + 3/\tau_e)^{-1}}{1 + \omega^2 (6D_o + 3/\tau_e)^{-2}} \qquad (15)$$

where τ_e is the lifetime of the methyl orientation in each of the three stable conformers. Equation (15) is qualitatively similar to the free internal diffusion result [equation (8)]. In the extreme narrowing limit we have

$$J(\omega) \longrightarrow \left[1 - \frac{3\rho}{1 + \rho} \sin^2 \beta + \frac{9}{4} \frac{\rho}{1 + \rho} \sin^4 \beta \right] (6D_o)^{-1} \qquad (16)$$

where in this case $\rho = (2D_o\tau_e)^{-1}$. The choice of which model is more appropriate in a given case is considered in Section 3.8. The case of three unequally stable states is discussed in Section 3.6.

Recent studies have indicated that a number of biologically interesting molecules are bistable. For example, the phenylalanine and tyrosine residues of proteins undergo 180° flips which can alter the relaxation parameters (Section 4.3.1), although generally the rate is too slow to have a significant effect on T_1. The nicotinamide ring of NAD$^+$ has also been found to jump slowly between *syn* and *anti* conformations (Zens *et al.*, 1975). Alternatively the five-membered pyrrolidine ring of proline (Torchia, 1971; Abraham and

Thomas, 1964; Pogliani *et al.*, 1975) and the six-membered tetrahydro-
pyrazine ring of tetrahydrofolate (Poe and Hoogsteen, 1978) also appear to
be bistable and the interconversion rates in the former have been shown to
be sufficiently rapid to alter the T_1 values (London, 1978). When D_o is the
overall (isotropic) diffusion coefficient, τ_A and τ_B are the lifetimes of the two
stable states, the angle β is defined by the relaxation vector and the axis
about which the vector jumps, and θ is half of the jump range (i.e., the jump
takes the C–H vector between $+\theta$ and $-\theta$), the spectral density is given by

$$J(\omega) = (1 - C)\frac{(6D_o)^{-1}}{1 + \omega^2(6D_o)^{-2}} + C\frac{(6D_o + 1/\tau_c)^{-1}}{1 + \omega^2(6D_o + 1/\tau_c)^{-2}} \qquad (17)$$

where

$$\frac{1}{\tau_c} = \frac{1}{\tau_A} + \frac{1}{\tau_B} \qquad (18)$$

and

$$C = \frac{3\tau_A\tau_B}{(\tau_A + \tau_B)^2}\,[\sin^2\beta\,(1 - \cos 2\theta)] \cdot [2 - \sin^2\beta\,(1 - \cos 2\theta)] \qquad (19)$$

Models corresponding to the limit $\tau_A = \tau_B$ have also been derived (Jones,
1977; Ghesquiere *et al.*, 1977). Equation (17) leads to several interesting
predictions regarding spin-lattice relaxation behavior. (1) The effects of
internal jumping are most pronounced for the case $\beta = 90°$, $\theta = 45°$ (Figure
7). As in the free internal diffusion case, T_1 can either increase or decrease as
a result of internal jumps between two stable states, the result depending on
the diffusion coefficients involved. The maximum increase in NT_1 (N = the
number of directly bonded protons) is 4.0 in sharp contrast to the free
internal diffusion case for which, depending on the particular values of β, D_o,
and D_i the relaxation times can increase without limit (in the absence of
other relaxation mechanisms becoming dominant). (2) Internal jumps
characterized by the parameters $\beta = 90°$, $\theta = 90°$ produce no effect since
such motion inverts the (unsigned) C–H vector leaving it with the same
orientation relative to the magnetic field (Figure 8). (3) The effect of the
internal jump on the relaxation rates is maximal for the case $\tau_A = \tau_B$ (Figure
9). It is interesting to note that in the limit $\tau_c \ll \tau_o = 1/(6D_o)$ so that the
internal motion is rapid relative to overall molecular tumbling, the effect of
the internal motion is still negligible if τ_A and τ_B are very different. This
reflects a decrease in the coefficient C (equation 19). (4) Although the jump is
uniquely defined, the choice of angles β and θ describing the jump is not, so
that any set of angles with a constant value of $\sin^2\beta\,(1 - \cos 2\theta)$ gives the
same result. This is discussed in greater detail by London (1978).

Other models corresponding to jumps between varying numbers of allowed states have appeared in the literature (Noack, 1971; Anderson, 1973;
Hubbard and Johnson, 1975; Wright *et al.*, 1979; King *et al.*, 1978). The

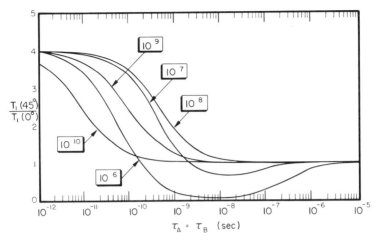

Figure 7. Ratio of the spin-lattice relaxation times calculated for a two-state model [equations (17)–(19)] for $\theta = 45°$ to the values for $\theta = 0°$ (i.e., no internal motion). Values of the overall isotropic diffusion coefficients are indicated for the individual curves, $\beta = 90°$ in all cases. Results are plotted as a function of $\tau_A = \tau_B$. Internal jumping which is rapid compared with the rate of overall motion increases T_1 by a maximum value of 4.0 for these parameters ($\beta = 90°$, $\theta = 45°$) [reprinted with permission from London (1978), *J. Am. Chem. Soc.*, **100**, 2678, copyright by the American Chemical Society].

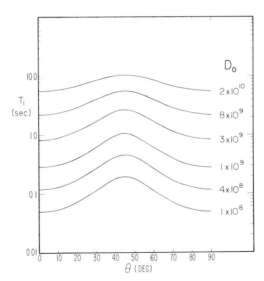

Figure 8. Spin-lattice relaxation times T_1 calculated for a two-state jump model corresponding to $\beta = 90°$, $\tau_A = \tau_B = 10^{-11}$ s, and the D_o values indicated [reprinted with permission from London (1978), *J. Am. Chem. Soc.*, **100**, 2678, copyright by the American Chemical Society].

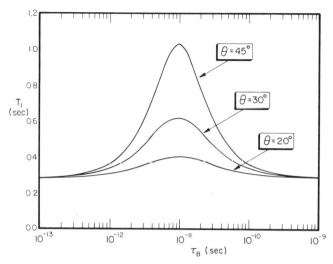

Figure 9. Spin-lattice relaxation times T_1 calculated for a two-state internal jump model with τ_A fixed at 10^{-11} s and τ_B varied from 10^{-13} to 10^{-9} s, $D_0 = 10^9$ s^{-1}, $\beta = 90°$, A and the θ values indicated. Internal jumping has a maximum effect on the spin-lattice relaxation time if the condition $\tau_A = \tau_B$ is met [reprinted with permission from London (1978), *J. Am. Chem. Soc.*, **100**, 2678, copyright by the American Chemical Society].

two-state model has been considered in detail since several of the features associated with such models are easily calculated and physically interpreted in this case.

3.6. Multiple Internal Rotations or Jumps

A characteristic pattern of monotonic increases in the NT_1 values for carbons further from the peptide backbone is observed for peptide residues with long, aliphatic side chains such as lysine, arginine, valine, leucine, etc. (Keim *et al.*, 1973a, b, 1974; Deslauriers and Smith, 1976; Howarth and Lilley, 1978). In most protein spectra, even those of relatively large molecular weight, a sharp peak corresponding to lysine C_ϵ can be observed although nearly all other methylene groups give much broader resonances (Allerhand *et al.*, 1970, 1971a; Conti and Paci, 1971; Nigen *et al.*, 1973). Both of these observations can be described by models which assume slow motion of the peptide backbone and multiple internal diffusive rotations or jumps about successive side-chain bonds. The simplest approach to this problem is generalization of the free internal diffusion model discussed in Section 3.2 to the case of multiple internal rotations (Wallach, 1967; Levine *et al.*, 1973) (Figure 10). In terms of the B matrix formalism (London and

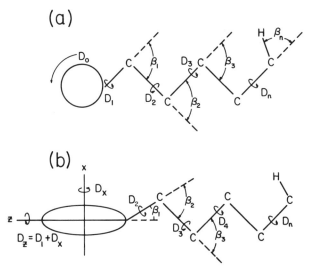

Figure 10. (*a*) Multiple internal diffusion model of Wallach (1967). Overall motion with diffusion coefficient D_o is assumed isotropic. Internal diffusion coefficients are $D_1, D_2, \ldots, D_n$. Angles between successive axes of internal diffusion are $\beta_1, \beta_2, \ldots, \beta_{n-1}$; the angle between the final diffusion axis and the relaxation vector is β_n. (*b*) Multiple internal diffusion model assuming overall anisotropic, axially symmetric motion (Levine *et al.*, 1974; London and Avitabile, 1976). The calculation is physically analogous and mathematically equivalent to (*a*) with $D_o = D_x$, $D_1 = D_z - D_x$ for the first diffusion coefficient and β_1 the angle between the axial z axis and the first internal diffusion axis. The first actual internal diffusion process is therefore described by D_2. The mathematical equivalence noted here corresponds to that illustrated in Figures 2*a,b*.

Avitabile, 1976), the spectral density for the nth carbon, $J_n(\omega)$, is given by

$$J_n(\omega) = \sum_{a,b,c,\ldots=0}^{2} B_{ab}(\beta_1)\, B_{bc}(\beta_2) \cdots B_{mn}(\beta_{n-1}) \beta_{no}(\beta_n)\, \frac{\tau}{1 + \omega^2\tau^2} \quad (20)$$

where

$$\tau = (6D_o + a^2D_1 + b^2D_2 + \cdots + n^2D_n)^{-1}$$

In the above, β_i are the angles between successive internal rotation axes (generally the carbon–carbon bonds), β_n is the angle between the final internal rotation axis and the internuclear C–H vector, D_o is the diffusion coefficient for overall isotropic molecular diffusion, and D_i are the internal diffusion coefficients about successive axes. Equation (20) also describes the motion if the overall molecular diffusion is assumed to be anisotropic but axially symmetric. In this case, D_o becomes $D_x = D_y$, the first internal diffusion coefficient D_1 is defined so that $D_z = D_1 + D_x$, and D_2 corresponds to the first actual internal diffusion process (Figure 10*b*) (Woessner *et al.*,

1969; Levine *et al.*, 1974; London and Avitabile, 1976). Thus, τ is given by:

$$\tau = (6D_x + a^2 (D_z - D_x) + b^2 D_2 + \cdots + n^2 D_n)^{-1} \qquad (21)$$

For the case of an axially symmetric ellipsoid with one degree of internal motion, the result of Woessner *et al.* (1969) can be obtained using two B matrices and making the identifications $\alpha = \beta_1$, $\Delta = \beta_2$, $D = D_2$, $R_2 = D_x$, and $R_1 = D_z$. The primary feature of the approach outlined above is a prediction of the cumulative effect of successive internal rotations on the relaxation behavior. Thus, even if all of the D_i are equal, the sixth carbon, for example, will exhibit a longer T_1 value than the second. This is illustrated in Figure 11 which gives a set of D_i values and the corresponding calculated T_1 values. The limiting behavior of the multiple free internal diffusion model for the case of an infinite chain with β and D_i all identical has recently been treated by Kuo *et al.* (1979), and the results applied to ^{2}H relaxation. For $\beta \neq 0$ the limiting behavior is independent of β (although the rate at which this limit is approached does depend on β) and is given by

$$J_n(\omega) \xrightarrow[n \to \infty]{} \frac{\tau}{1 + \omega^2 \tau^2}; \ \tau = (6D_o + 2nD_i)^{-1}$$

where the internal diffusion coefficients D_i about all bonds are assumed equal.

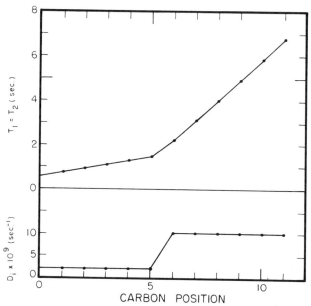

Figure 11. T_1 and T_2 values calculated using the multiple free internal diffusion model (Figures 10a) with $\beta_i = 70.5°$ and the diffusion coefficients indicated: $D_i = 2 \times 10^9 \ s^{-1}$ ($i = 0$ to 5), $D_i = 10^{10} \ s^{-1}$ ($i = 6$ to 11). Cumulative effect of the multiple internal motions is indicated by the fact that $T_1 = T_2$ increases although D_i remains constant for the first five carbons. An abrupt jump in the D_i values leads to an increased slope in the T_1 curve.

At this point, we note that an approximation to equation (20) originally suggested by Wallach and widely quoted in the literature is based on the assumption that if $D_1, D_2, D_3, \ldots \gg D_0$, all terms in the sum with $a, b, c, \ldots \neq 0$ will be negligibly small. In this case, only the B_{00} terms contribute. For all of the β_i equal, as is true for a polymethylene chain, equation (20) reduces to

$$J_n(\omega) \longrightarrow B_{00}(\beta)^n \frac{(6D_0)^{-1}}{1 + \omega^2(6D_0)^{-2}} = \left(\frac{3\cos^2\beta - 1}{2}\right)^{2n} \frac{(6D_0)^{-1}}{1 + \omega^2(6D_0)^{-2}} \quad (22)$$

This approximation is a very poor one. There are no experimental systems undergoing multiple internal rotations which exhibit the enormous T_1 increases predicted by equation (22). This is not a problem with the model, since calculations corresponding to a wide range of internal diffusion coefficients indicate that equation (22) is a very poor approximation to equation (20).

Beyond the calculation problem, the model assuming multiple free internal diffusion processes clearly runs into problems for more than two to three such rotations since it fails to exclude molecular conformations in which remote atoms occupy the same space. Another difficulty is the fact that particular stable conformations, for example, the *trans* conformation, are not weighted any more heavily than eclipsed conformations. A solution to the second problem, which also partially deals with the first problem, is to adopt a jump formalism in which the bonds jump among the allowed *gauche* and *trans* states (Anderson, 1973; London and Avitabile, 1977b; Tsutsumi, 1979). For example, London and Avitabile (1977b) have assumed that due to the steric barrier corresponding to the eclipsed conformation for a polymethylene chain, no direct $g^+ \leftrightarrow g^-$ jumps occur so that the motion is described by

$$g^+ \leftrightarrow t \leftrightarrow g^- \quad (23)$$

where the g^+, t, and g^- orientations are related by 120° jumps about the carbon–carbon bonds. The parameters β and D_i used to define the motion in the free internal diffusion model are replaced by the parameters β, τ_t, and σ, where τ_t is the lifetime of the *trans* conformation and σ is the equilibrium *gauche/trans* ratio:

$$\sigma = \frac{[g^+] + [g^-]}{2[t]} \quad (24)$$

It is interesting to note that in this case, the relaxation rates will depend on a parameter σ, which is a measure of an equilibrium property of the system. For this reason, the measured relaxation rates will be dependent on the ordering as well as the dynamics of the jump (London and Avitabile, 1977b). For values of $\sigma < 1.0$, the chain favors an extended all *trans* conformation which of course reduces the problem of spatial overlap of remote atoms. An interesting consequence of this model is that the characteristics associated

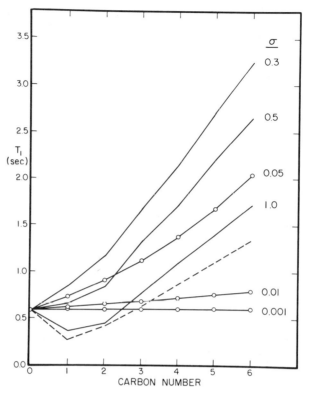

Figure 12. Spin-lattice relaxation times T_1 calculated for a six carbon chain undergoing *gauche–trans* isomerism as described in Section 3.6. Overall diffusion is assumed isotropic with diffusion coefficient $D_o = 10^6$ s^{-1}, the trans lifetimes are all assumed equal, $\tau_t = 10^{-10}$ s. The broken line corresponds to results for the multiple free internal diffusion model (Figure 10*a*) in which $D_i = \frac{1}{6} \tau_t (1 + \frac{1}{2} \sigma)$ and $\sigma = 1.0$. σ is the mean *gauche/trans* ratio, that is, $\sigma = ([g^+] + [g^-])/(2[t])$ [reprinted with permission from London and Avitabile (1977b), *J. Am. Chem. Soc.*, **99**, 7765, copyright by the American Chemical Society].

with the motion of the slowly tumbling end of the molecule (C$_0$), for example, reduced NOE value and frequency dependence of the spin-lattice relaxation, will be transmitted farther along the chain in sharp contrast to the multiple unrestricted internal diffusion model. In the limit $\sigma \to 0$, all atoms will have the same relaxation parameters as C$_0$ (Figure 12).

A final problem which arises in connection with the treatment of multiple internal jumps is related to the fact that such motions are not independent as assumed in the above calculation, but are likely to be correlated. This behavior has been discussed most extensively in connection with lipid fatty acyl chains which undergo kink formation and diffusion, where a kink is defined as a g^+tg^- or g^-tg^+ sequence (Pechhold, 1968; Traüble, 1971; Horwitz, 1972). In general, treatments incorporating such effects are likely to introduce more parameters than can generally be deduced from experiments. London and Avitabile (1977b) have suggested the use of a correlation parameter to qualitatively gauge such effects. A more detailed weighting procedure has been given by Wittebort and Szabo (1978) based on the tetrahedral lattice model of Monnerie and coworkers (Monnerie and Geny, 1969; Valeur *et al.*, 1975). According to this model, the methylene chain is constrained to lie on a tetrahedral lattice. Configurations in which two atoms occupy the same position in space are eliminated from the calculation, thereby dealing with the excluded volume problem. Further, this model allows the effects of concerted jumps involving several bonds to be treated, in contrast to the above models which assume independent motions about each bond. Of course, the approach of enumerating each of the allowed configurations can become cumbersome for longer chains.

3.7. Application of Dynamic Models to Experimental Problems

As noted at the beginning of Section 3, the set of relaxation parameters does not uniquely define the motion of a complex system and hence the choice of an appropriate dynamic model becomes critical. Although several of the models discussed contain a number of adjustable parameters, the parameters have been introduced in a very specific way so that it is not possible to apply any model to any given set of data. The approach which generally has proven most fruitful is the initial selection of a particular model based on the rotational possibilities of the system and inherent physical constraints, as well as on other experimental or theoretical data. For example, it is obviously a gross oversimplification to try to interpret the motion of a long hydrocarbon chain using a two-state model. The success of the chosen model in explaining the data with physically reasonable parameters supports the validity of the interpretation. A further procedural rule is the initial selection of the simplest physically reasonable model available. Interesting conclusions can be derived from the failure of a given model to fit the data adequately, since this indicates that at least one of the assumptions inherent in the model is inapplicable. Thus, Deslauriers and coworkers have shown unequivocally that differences in the relaxation times of various peptide carbons cannot be explained by a model which assumes rigid, anisotropic motion (Section 4.1). This result necessitates the conclusions of significant internal motion in these molecules.

As another example, the model used for multiple unrestricted internal rotations cannot predict marked decreases in NT_1 values along a chain (assuming the overall motion is in the extreme narrowing limit), so that this result is inapplicable, for example, to the case of phytol which exhibits a pronounced nonmonotonic NT_1 gradient (Goodman *et al.*, 1973). Therefore, the T_1 behavior does not reflect a reduced rate of *internal* diffusion about some of the bonds due to the presence of the methyl groups. In this case, the failure of the multiple free internal diffusion model probably reflects the inadequacy of the assumption that one end of the molecule diffuses completely independently.

The use of dynamic models has also been valuable in the interpretation of polymer motion. Models for defect diffusion discussed by Valeur *et al.* (1975) and by Jones and Stockmayer (1977) which assume an infinite chain length lead to nonexponential autocorrelation functions and to corresponding spectral densities which are inconsistent with many observations. The failure of these models implies greater local freedom of motion. Modifications of this approach which limit the dynamic coupling among more distant bonds provide a reasonable fit of the data and can be used to estimate the extent of such long-range dynamic coupling.

A difficult problem to resolve is whether methyl-group motion is described more adequately by internal diffusion [equation (8)], or as a jump process [equation (15)]. An interesting approach to this problem has been taken by Ladner *et al.* (1976) in a study of the spin-lattice relaxation rates of the methyl groups in 7,12-dimethylbenz[a]anthracene (Figure 13). In this molecule, the 7-methyl group experiences a sixfold rotational barrier. The spin-lattice relaxation time is relatively long for this group indicating a large value of ρ [equations (5) and (16)], and the activation energy for internal motion is low (<0.4 kcal/mol) suggesting that the internal motion is reasonably described as a free internal diffusion process. Ladner *et al.* note that a highly sterically hindered group will reorient rapidly if the unfavorable interactions are comparable for all rotameric conformers, as in this sixfold barrier case. In contrast, the spin-lattice relaxation time of the 12-methyl group, which is subject to a high threefold barrier to internal rotation, exhibits a much shorter spin-lattice relaxation time and a higher activation energy (>4.4 kcal/mol). In this case, the data indicate that the internal reorientation of the

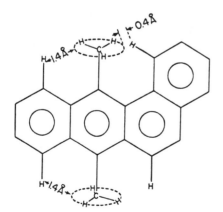

Figure 13. Structure of 7,12-dimethylbenz[a]-anthracene showing the distances of closest approach between the methyl protons and the adjacent ring protons [reprinted with permission from Ladner *et al.* (1976), *J. Phys. Chem.*, **80**, 1783, copyright by the American Chemical Society].

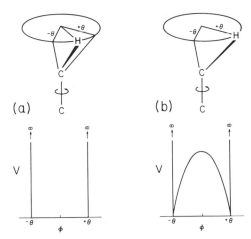

Figure 14. Models for restricted internal motion and the corresponding potential energy surfaces. (a) Diffusion about a C—C bond restricted by a square-well potential limiting the range of motion to $(-\theta, +\theta)$ (as in Figure 4a). (b) Motion about the C—C bond restricted to two allowed conformations with azimuthal angle $+\theta$ or $-\theta$. The corresponding potential-energy surface contains a barrier sufficiently high so that the two-state approximation is valid, but low enough to permit rapid interconversion between the two states.

12-methyl group occurs in distinct jumps. Thus, if the internal diffusion rate for the 7-methyl group is D_i, and the lifetime of the 12-methyl group in one of the allowed rotamers is τ_e, the condition $D_i \gg (\tau_e)^{-1}$ is apparently satisfied. This study illustrates the value of recognition of the physical constraints involved, as well as evaluation of the diffusion rates and activation energies measured, in the choice of an appropriate dynamic model.

As a final example, restricted internal motion occurring within a protein or peptide might be characterized by either of the two potential energy diagrams illustrated in Figure 14. The square-well potential corresponds to a limit on the allowed range of motion, while the humped potential energy surface leads to a bistable system whose motion is characterized as a jump process. In Figure 15, the corresponding spin-lattice relaxation behavior as a function of the half-range θ is plotted for the parameters indicated. It is apparent that the obvious differences provide a possible basis for discriminating between these models. For low-amplitude motions, the bistable potential will lead to much larger increases in the T_1 values. This analysis can be applied, for example, to the relaxation behavior of proline to deduce that in fact an appreciable barrier must exist for the ring puckering. Thus, if the square-well potential model were used, the required range of motion would be physically unacceptable (London and Avitabile, 1978).

In view of the lack of uniqueness of the dynamic models, it is of value to make as many relaxation measurements as are feasible. Frequency-

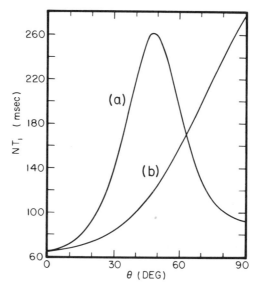

Figure 15. Calculated NT_1 values for the two models illustrated in Figure 14. (*a*) The two-state model with parameters $D_o = 10^7$ s^{-1}, $\beta = 70.53°$, and $\tau_A = \tau_B = 1.67 \times 10^{-12}$ s. (*b*) The restricted diffusion model with the same values of D_o and β, and $D_i = 10^{11}$ s^{-1}. We note that the choice of $\tau_A = \tau_B = \frac{1}{6} D_i$ is arbitrary. However, the parameters have been chosen so that D_i, $(\tau_A)^{-1}$, $(\tau_B)^{-1}$ $>> D_o$ so that the calculated T_1 values are fairly insensitive to the exact values of the internal motion parameters [reprinted with permission from London and Avitabile (1978), *J. Am. Chem. Soc.*, **100**, 7159, copyright by the American Chemical Society].

dependent data are of obvious value, and in a number of interesting cases have shown that the simplest models for motion appear to break down (Wilbur *et al.*, 1976; Jones *et al.*, 1976; Llinas *et al.*, 1977; Visscher and Gurd, 1975; Howarth, 1978). Data corresponding to spectral densities at different frequencies can also be obtained by monitoring the ^{1}H satellites in ^{13}C enriched molecules (London *et al.*, 1977b). $T_{1\rho}$ measurements have also been applied recently to the study of protein dynamics (James *et al.*, 1978; Bleich and Glasel, 1978). Measurements of relaxation rates as a function of temperature can provide activation energies which are of value, as in the illustration of methyl rotation above.

An additional means of characterizing molecular motion is available if several interactions leading to nuclear relaxation are significant. In general, these interactions are not independent, the relationships being described by cross-correlation functions. These functions depend on the dynamics of the systems, much as the autocorrelation functions discussed above, and can be calculated assuming different dynamic models. A complete theory for nuclear relaxation, as given for the case of ^{13}C by Werbelow and Grant (1975a, b; 1977), includes both the cross- and autocorrelation functions. Inclusion of cross-correlation effects leads to a prediction of nonexponential spin-lattice relaxation, which is usually difficult to observe. Deviations are generally largest for highly anisotropic motion, and have been observed, for example, for the spin-lattice relaxation of methionine methyl groups in tetragastrin (Cutnell and Glasel, 1976a, b). In general, sensitivity problems associated with studies of proteins and peptides limit the applicability of this type of measurement, and the initial part of the recovery curve measured in a 180°–τ–90° experiment is equal to that calculated neglecting cross-correlation effects. More recently, it has been noted that measurements of ^{13}C linewidths in ^{1}H coupled spectra provide a convenient means for evaluation of certain cross-correlation functions (Prestegard and Grant, 1978). Application to the methylene resonances of carboxymethylated human carbonic anhydrase B has provided information on the internal motion of the carboxymethyl group [Uiterkamp et al., 1978 (Section 4.5)]. It will be particularly interesting to determine whether such measurements can supply the critical information necessary to distinguish between alternative but similar dynamic models.

3.8. Effective Correlation Times and Statistical Distributions

Up to this point, the approaches outlined in this section have been based on the construction of specific physical models for anisotropic or internal motion and the calculation of relaxation parameters from these models. As noted in Section 3.7, in many cases of interest various models will fail to provide an adequate fit of the experimental data, indicating that the model does not provide a reasonable physical picture of the dynamic processes involved. Several alternative approaches to the interpretation of relaxation data are most useful when the complexity of the molecule or the variety of possible dynamic processes suggests that application of any of the available models will constitute a drastic oversimplification of the problem. Two approaches which are frequently used in such situations are descriptions in terms of effective correlation times (Lyerla et al., 1974) and statistical distributions of correlation times (Noack, 1971). Both approaches lead to spectral density terms of the form of equation (2). In the case of a statistical distribution, however, the coefficients of these terms are chosen to optimize the fit of the data and do not include physical constraints such as β or θ which characterize the motion. These approaches have been particularly useful in the interpretation of data for nonglobular biopolymers such as elastin (Lyerla and Torchia, 1975), collagen (Torchia and VanderHart, 1976), and simple polypeptides (Saito et al., 1978).

A simple, qualitative description of the relative motion of a complex system is obtained by deducing effective correlation times for each nucleus, τ_{eff}^i, based on T_1 data and using the isotropic motion spectral density relations. Such effective correlation times actually contain amplitude and isomer probability factors (e.g., London and Avitabile, 1977b) so that they are not in general simply related to rates of motion. However, a qualitative description of the relative mobilities of the nuclei, that is, $1/\tau_{eff}^{i}$, can be of great value in providing a rough dynamic picture of a complex system. Since the isotropic T_1 relation is a double-value function of τ, NOE data are often used to determine which side of the T_1 minimum is applicable (Allerhand and Oldfield, 1973; Fossel *et al.*, 1974). Using the correlation time corresponding to the most immobilized nucleus, τ_o, an effective correlation time for internal motion of the *i*th nucleus τ_{int}^{i}, can be calculated (Lyerla *et al.*, 1974):

$$\frac{1}{\tau_{int}^{i}} = \frac{1}{\tau_{eff}^{i}} - \frac{1}{\tau_o} \tag{25}$$

One difficulty with this approach is that frequently no nuclei are completely immobilized in the macromolecule. A similar problem is encountered in the application of the multiple internal diffusion models discussed in Section 3.6. In addition, the approach tends to overemphasize overall motion (τ_o) in the evaluation of internal motion correlation times.

Statistical distributions are of particular value in the interpretation of the dynamics of polymers (Connor, 1964) and small molecules which are part of aggregated systems (Levy *et al.*, 1978). In the case of polymers, the dynamic range of the motion is large since the motions which dominate T_1 and T_2 frequently differ by several orders of magnitude. This, in turn, requires a very broad distribution of correlation times and has led to the use of logarithmic distributions such as the log normal (Connor, 1964) and the log χ^2 (Schaefer, 1973) distributions. These distributions are defined by a mean correlation time, $\bar{\tau}$, and a width parameter for the distribution. The simplicity of a two-parameter fit leads to an overdetermined problem if T_1, T_2, and NOE values can all be measured. However, this simplicity is based on the assumptions inherent in the choice of the distribution function. For example, if the actual motion of the polymer is characterized by several modes of rapid internal motion as well as a slow rate of overall tumbling, the true distribution of correlation times might well be bimodal, in contrast with the assumption of the log χ^2 model. In particular, the two main features of polymer relaxation: submaximal ^{13}C NOE values and $T_1 \gg T_2$, also characterize a rotating methyl group of a globular protein in which two correlation times, τ_o and τ_i, clearly provide the most valid physical description. It must be noted, however, that two correlation times typically are inadequate to describe polymer motion (Schaefer and Natusch, 1972).

4. ILLUSTRATIVE APPLICATIONS

The dynamic models discussed in Section 3, as well as the formalism for chemical exchange processes, have been used to interpret the dynamic behavior of a wide variety of proteins and peptides. Based on the data presently available, it appears that certain types of motion are a general characteristic of particular amino acid side chains, but that the rates and the ranges of motion are in general strongly dependent on interresidue interactions. In this section, representative experimental NMR data for proteins and peptides are reviewed in light of the available formalism for treating dynamic processes.

4.1. The Polypeptide Backbone

Dynamic analysis of the peptide backbone can be extremely informative, but also extremely complex since overall, possibly anisotropic motion as well as internal, "segmental" or librational motion must be considered. The relative importance of these dynamic processes in determining the observed relaxation parameters varies widely among the systems which have been studied. Results are most conveniently interpreted by separately considering three classes of molecules: globular proteins, nonglobular proteins and polypeptides, and low-molecular-weight peptides ($\lesssim 20$ residues).

Globular proteins provide the simplest systems for analysis. Internal motion of the peptide backbone which is slow compared with the rate of overall molecular tumbling, such as that which mediates slow H–D exchange of amide protons, will not affect the T_1 or NOE values. Rapid, low-amplitude librational motion is also not expected to have a large effect on the α-carbon relaxation rates (Section 3.4). This is consistent with data of Visscher and Gurd (1975) and Bauer et al. (1975), who find good agreement between rotational correlation times calculated on the basis of α carbon spin-lattice relaxation times and values based on depolarized light scattering experiments. Similar conclusions have been reached by Hull and Sykes (1975b). The above conclusion differs somewhat from that of Howarth (1978) and DeWitt et al. (1978) who suggest that librational motions with low amplitudes will lead to larger changes in the T_1 values. These analyses were completed using earlier theories of Woessner (1962a) and van Putte (1970), rather than the more recently developed formalism of Section 3.4. The dependence of the van Putte model on the range differs significantly from that illustrated in Figures 5 and 6; T_1 increases dramatically as the range parameter (labeled β_0 in this case) decreases toward $0°$. This behavior reflects the absence of a contribution due to overall molecular tumbling which is significant for proteins and which becomes dominant as the range of internal motion approaches zero. Nevertheless, low-amplitude librational motion may be significant for other nuclei such as ^1H if both the orientation and the length of the internuclear vector change as a result of the motion.

Rotational correlation times for globular proteins calculated on the basis of the assumption of isotropic motion are generally in fairly good agreement with values calculated using a Stokes–Einstein equation:

$$\tau = \frac{4}{3}\pi r^3 \frac{\eta}{kT} \sim \left(\frac{\text{MW}}{N_a\,\rho}\right)\frac{\eta}{kT} \tag{26}$$

where η is the viscosity of the medium, r is the molecular radius, MW is the molecular weight, N_a is Avogadro's number, ρ is the protein density (typically ~ 1.4 g/cm^3), k is Boltzmann's constant, and T is the absolute temperature. The expected proportionality of rotational correlation time to molecular weight has recently been nicely demonstrated by James *et al.* (1978), although the agreement with the Stokes–Einstein relation is poorer than indicated in Figure 6 of this reference due to a calculational error leading to τ values $\frac{4}{3}\pi$ too large.

Relatively little work has been done in characterizing the anisotropic motion of globular proteins. Wilbur *et al.* (1976) studied the spin lattice relaxation of the α carbons of four proteins: myoglobin, hemoglobin, hen egg-white lysozyme, and bovine serum albumin, at two different NMR frequencies. For the more spherical proteins hemoglobin and myoglobin, calculated rotational correlation times at the two frequencies were found to be in good agreement, while for the relatively nonspherical proteins hen egg-white lysozyme and bovine serum albumin, the agreement was poor. An attempt to fit the α carbon relaxation in the latter two cases using the axially symmetric ellipsoid model (Section 3.2) for motional anisotropy did not significantly improve the agreement. These data are difficult to interpret due to the many α carbons corresponding to various values of β, which contribute to the envelope of resonances observed. More recent calculations suggest that the discrepancies between the correlation times obtained for lysozyme at the two frequencies can be resolved by using a longer C—H bond length, rather than by invoking motional anisotropy (Dill and Allerhand, 1979).

Nonglobular proteins exhibit dynamic characteristics similar to polymers, and the extensive literature in this area will not be reviewed here. Differences relative to globular proteins reflect two characteristics: (1) the absence of many interresidue contacts may be consistent with relatively large-amplitude segmental motion which can contribute significantly to the T_1 and NOE values; (2) the extreme structural anisotropy is generally accompanied by extreme dynamic anisotropy. Motion perpendicular to the long axis is relatively slow and may not contribute significantly to T_1, but dominate the observed linewidth. In extreme cases such as collagen fibrils studied by Torchia and VanderHart (1976), motion perpendicular to the helical axis is sufficiently slow so that static dipolar contributions to the linewidth are significant, and techniques for obtaining high resolution in the solid state, such as high-power proton decoupling, have been used. A few representative studies are considered below.

Effects of segmental motion have been observed in poly(γ-benzyl L-glutamate) (Allerhand and Oldfield, 1973). As the proportions of chloroform and trifluoroacetic acid in the solvent are varied, the protein undergoes a helix–random coil transition. Analysis of ^{13}C T_1 and NOE data indicates a 30-fold increase in the effective correlation time for C$_\alpha$ in the helical conformation, attributable to a loss of segmental motion.

Studies of the polytetrapeptide of tropoelastin, (Val—Pro—Gly—Gly)$_n$, indicate an abrupt increase in the effective correlation time deduced from ^{13}C T_1 measurements as the temperature is increased above 60°C (Urry et al., 1978). This inverse temperature transition is attributed to a loss of segmental motion which is correlated with Val$_1$ NH—GlyCO$_4$ hydrogen-bond formation and a hydrophobic Val$_1$ C$_\gamma$—Pro$_2$ C$_\delta$ interaction. The sensitivity of the relaxation measurements to these structural transitions underlines the value of the approach even when the complexity of the dynamics limits the analysis to consideration of effective correlation times.

An interesting analog of the helix–coil transition noted above is observed for the α1-CB2 collagen peptide (Torchia et al., 1975). In this case, a triple-stranded helix is formed at low temperature with dimensions approximating a symmetric ellipsoid: 11.5–14.5 Å × 110–120 Å, (MW ~10,000). In the triple helix, linewidth, T_1, and NOE data for the glycine C–H vectors have been interpreted using values of β deduced from the crystal structure of a related peptide, poly(Gly—Pro—Pro) which forms a triple helix in the solid state. Agreement of calculated parameters with experiment was improved slightly using the dimensions of a hydrated polymer, although uncertainties in the measured T_1 values preclude a definite conclusion. In the random-coil conformation obtained at 30°C, relaxation parameters reflect considerable backbone mobility. In reconstituted collagen fibrils (Torchia and VanderHart, 1976), the helical form prevails, and spin-lattice relaxation is dominated by motion about the helical axis. Azimuthal reorientation in the range 10^6–10^8 s^{-1} and a jump amplitude, estimated using a two-state model, of $\geq$ 30° were determined. These results lead to the remarkable conclusion that intermolecular forces in the reconstituted collagen fibrils do not depend strongly on the azimuthal angle and, hence, do not arise from a unique set of interactions between the side chains.

The relative importance of overall and segmental motion in small peptides is of considerable interest from a conformational standpoint. If the latter is important, relaxation measurements can be used to probe secondary structure such as β and γ turns which have frequently been postulated for these systems. Support for the importance of segmental motion in determining the observed T_1 and NOE values comes from a study of gramicidin A (Fossel et al., 1974). Measurements of gramicidin A as a monomer (in dimethyl sulfoxide) and as a dimer (in methanol) indicate a sixfold increase in the effective rotational correlation time of the dimer relative to the monomer. This increase is considerably greater than that expected on the basis of the increased molecular volume (equation 26), and is interpreted to reflect an increased rigidity of the peptide backbone of the dimer. Thus, in the monomer, internal motion of the peptide backbone contributes significantly to the observed relaxation of the α carbons. [It is interesting to note that in both this experiment and the poly(γ-benzyl L-glutamate) study noted above, changes in the effective correlation times are deduced largely from the change in the NOE values, with the ^{13}C T_1 values changing only slightly due to movement from one side of the T_1 minimum to the other.]

   In contrast to the gramicidin A study, evidence for internal motion of the peptide backbone of short peptides on a time scale faster than overall tumbling is equivocal in most of the reported studies. Two observations tend to suggest that, in general, internal motion must be slow compared with overall molecular tumbling. (1) Rotational correlation times deduced from α-carbon NT_1 values for a number of peptide hormones including angiotensin II (Deslauriers *et al.*, 1975; 1977b), methionine enkephalin (Tancréde *et al.*, 1978), and luteinizing hormone–releasing hormone (LH–RH) (Deslauriers and Somorjai, 1976), are in close agreement with values predicted using the Stokes–Einstein equation (26). (2) NT_1 values for the α carbons (excluding terminal residues) are generally very similar. The latter observation can be interpreted using the multiple uncorrelated internal diffusion model discussed in Section 3.6 (Deslauriers and Somorjai, 1976). Assuming the center of mass to be located near the middle of the peptide chain, this model predicts a significant NT_1 gradient if internal motion is more rapid than overall tumbling (Figure 11). Since this gradient is not observed, internal motion is concluded to be slow.

   Neither of the above observations is conclusive, however. The Stokes–Einstein calculation involves several rather important assumptions which are discussed in the references noted above as well as by Deslauriers and Smith (1976). The second observation is dependent on the applicability of the multiple diffusion model. At this point, a distinction between "internal" and "segmental" motion may be of value. Models for internal motion assume that at least one atom in the molecule moves independently of the others. This approximation is probably not unreasonable for the short side chains anchored to the polypeptide backbone of a globular protein, and also gives reasonable values for internal diffusion rates of peptide side chains. Alternatively, nonglobular proteins or peptides may undergo relative motion in which no particular atom is moving independently. This problem has not yet been dealt with in a satisfactory way. A qualitative approach might be based on a set of calculations in which *each* of the atoms in the molecule is chosen to diffuse independently and the remaining atoms diffuse internally using the multiple internal diffusion model. The calculated relaxation parameters would then reflect a weighted average of each of these calculations, although it is quantitatively unclear how to determine the weights used. On the basis of such a model, a significant degree of segmental motion may not be inconsistent with equality of the different α-carbon NT_1 values. From a practical standpoint, the complex motion of the peptide backbone in nonglobular systems can best be treated using effective correlation times or distributions of correlation times (Section 3.8) or restricted amplitude internal motion models (Section 3.4).

   Internal motion of the peptide backbone is of general importance in two cases. (1) Relaxation data for the terminal α carbons and in some cases the α carbons of the penultimate residues as well, indicate a significant degree of internal motion. This result is not surprising, since these residues are functionally equivalent to short side chains attached to the peptide backbone. In this context, it may be noted that the increased NT_1 values for the α carbons of terminal residues are more pronounced for lower molecular weight residues. For example, in LH–RH, the Gly C_α NT_1 is 2.4 times that of the shortest value (corresponding to Trp^3 C_α), while for the terminal pyroglutamyl residue the ratio is only 1.8. (2) Nonterminal glycine residues appear to exhibit increased flexibility based on NT_1 measurements (Deslauriers *et al.*, 1977a; Combrisson *et al.*, 1976; Tancréde *et al.*, 1978). In view of the above discussion on segmental motion, it is interesting to note that increased NT_1 values for nonterminal glycine residues can lead to nonmonotonic NT_1 gradients along the peptide backbone. This pattern is inconsistent with a multiple, unrestricted internal diffusion model, but not with a segmental motion analysis as indicated above.

Another approach toward relating relaxation measurements to peptide conformation has been explored extensively by Deslauriers and coworkers. Particular molecular conformations will presumably exhibit differing motional anisotropy. Application of the model of Section 3.2 for an asymmetric ellipsoid to the ^{13}C NT_1 data for the backbone carbons of a peptide might therefore reveal the predominant molecular conformation. Studies of amino acids and diketopiperazines (Somorjai and Deslauriers, 1976; Deslauriers and Smith, 1977) indicate that in some cases only, data corresponding to carbons not expected to undergo significant internal motion can be fitted using the model of Section 3.2. However, in several of these cases, the degree of motional anisotropy required to achieve a satisfactory fit of the data is physically unreasonable. Optimal fits of the data for cyclic peptides do indicate, however, that overall molecular shape is the deciding factor in determining the anisotropy of the motion.

Analysis of the relaxation behavior for the linear peptide hormones angiotensin II (Deslauriers *et al.*, 1975; Deslauriers *et al.*, 1977b), melanostatin (Deslauriers *et al.*, 1977c) and methionine-enkephalin (Tancréde *et al.*, 1978) indicates that an anisotropic, rigid-body model of overall tumbling cannot explain the relaxation results. The interpretation of ^{13}C spin lattice relaxation data for peptide hormones does not require a detailed knowledge of molecular conformation and, conversely, cannot decide between different proposed overall molecular conformations. The correlation times calculated for internal motion are, in general, not very sensitive to overall anisotropic motion. In contrast, it is concluded that the sensitivity of ^{13}C T_1 values to internal motion indicates that side-chain–side-chain interactions which influence this motion can profitably be studied by ^{13}C NMR.

The use of motional anisotropy deduced from relaxation measurements is further limited by the tendency of many peptides to aggregate. Intermolecular association for angiotensin (Ferretti and Marshall, 1978), gramicidin A (Fossel *et al.*, 1974), and the enkephalins (Khaled *et al.*, 1977) have all been reported. In addition to altering the effective molecular dimensions, such association may also affect the dynamic properties by altering the molecular conformation. For example, in acetyl proline N-methylamide studied in chloroform, both the *cis/trans* ratio (Higashijima *et al.*, 1977) and the γ-turn probability (London, 1979) are dependent on concentration.

Despite the difficulties associated with elucidating overall molecular conformation from ^{13}C relaxation data, these data have provided a clear picture of the flexibility of these peptides. As noted in Section 2, conformational information has been successfully provided by ^1H T_1 and NOE measurements.

4.2. Aliphatic Residues

4.2.1. Methionine. The ease of labeling the methionine methyl group has resulted in a relatively large amount of data for this amino acid, so we begin this section with a detailed analysis of the dynamic characteristics of methionine residues. All relaxation data thus far obtained for methionine-containing peptides, for example, Gly—Gly—Met—Gly—Gly (Keim *et al.* 1974); Met-enkephalin (Bleich *et al.*, 1976a, Combrisson *et al.*, 1976, Tancréde *et al.*, 1978) and tetragastin (Bleich *et al.*, 1976b); indicate a monotonic increase in NT_1 for methionine carbons further from the peptide backbone. These data can be explained by a model assuming multiple unrestricted internal rotations about the various C—C and C—S bonds (Section 3.6). In the absence of additional data, more detailed conclusions are presently unwarranted. This behavior is typical for the aliphatic residues (Keim *et al.*, 1973a, b; 1974; Deslauriers and Smith, 1976).

The ^{13}C relaxation rates of [methyl-^{13}C] methionine-labeled sperm whale myoglobin (Jones *et al.*, 1976), myelin basic protein (Deber *et al.*, 1978), and dihydrofolate reductase (Blakley *et al.*, 1978; London and Avitabile, 1978) have been determined. In all cases, comparison with relaxation rates corresponding to immobilized carbons indicates a significant degree of internal motion. Data for sperm whale myoglobin was fitted assuming free internal diffusion about the S—CH_3 bond and overall isotropic diffusion (Section 3.3). This model gives good agreement for the NT_1 and NOE values (Jones *et al.*, 1976), but fails to quantitatively predict the dependence of T_1 on frequency. Further, the studies of Wilbur *et al.* (1976) discussed in Section 4.1 indicate that the discrepancies cannot be explained on the basis of overall anisotropic motion of the protein.

In the case of dihydrofolate reductase, resonances corresponding to all seven of the methionine residues are not resolved at 100 MHz, although additional resonances can be resolved in some of the binary and ternary complexes studied by Blakley *et al.* (1978). An overall, isotropic rotational correlation time of 2×10^{-8} s was determined for the *S. faecium* enzyme from data obtained for the [guanido-^{13}C] arginine-labeled enzyme (M.W. = 20,000) (Cocco *et al.*, 1978). Using this value, the free internal diffusion model (Section 3.3) did not produce an adequate fit of the data. In particular, the calculated linewidths are significantly broader than those observed for any of the resolved or partially resolved peaks, and the calculated NT_1 values were too short in some cases, regardless of the internal diffusion coefficient used in the calculation. Some improvement can be obtained by assuming that β deviates significantly from the tetrahedral angle, as has been suggested recently (Blunt and Stothers, 1977). In the present case, the dispersion in T_1 values, linewidths, and NOE values corresponding to the various methionine methyl resonances would require an unlikely dispersion in β values. A more probable explanation for the discrepancy is additional internal motion, most likely about the CH_2—S bond. A model assuming free, uncorrelated diffusion about the CH_2—S and S—CH_3 bonds appears to be adequate to describe the relaxation parameters of the sharpest resonances only (Blakley et al., 1978). An interesting characteristic of the various methionine methyl resonances is the parallel behavior of NT_1, NOE, and $1/\nu$ (where ν is the linewidth). Thus, the peaks with the longest T_1 values also exhibit the largest NOE values and the narrowest lines. This trend cannot be reproduced by varying the diffusion coefficients within the context of a model assuming either one or two unrestricted internal diffusion processes. Thus, more rapid internal motion generally increases NT_1, but decreases the theoretical NOE values (Doddrell et al., 1972; London and Avitabile, 1976). In contrast, a model assuming variation in the amplitude of motion does predict the trend noted above. Thus, greater amplitude of internal motion is associatied with longer T_1 and T_2 values and a larger NOE (Section 3.4; London and Avitabile, 1978). This suggests that differences in relaxation parameters among the various methionine residues are related to variations in the amplitude rather than the rates of internal motion. Since the predicted T_1 values assuming one internal diffusion process were in general too short, and the values assuming two internal diffusion processes were in general too long, a reasonable model is based on the assumption of free methyl rotation, that is, motion about the S—CH_3 bond, and restricted amplitude diffusion about the CH_2—S bond (Figure 16). This is a physically reasonable model since methyl rotation requires no additional volume and should therefore not be restricted by interresidue interactions. In contrast, diffusion about the CH_2—S bond does require a considerable increase in volume over a nonrotating residue (Figure 16). In dihydrofolate reductase, it was found that the amplitude of rapid internal diffusion ranged from ~90–180° (London and Avitabile, 1978). It should be emphasized that complete 360° rotations about the CH_2—S bonds are possible on a time scale long compared with the overall tumbling rate of the protein ($\tau > 2 \times 10^{-8}$ s). The particular ranges deduced for internal motion will, of course, be dependent on the particular model applied to interpret the data, the important conclusion being that variations in the amplitude of motion, rather than the rates, among the various methionine residues are required to explain the data. This dynamic model also has been successfully applied to *iso*leucine relaxation data of myoglobin (Wittebort et al., 1979).

A question of practical importance concerns the relationship of internal

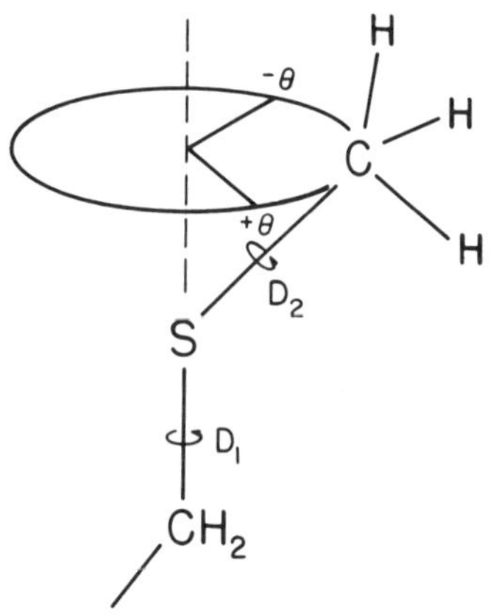

Figure 16. Model used to describe the methionine methyl relaxation data of [methyl-^{13}C]methionine labeled dihydrofolate reductase from *S. Faecium*. Motion about the CH_2—S bond is restricted by boundary conditions as in Figure 4a, but diffusion of the methyl group about the S—CH_3 bond is unrestricted.

mobility to chemical shift inequivalence resulting from protein folding. Resolution of single carbon resonances requires significant shift contributions resulting from perturbations produced by nearby residues; but, to some extent, close interresidue contacts are incompatible with internal motion of the residues. This point is illustrated by the observation that protein denaturation abolishes most or all of the chemical shift inequivalence for a given class of carbons (Figure 17). Similarly, protein denaturation leads to increases in the mobilities of the residues as indicated, for example, by sharpening of the resonances and increases of the spin lattice relaxation times (Allerhand *et al.*, 1971a; Glushko *et al.*, 1972; Allerhand, 1978). Internal rotation of methyl groups does not require additional volume and is therefore compatible with close packing of the protein residues. In contrast, rotation about the CH_2—S bond is less consistent with the packing constraints. It is therefore not surprising that the restricted amplitude diffusion about this bond deduced from the relaxation data is consistent with the chemical shift dispersion of the methionine residues in the native enzyme (Figure 17c). Further, it can be noted that resonances for which the amplitude of internal motion about the CH_2—S bond is calculated to be $\geq 180°$ exhibit shifts close to the value for the denatured enzyme (Blakley *et al.*, 1978). In general, the data of Blakley *et al.* suggest a rough correlation between the extent of residue immobilization, as indicated by the amplitude restriction, and the chemical shift relative to the denatured enzyme resonance. For example, the resonances with the shortest NT_1 values tended to be furthest upfield in the binary enzyme–methotrexate complex (Table 2 in Blakley *et al.*, 1978). Despite this correlation, the need for care in associating shifts with motional restriction is obvious since particular interactions may, for example, greatly restrict the motion while producing a minimal chemical shift.

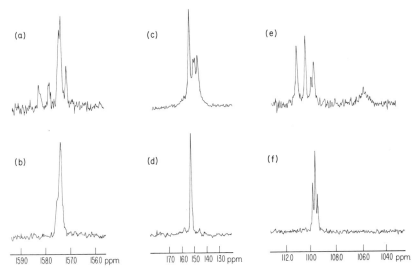

Figure 17. Proton decoupled ^{13}C spectra for the native and urea denatured forms of *S. faecium* dihydrofolate reductase. (a) Guanido carbons of the eight arginine residues in the native enzyme. (*b*) arginine guanido carbons of the denatured enzyme (chemical shift scale has been modified due to isotope effects resulting from differences in the D_2O/H_2O ratio). (*c*) Methyl carbons of the seven methionine residues in the native enzyme. (*d*) Methionine methyl carbons of the denatured enzyme. (*e*) Quaternary C_γ carbons of the four tryptophan residues of the native enzyme. (f) Tryptophan quaternary carbons of the denatured enzyme. Data for *a* and *b* from Cocco *et al.*, 1978; data for *c* and *d* from Blakley *et al.*, 1978; data for *e* and *f* from Groff *et al.*, unpublished.

4.2.2. Arginine. Relaxation studies in arginine-containing peptides, for example, Gly—Gly—Arg—Gly—Gly (Keim *et al.*, 1974), bradykinin (London *et al.*, 1978), angiotensin derivatives (Deslauriers *et al.*, 1975) and luteinizing hormone-releasing hormone (LH–RH) (Deslauriers and Somorjai, 1976), all indicate a monotonic increase in the ^{13}C NT_1 values proceeding away from the peptide backbone. Similarly, the data can be fit by a model assuming multiple unrestricted internal rotations about the side-chain C—C bonds. Using this model, correlation times of $\sim 10^{-10}$ s for internal diffusion were determined in LH–RH and bradykinin.

The guanido carbons are most suitable for observation in proteins due to the absence of directly bonded protons. Relaxation rates for the Arg guanido carbons in cyanoferricytochrome *c* and hen egg-white lysozyme (Oldfield *et al.*, 1975), and in *S. faecium* dihydrofolate reductase (Cocco *et al.*, 1978) have all indicated significant differences in the relaxation rates obtained in H_2O and D_2O. This effect is probably dominated by the five exchangeable guanido protons. A substantial range in the T_1 values corresponding to the various arginine residues of hen egg-white lysozyme was found, and interpreted to reflect a significant degree of internal, segmental motion, analogous to that noted above for the peptides. This result may reflect the fact that many of the charged arginine residues lie on the surface of the protein and are therefore subject to fewer motional constraints than internal residues.

Relaxation studies of the arginine residues of *S. faecium* dihydrofolate reductase by Cocco *et al.* (1978) suggest a separation of the residues into two classes. Residues of the first class, which exhibit chemical shifts close to the value for the denatured enzyme, exhibit T_1 values which increase with increasing temperature. This result indicates motion in the extreme narrowing limit ($\omega\tau < 1$), so that these residues are probably undergoing significant segmental motion analogous to that observed for the arginine residues in lysozyme. This conclusion is also consistent with the chemical shifts which would be expected for solvent exposed residues located on the surface of the protein. In contrast to the lysozyme results, the NOE values for these residues are small (~ 1.2), suggesting that, in this case, the internal motion is still fairly restricted. In contrast, a second class of residues, which were shifted significantly from the position of the denatured enzyme, exhibited decreasing T_1 values with increasing temperature and small NOE values. This indicates motion in the slow tumbling range, which would characterize the entire macromolecule. Ascribing the differences in the T_1 values for these resonances measured in D_2O compared with H_2O to the five exchangeable guanido protons, a rotational correlation time of 2×10^{-8} s was calculated for the protein (M.W. $= 20,000$). As in the methionine studies discussed above, a reasonable assumption is that the same interresidue interactions which lead to the chemical shift inequivalence are responsible for the motional restriction. Two of the three residues corresponding to class 2 are shifted downfield relative to the denatured enzyme, or equivalently the class 1 residues. Since deprotonation of the guanido group at high pH (pK ~ 12.5) produces a downfield shift (Walker and London, unpublished observations), the interresidue interactions may involve a salt linkage/hydrogen bond in which the guanido group donates a proton.

It is apparent from the above discussion that two aliphatic residues with very different functional groups exhibit closely analogous dynamic behavior in both peptides and proteins. Summarizing the above conclusions: (1) monotonic NT_1 gradients are observed for both residues in peptides suggesting multiple internal rotations about all bonds; (2) spin lattice relaxation times for both types of residues show a considerable variation in the proteins studied; (3) the residues exhibiting the longest T_1 values and inferred to have the greatest internal mobility also exhibit shifts closest to the values for the denatured enzyme; (4) conversely, residues whose motion is more restricted as indicated by the relaxation parameters are more likely to be shifted significantly from the position of the denatured enzyme; (5) internal motion which does not require much additional volume, such as methyl group rotation in methionine, occurs over a time scale short compared with overall molecular tumbling; (6) internal motion requiring greater additional volume is more likely to occur over a shower time scale and/or over a faster time scale with restricted amplitude.

4.3. Aromatic Residues

4.3.1. Phenylalanine and Tyrosine. In contrast to the aliphatic residues considered above, motion of the bulkier aromatic side chains tends to be more hindered. A survey of the values for the ratio $NT_1(C_\alpha)/NT_1(C_\beta)$, which is a qualitative reflection of the internal diffusion coefficient about the C_α—C_β bond, indicates that the internal motion about this bond is frequently, but not always slower for the aromatic residues. Internal motion about the C_β—C_γ bond has proved to be of particular interest. In phenylalanine residues, the C-4—H bond of the benzyl ring is parallel to the C_β—C_γ bond so that internal diffusion about the latter does not affect the relaxation parameters of C-4. Thus, in general the ratio $NT_1(C_\gamma)/NT_1(C-4) = 1$ for phenylalanine residues. This relationship is reasonably well satisfied in most, but not all peptides studied. Values of 0.96 (at pH 4.1) and 1.1 (at pH 1.1) in [Pro[3], Pro[5]]-angiotensin (Deslauriers *et al.*, 1976a), 1.1 in gramicidin S (Allerhand and Komoroski, 1973), 1.2 in tetragastrin (Bleich *et al.*, 1976b), but 1.4 in [Ile[5]]-angiotensin (Deslauriers *et al.*, 1975), and 0.83 in Met-enkephalin in dimethyl sulfoxide (Bleich *et al.*, 1976a) have been obtained. These discrepancies may reflect contributions of nonbonded protons to the relaxation rates, anisotropic motion effects, or experimental error. If the internal motion about the C_β—C_γ bond is rapid relative to overall molecular tumbling, the NT_1 values for the *ortho* and *meta* carbons will be greater than for C_γ and C-4, assuming overall motion in the extreme narrowing limit. A particularly interesting characteristic of the peptides which have been studied is the significant difference in the diffusion about the C_β—C_γ bond between phenylalanine and tyrosine residues. Rapid internal motion is observed for Phe[4] but not Tyr[1] of Met-enkephalin (Bleich *et al.*, 1976a; Combrisson *et al.*, 1976; Tancréde *et al.*, 1978) and for Phe[8] but not Tyr[4] of [Ile[5]]-angiotensin (Deslauriers *et al.*, 1975). Rapid internal motion about the C_β—C_γ bond of phenylalanine residues is also observed in bradykinin (London *et al.*, 1978), gramicidin S (Allerhand and Komoroski, 1973), and tetragastrin (Bleich *et al.*, 1976b). Slow internal motion (relative to overall molecular tumbling) is observed for the tyrosine residues of LH–RH (Deslauriers and Somorjai, 1976), oxytocin (Walter *et al.*, 1974), and [Pro[3], Gly[4]]-oxytocin (Deslauriers *et al.*, 1978). These results suggest that the tyrosine hydroxyl group locks the aromatic side chain into a hydrogen-bonded solvent structure, significantly reducing the rate of internal diffusion about C_β—C_γ.

Studies of the ^{13}C spin-lattice relaxation rates of aromatic side chains in globular proteins have indicated that any internal motion is slow relative to overall molecular tumbling (Visscher and Gurd, 1975; Glushko *et al.*, 1972;

Oldfield *et al.*, 1975; Hunkapiller *et al.*, 1973; Sternlicht *et al.*, 1973; Browne *et al.*, 1973; London *et al.*, 1975; Wüthrich and Baumann, 1976). A more recent deduction of considerable interest is the conclusion of Campbell *et al.* (1975c) that there is considerable motional freedom for these residues at rates frequently exceeding 10^4 s^{-1}. This conclusion is based on the observation that the *ortho* and *meta* protons (and in some cases ^{13}C resonances as well) in most of the proteins studied give rise to only a single resonance each, although if these residues were held rigidly in the protein, the environments of the four protons (or carbons) should in general be different resulting in four separate observable peaks per phenylalanine or tyrosine residue. Campbell *et al.* further showed that the addition of lanthanide ions to hen egg-white lysozyme did not remove this degeneracy. Thus, the two *ortho* (or *meta*) protons still behave as if they have the same orientation relative to the bound lanthanide ion, although substantial differences are predicted on the basis of the crystal structure. It was therefore postulated that internal motion about the C_β–C_γ bond on a time scale rapid compared with estimated chemical shift differences in hertz is required to produce the observed shift equivalence. It was further noted that close interresidue contacts make free internal diffusion about this bond unlikely, so that the averaging process more probably reflects 180° flips about the C_β—C_γ bond:

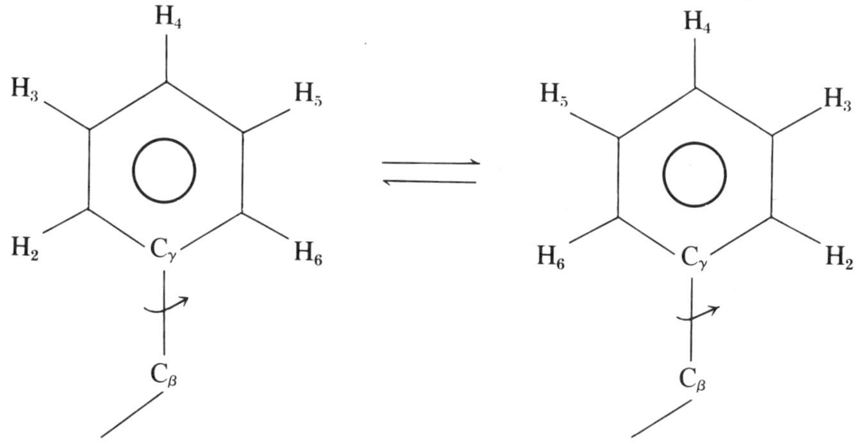

Theoretical calculations of Gelin and Karplus (1975) indicate barriers to this motion frequently exceeding 100 kcal/mol based on the crystal structure of bovine pancreatic trypsin inhibitor (BPTI). It must therefore be concluded that internal 180° flips at the required rate to produce chemical shift equivalence reflect a dramatic lowering of the rotational energy barrier due to relaxation of the protein structure around the partially rotated residue. Calculations assuming this effect yield reasonable energy barriers for this process. Since the ^{13}C spin lattice relaxation rates of most aromatic carbons are close to the rates for the peptide backbone α carbons, an upper limit for the rate of these 180° flips is 10^7 s^{-1} for low-molecular-weight proteins, M.W. ~20,000. In addition to lysozyme, evidence for rapid 180° flips of phenylalanine and tyrosine residues has been observed for BPTI (Wüthrich and Wagner, 1975), parvalbumin (Cave *et al.*, 1976), dihydrofolate reductase (Feeney *et al.*, 1977; Kimber *et al.*, 1977), and alkaline phosphatase (Hull and Sykes, 1975b). Although in the majority of cases the flipping is sufficiently rapid to eliminate chemical shift inequivalence, this is not always the case. Observation of tyrosine resonances in ferrocytochrome c over a temperature range of nearly 100°C indicates a gradual transition from a slow exchange to a rapid exchange limit with the rate of internal motion varying by about 4 orders of magnitude (Figure 18). The thermodynamic parameters deduced from this study are $\Delta H = 23$ kcal/mol and $\Delta S = 2.29 \times 10^{-2}$ kcal/°K at 25°C (Campbell *et al.*, 1976). An interesting feature of the data is the linearity of the Arrhenius plot for the flip rate as a function of $10^3/T$ (Figure 18), which indicates a single activation energy over the entire temperature range. Since the activation energy is a reflection of the interresidue interac-

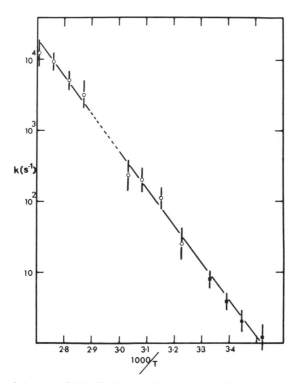

Figure 18. Plot of the rate of 180° flipping, k, of a tyrosine residue in ferrocytochrome c as a function of $1000/T$ with k values corresponding to the open circles based on spectral simulation and k values corresponding to the closed squares deduced from cross saturation experiments (from Campbell *et al.*, 1976).

tions, the results indicate a high degree of structural integrity over a wide temperature range. Campbell *et al.* further note a similar temperature dependence for the tyrosine flipping motion in ferrocytochrome *c* from other sources. Thus, the internal mobility of this residue is conserved despite many differences in the primary structures of the proteins.

Studies of 5-fluorotyrosine labeled alkaline phosphatase (Hull and Sykes, 1974, 1975a, b) have yielded results analogous to those discussed above. Observation of the 11 fluorine resonances in [5-fluorotyrosine] alkaline phosphatase indicates rapid internal flipping with rates in the range 10^6–10^8 s^{-1} for nine of the tyrosines and 10^2–10^5 s^{-1} for two others. In this case, internal flip rates are calculated based on chemical shift anisotropy contributions to the fluorine linewidth since the ^{19}F–^1H dipolar interactions with H_6 is not sensitive to this motion, being characterized by a value of $\beta \sim 0$. A particularly interesting result is a correlation between the fluorine linewidths and chemical shifts: tyrosine resonances which resonate closest to the position of fluorotyrosine in the denatured enzyme exhibit chemical shift anisotropy linewidths significantly less than those observed for the low-field tyrosines. This effect has been interpreted to reflect greater mobility about the C_β—C_γ bond for resonances located on the surface of the protein, a result analogous to that obtained for aliphatic residues discussed above. In view of the possibility of structural perturbations of the enzyme arising from the presence of the fluorinated tyrosine residues, it would be interesting to examine the C_β—C_γ flip rate of both a tyrosine and a fluorotyrosine residue in a given enzyme position, particularly if the tyrosine flip rate is slow as in the case of ferrocytochrome *c* discussed above.

Evidence for internal motion of phenylalanine residues on a time scale shorter than that characterizing overall protein tumbling has been reported for the triple stranded helical form of α1–CB2 collagen peptide discussed in Section 4.1 (Torchia *et al.*, 1975), and muscle calcium parvalbumin (Opella *et al.*, 1974; Nelson *et al.*, 1976). Although this motion has generally been considered in terms of the unrestricted internal diffusion model, the dynamics are probably better described as a low-amplitude libration punctuated by the 180° flips discussed above. This analysis is supported in the case of the collagen peptide by both theoretical potential energy calculations and X-ray data indicating a bistable potential for the phenylalanine residue (Torchia *et al.*, 1975). The effects of restricted amplitude internal diffusion have been described in Section 3.4. Although such motion is in general relatively ineffective in altering the relaxation parameters, dynamic simulation of bovine pancreatic trypsin inhibitor by Karplus and McCammon (1979) indicates that tyrosine ring orientation fluctuations of $\pm$ 30° from the average occur over a time scale of 100 ps. Internal diffusion characterized by this amplitude might be sufficient to cause measurable changes in the relaxation times. In addition, the 180° flips can also alter the relaxation times. This problem has been discussed from the standpoint of phenyl group motion in polystyrene by Jones (1977). The 180° flip motion is described by equations (17)–(19) for $\tau_A = \tau_B$ and $\theta = 90°$ (corresponding to 180° flips between $+\theta$ and $-\theta$). In this limit, $C \to B_{10}$ and equation (17) becomes

$$J(\omega) = (1 - B_{10})\frac{(6D_o)^{-1}}{1 + \omega^2(6D_o)^{-2}} + B_{10}\frac{(6D_o + 1/\tau_c)^{-1}}{1 + \omega^2(6D_o + 1/\tau_c)^{-2}} \qquad (27)$$

Equation (27) is clearly analogous to the free internal diffusion case [equation (8)]. It has the property that for $\beta = 0°$, the second term on the right vanishes since an instantaneous 180° flip of the dipolar interaction vector will not affect the relaxation. This is not the case, however, for $\beta = 60°$, corresponding to the *meta* and *ortho* protons of phenylalanine or tyrosine. This result differs quantitatively from the free internal diffusion case. For example, in the asymptotic limit of very fast internal motion, for $\beta = 60°$, T_1 increases by a factor of 64 based on equation (9) but by only a factor of 2.3 based on the above relation. In general, internal 180° flips of phenylalanine or tyrosine residues are very much less effective than unrestricted internal diffusion in altering the spin-lattice relaxation rates. Geometric considerations suggest that the internal 180° flip model may be more generally applicable for the phenylalanine and tyrosine residues, even in peptides.

In addition to the ring flipping motion discussed above, other slow equilibria have been monitored by resonances of aromatic side chains. An interesting example is the high-temperature–low-temperature equilibrium of transaminated basic pancreatic trypsin inhibitor (Brown *et al.*, 1978). In this derivative, the Tyr[23] H-3 and H-5 proton resonances which give rise to a doublet due to coupling with H-2 and H-6, respectively, are further split due to a slow conformational equilibrium. "High-temperature" and "low-temperature" enzyme conformations have been identified, with the ratio varying from 4:1 at 50°C to 1:4 at 5°C. In both forms, rapid flipping about C_β—C_γ bond occurs so that the H-3 and H-5 resonances are equivalent. Since no temperature dependence was observed for other nonlabile protons of the transaminated inhibitor, it was concluded that the equilibrium being monitored involved only the local environment of Tyr[23]. A slow exchange process monitored by a heme methyl resonance in cytochrome c has also been postulated to reflect a slow dynamic equilibrium of a nearby Phe residue (Burns and La Mar, 1979). Slow conformational equilibria leading to split ^{13}C resonances of tryptophan have also been observed and are discussed in the following section.

4.3.2. Tryptophan. The behavior of tryptophan spin lattice relaxation rates of tryptophan-containing peptides indicates internal motion with rates slower than that of the peptide backbone in LH–RH (Deslauriers and Somorjai, 1976) and tetragastrin (Bleich *et al.*, 1976b). Measurements of ^{13}C spin lattice relaxation rates in proteins indicate that internal motion must be slow compared with overall protein tumbling, analogous to the results for phenylalanine and tyrosine (Visscher and Gurd, 1975; Oldfield *et al.*, 1975). The asymmetry of the indole sidechain makes the possibility of rapid flipping of the indole ring about the C_β—C_γ bond unlikely, since a much greater deformation of the protein structure would be required in the intermediate, partially rotated, state. Thus, in making the initial observations on aromatic ring flips, Campbell *et al.* (1975c) concluded that tryptophan residues would be very unlikely to have the rapid internal mobility deduced for phenylalanine and tyrosine residues. Therefore, if different rotational conformers of the tryptophan sidechain are significantly populated, the interconversion between these forms might be slow on the NMR time scale, that is, $\tau \Delta \nu \gg 1$. In fact, such a slow exchange has recently been observed for the ^{13}C resonances of [γ-^{13}C tryptophan] dihydrofolate reductase derived from *S. faecium* (London *et al.*, 1979a). In this enzyme, intensity data indicate that two resonances with a peak separation of 0.2 ppm correspond to a single tryptophan residue (Figure 19). The two resonances, labeled 3a and 3b in Figure 19 have an intensity ratio of ~2:3 at 15°C so that the corresponding protein conformations are not equally stable. Simulation of the spectrum assuming a two-site chemical exchange formalism gives a reasonable fit of the 15°C data for the parameters $p_A = 0.4$, $p_B = 0.6$, $\tau_A = 0.5$ s, $\tau_B = 0.75$ s, and $T_{2A} = T_{2B} = 0.16$ s, where p_A and p_B are the probabilities of the two conformations labeled 3a and 3b, τ_A and τ_B are the lifetimes of the two conformations, and $T_{2A} = T_{2B}$ are the spin–spin relaxation times in the absence of exchange which correspond to linewidths of 2 Hz. The lifetimes are minimum values since slower rates of exchange are also consistent with the observed spectrum. The microscopic nature of this dynamic process cannot be directly deduced from this data alone. Two possibilities consistent with the slow exchange rate include a 180° flip of the indole side chain about the C_β—C_γ bond as in the cases of Phe and Tyr discussed above, or a slow cis/trans isomerism of a nearby X-Pro peptide bond (Section 4.4.2). Recent studies indicate that the relative stability of the two conformers is altered by KCl or the inhibitor methotrexate, both of which have a slightly higher affinity for the A conformer (Groff et al., in preparation).

In addition to the observation of two resonances corresponding to a single tryptophan residue, the ^{13}C NMR spectrum of [γ-^{13}C tryptophan] dihydrofolate reductase illustrates another unusual dynamic property. The tryptophan resonance furthest upfield is substantially broader than the remaining resonances at 15°C and broadens further and shifts downfield with increasing temperature (Figure 19). This behavior indicates a slow exchange process for this tryptophan residue (lifetime based on linewidth ~0.03 s at 15°C), since all other broadening mechanisms likely to be important exhibit a temperature dependence opposite to that which is observed. Based on a quantitative analysis of the broadening and shift behavior of this resonance with temperature, as well as on frequency-dependent data, it can be concluded that the spectral properties reflect a slow chemical exchange process with a conformation in which the tryptophan residue in question exhibits a shift closer to the remaining tryptophan residues and to the denatured enzyme position as well (Groff *et al.*, in preparation). The probability of the

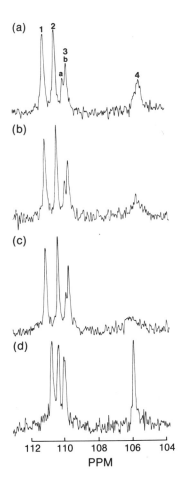

Figure 19. Proton decoupled ^{13}C NMR spectra of the tryptophan C_γ region of [γ - ^{13}C (90 atom %)] Trp-labeled dihydrofolate reductase (1 mM in 0.05 M phosphate buffer, pH 7.3, 0.5 M KCl, 0.02% sodium azide, 1 mM EDTA, and 10% D$_2$O for the NMR lock) obtained at: (a) 5°C; (b) 15°C; (c) 25°C. A spectrum obtained in the presence of a saturating concentration of 3',5'-dichloromethotrexate at 15° is also shown (d) (from London et al., 1979b).

latter (p_A) is very small compared with the probability of the conformation in which the tryptophan has the large upfield shift (p_B). However, p_A increases significantly with increasing temperature. One explanation for the observed exchange effects is a relaxation of the protein structure or "breathing" so that in the more open state A the tryptophan residue is more exposed to solvent and consequently exhibits a more typical shift. The increase in p_A with temperature indicates a correspondingly larger fraction of the relaxed structure in solution. If the above explanation is correct, observation of analogous effects should be possible in many other systems, in particular for resonances which are significantly shifted relative to the solvent exposed position since slow exchange effects are more easily monitored for a larger $\Delta\omega$ value. A parallel observation has been made in [5-fluorotyrosine] alkaline phosphatase; the tyrosine residues with ^{19}F shifts closest to the frequency for the denatured enzyme exhibit the sharpest resonances while the fluorotyrosine resonances #1 and #3 with shifts furthest from the denatured enzyme value are exchange broadened (Hull and Sykes, 1975b). A similar correlation between chemical shift and linewidth is observed in the ^{19}F NMR spectrum of [3-fluorotyrosine]-labeled *lac* repressor (Lu *et al.*, 1976). However, the broadening mechanism in this case has not been demonstrated to arise from chemical exchange effects, as in the example of $[\gamma\text{-}^{13}C]$ tryptophan-labeled dihydrofolate reductase discussed above. We emphasize that two effects can lead to a linewidth–chemical-shift correlation: (1) narrowed resonances of surface residues discussed in Section 4.2; (2) broadened resonances of interior residues due to chemical exchange, as discussed above for tryptophan. In the first case, internal motion rapid relative to overall protein tumbling leads to narrow resonances for residues with shifts closest to the denatured enzyme value. In the second case, resonances which are shifted significantly from the denatured enzyme value will exhibit broadening due to slow breathing motion of the protein. Broadening due to other slow exchange processes is also more likely for interior residues with restricted motion. The two effects noted above can readily be separated by a temperature-dependent linewidth study, as depicted in Figure 19.

We note finally that in the presence of a variety of inhibitors such as methotrexate, the upfield tryptophan resonance is significantly narrowed, indicating a decrease in the rate of chemical exchange (Figure 19). Presumably, the inhibitor locks the protein into a particular folded conformation, and the binding energy tends to reduce the rate of relaxation to a more open, solvent accessible structure. Analogous sharpening of tryptophan resonances has been observed for the Trp[63] in lysozyme (Dobson and Williams, 1975) and for all of the ^{19}F resonances of [6-fluorotryptophan]-dihydrofolate reductase in the presence of inhibitors and substrates (Kimber *et al.*, 1977). The latter is a less specific effect involving all of the four tryptophan residues. It was interpreted to indicate that in the absence of ligands, the enzyme exists in a number of interconverting configurations, while the ligand complexes are essentially in a single conformation (Kimber *et al.*, 1977; Roberts *et al.*, 1977).

4.4. Proline

4.4.1. Ring Puckering.   The dynamics of proline residues have attracted considerable interest due to: (1) differences in the NT_1 values of the pyrrolidine ring carbons indicating significant internal motion of the ring, and (2) the relatively high *cis/trans* ratio making possible kinetic studies of the *cis* ↔ *trans* interconversion rates. Calculations of Somorjai and Deslauriers (1976) and Deslauriers and Smith (1977) demonstrate unequivocally that the NT_1 differences cannot be explained using only the assumption of rigid, anisotropic motion. This result is further supported by the fact that qualitatively similar results for proline are obtained in many different peptides for which anisotropic characteristics of the motion are markedly different. Semiquantitative approaches have been used to estimate the internal mobilities of the various ring carbons (Torchia and Lyerla, 1974; Deslauriers and Somorjai, 1976), but these treatments do not provide insight into the amplitude of motion required to produce the observed effects on T_1. Interpretation of the data using a more detailed model presents two problems: (1) why does the low-amplitude motion required by the constraints of the ring geometry have such a significant effect on the NT_1 values, particularly in light of the general results of Section 3.4 on low-amplitude internal diffusion; (2) can a reasonably simple model be used given the large number of possible puckered conformations of the ring? An answer to the second question is suggested by theoretical conformational energy calculations of DeTar and Luthra (1977) and Venkatachalam *et al.* (1974, 1975) indicating that the ring is essentially bistable (Figure 20). Thus, in contrast with a pseudorotation description of

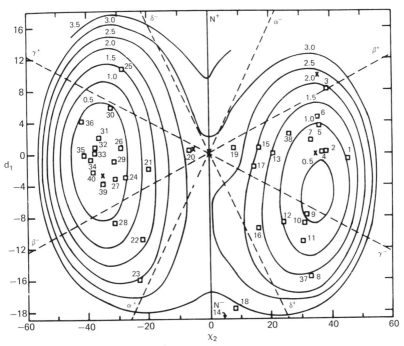

Figure 20. Energy contour plot for ring conformations of s-*trans*-Ac—Pro—OCH₃ defined by torsion χ_2 and d_1 (Detar and Luthra, 1977). There are two regions of minimum energy. For both, the inner-most contour is 0.5 kcal/mol above the global minimum which is located at $\chi_2 = -36$, $d_1 = -1$; the minimum at $\chi_2 = 36$, $d_1 = -5$ is 0.3 kcal/mol above the global minimum. Successive contours are at 1, 1.5, 2, 2.5, and 3 kcal. The region between the two 2.5 kcal contours includes the saddle point of the pass, whose minimum height is 2.7 kcal. The diagonal lines are loci of envelope forms with the indicated atom up (+) or down (−) with respect to the average plane. The numbered squares show conformations for prolone rings studied by X-ray crystallography [reprinted with permission from DeTar and Luthra (1977), *J. Am. Chem. Soc.*, **99**, 1232, copyright by the American Chemical Society].

cyclopentane-like molecules in which all carbons are equally likely to assume a puckered orientation, DeTar and Luthra find that as the ring is made less flexible by substitution, two of the available puckered conformations become significantly more stable. Further, the proline α carbon NT_1 values are typically similar to those of the remaining α carbons in the peptide, suggesting no significant internal motion for this carbon. The bistable nature of the system further explains the relative effectiveness of the internal motion on the relaxation parameters, as discussed below.

Proline relaxation data has been analyzed using the bistable model of Section 3.5 (London, 1978). If the overall motion of the peptide is assumed to be isotropic, and the $^{13}C-^{1}H$ dipolar interaction is assumed dominant as is typically the case (Smith *et al.*, 1975), the relaxation data can be used to determine whether an envelope exo $\leftrightarrow$ endo flip or a flip between half-chair conformations better describes the puckering (Figure 21). Thus, the first model predicts $NT_1^{\gamma} > NT_1^{\beta} = NT_1^{\delta} > NT_1^{\alpha}$ and the second predicts $NT_1^{\gamma} = NT_1^{\beta} > NT_1^{\delta} = NT_1^{\alpha}$. For most linear peptides, the motion appears to be intermediate between the two extreme cases with $NT_1^{\gamma} > NT_1^{\beta} > NT_1^{\delta} > NT_1^{\alpha}$. This pattern suggests that the two stable states correspond to half-

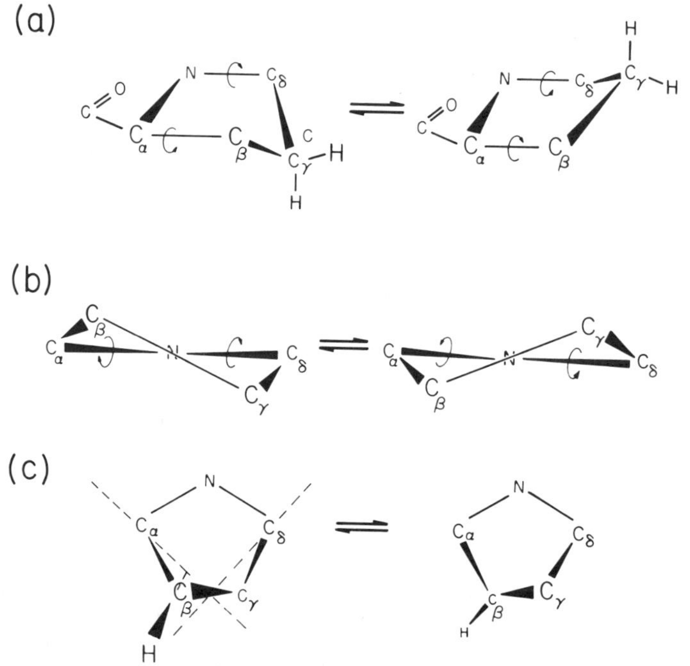

Figure 21. Models for ring puckering of proline: (*a*) Exo↔endo interconversion of C_{γ} with $\beta = 90°$ for C_{γ}, $70.5°$ for C_{β}, C_{δ}; (*b*) Interconversion between half-chair forms; (*c*) Same as *b* viewed perpendicular to the $C_{\delta}—N—C_{\alpha}$ plane [reprinted with permission from London (1978), *J. Am. Chem. Soc.*, **100**, 2678, copyright by the American Chemical Society].

chair conformations with C_γ further from the C_α—N_γ—C_δ plane than C_β, which also correspond to the most commonly observed conformations in a series of crystal structures analyzed by DeTar and Luthra (1977). Quantitative evaluation of the NT_1 behavior in terms of the bistable model leads to the following conclusions:

1. The nature of the restricted motion is crucial to an understanding of the relaxation data. Significantly increased NT_1 values reflect the bistable nature of the ring. Internal motion in which the C_γ–H vector can with equal probability lie anywhere within a given range (Figure 14a) require a physically unreasonable value for the allowed range of internal motion (London and Avitabile, 1978). Thus, the conformational energy map of Figure 20 manifests the theoretical potential energy hump of Figure 14b. Further, the height of the barrier leading to a two-state model is, from Figure 20, estimated to be ~3 kcal. Puckering ranges of 50–70° have been deduced for the proline C_γ–H vector (London, 1978).

2. Assignments of proline ^{13}C resonances can be based in part on the NT_1 relationships derived from the calculation. For example, the ratio $NT_1^\gamma / NT_1^\alpha$ has a maximum theoretical ratio of 4.0 (Figure 7) corresponding to a 90° range of motion. A more realistic range restricts the allowed ratio to a lower value. A ratio of 4.0 originally observed for the proline residue of the peptide Pro—Leu—Gly—NH_2 (Deslauriers et al., 1973) probably reflected an assignment error (Cutnell et al., 1975).

3. The calculation of Figure 9 indicates that in order to obtain a large $NT_1^\gamma / NT_1^\alpha$ ratio the probabilities of the two puckered conformations must be similar, that is, $\tau_A \sim \tau_B$. This condition fails to obtain in certain cyclic peptides (Fossel et al., 1975; Deslauriers et al., 1976b) and the $NT_1^\gamma / NT_1^\alpha$ ratio is greatly reduced. This conclusion is further supported by the analysis of ^1H–^1H scalar coupling constants which indicate a roughly equal mixture of two puckered conformations (Abraham and Thomas, 1964; Torchia, 1971; Ellenberger et al., 1974; Pogliani et al., 1975). It should be noted, however, that not all coupling data is consistent with this conclusion. Analysis of $^3J_{CC}$ coupling constants for Pro^3 in thyrotropin releasing hormone (Haar et al, 1975) suggests a strong preference for endo over exo ring conformations. In contrast, the relaxation data (Deslauriers et al., 1974) require a maximum difference between τ_A and τ_B of a factor of 3, with the latter value corresponding to a 90° range of motion. Jankowski et al. (1978) have analyzed the pH dependence of the $^3J_{HH}$ values of proline and find that while at acid and neutral pH a 50:50 exo:endo mixture prevails, the deprotonation of the nitrogen strongly favors the endo conformation leading to a 6:1 ratio. Deprotonation of the proline nitrogen does in fact decrease the $NT_1^\gamma / NT_1^\alpha$ ratio (Deslauriers et al., 1974), but the magnitude of the change is very much less than the magnitude predicted by Jankowski et al. These predictions are based on a

modified Karplus equation which is dependent on both the $^3J_{HH}$ and $^1J_{CH}$ couplings (Jankowski, 1977). Since, in fact, the relevant $^3J_{HH}$ values do not change appreciably in going to high pH, the conclusion is based largely on changes in the $^1J_{CH}$ values. Although there are a number of approximations inherent in both approaches, the discrepancy suggests possible difficulties with the modification of the Karplus equation proposed by Jankowski (1977).

4.4.2. *Cis* ↔ *Trans* **Isomerism.** In contrast with peptide bonds formed from amino acids in which the *trans* orientation, defined in terms of successive α carbons, is strongly predominant, the *cis* and *trans* conformations of peptide bonds formed from imino acids have similar stabilities:

trans   *cis*

As a result of the relatively high activation energy for *cis* ↔ *trans* isomerism, ΔH ~20 kcal/mol, interconversion of the two conformations is slow compared with the chemical shift differences for many of the resonances so that typically resonances corresponding to both conformations are observed (Thomas and Williams, 1972; Evans and Rabenstein, 1974; Fermandjian *et al.*, 1975). The dynamics of *cis* ↔ *trans* interconversion can be treated as a chemical exchange problem and the thermodynamic parameters and kinetics of the process quantitated (Maia *et al.*, 1976; Cheng and Bovey, 1977; Roques *et al.*, 1977). These results are of particular relevance to protein denaturation and renaturation studies. Brandts *et al.* (1975, 1977) have suggested that denaturation involves a rapid unfolding of the polypeptide chain followed by a slow *cis* ↔ *trans* isomerism. This model successfully resolves the conflict between thermodynamic parameters deduced from calorimetric studies which "see" only a two-state system and kinetic data which "see" both rapid and slow steps. Thus, the small enthalpy difference between *cis* and *trans* conformations makes the slow step nearly invisible to calorimetric study.

In addition to denaturation, renaturation kinetics, prolidase catalyzed hydrolysis of dipeptides having the sequence X—Pro has recently been studied by Lin and Brandts (1979). Two well-separated kinetic phases are observed, the amplitudes proportional to the *cis/trans* ratios of the X—Pro peptide bonds. It was concluded that the slow phase is rate limited by the *cis–trans* isomerization of the X—Pro peptide bond, and that the prolidase has an absolute requirement for the trans form of the peptide bond.

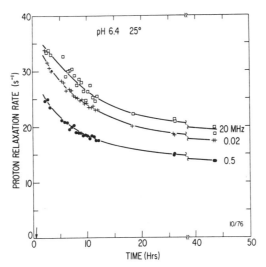

Figure 22. Time course of solvent proton relaxation rates at three values of the magnetic field, corresponding to proton Larmor frequencies of 0.02, 0.5, and 20 MHz, after addition of 0.8 mM Mn^{+2} to 0.4 mM apo-conconavalin A, pH 6.4 [reprinted with permission from Brown *et al.* (1977), *Biochemistry*, **16**, 3883, copyright by the American Chemical Society].

Cis ↔ *trans* isomerism also has been invoked recently to explain the very slow interconversion between folded states of concanavalin A (Brown *et al.*, 1977; Koenig *et al.*, 1978). Brown *et al.* have studied the magnetic field dependence of the spin-lattice relaxation rate of water in the presence of concanavalin A, Mn^{2+} and Ca^{2+}. Water T_1 values were found to undergo very gradual increases, for example, after the addition of Mn^{2+} to apo-concanavalin A (Figure 22). Similar time-dependent data obtained after adding Ca^{2+} to Mn^{2+}-concanavalin A were fit with a model postulating two conformational states with a small enthalpy difference and an energy barrier of 22 kcal/mol, determined from the temperature dependence of the isomerization rate. These parameters suggest a proline *cis* ↔ *trans* isomerism as discussed above. Subsequently, Reeke *et al.* (1978) have analyzed the X-ray structure of both the metallized and demetallized forms of concanavalin A and find a *cis* peptide bond in the first and a *trans* bond in the latter. This bond appears to correspond to Ala-207 rather than to Pro-206 as suggested by Brown *et al.* (1977).

The ubiquitous occurrence of proline residues in peptide hormones suggests that *cis* ↔ *trans* interconversion might also be related to the activities of these peptides. Recently an analog of bradykinin with full intrinsic activity, [Gly^6]-bradykinin, has been found to have a 40% *cis* probability about the Gly^6—Pro^7 peptide bond (London *et al.*, 1979b). This observation may provide a basis for studying the effects of proline *cis* ↔ *trans* isomerism on peptide activity.

4.5. Modified Enzymes

The complexity and sensitivity limitations of natural abundance NMR studies of proteins, and the difficulty of specific incorporation of labeled residues, make the use of labeled reagents an attractive approach in NMR investigations of proteins. Beyond these considerations, however, a strategically placed label may provide more or different information than could be obtained from direct studies of the residues themselves. Mobility of the carboxymethyl groups of derivatized cytochrome c (Nigen *et al.*, 1973; Eakin *et al.*, 1975), and carbonic anhydrase (Jeffers *et al.*, 1978; Uiterkamp *et al.*, 1978) has been studied by ^{13}C NMR. Correlation times characterizing the internal motion for these systems differ by several orders of magnitude and are sensitive to the presence of inhibitors and to changes in pH. In particular, reduction in the mobility of the His-200 carboxymethyl group of carbonic anhydrase near neutral pH relative to both high pH (Uiterkamp *et al.*, 1978) and low pH (Jeffers *et al.*, 1978), has been interpreted to reflect coordination of the carboxyl group in the zinc ion at the active site.

Relaxation rates of 90% ^{13}C enriched carbamyl groups attached to the N-terminal glycine and phenylalanine residues of insulin, as well as to the lysine B29 amino group of zinc and nickel insulin, have been determined by Led *et al.* (1975). Data were analyzed using either of two dynamic models for unrestricted internal rotation about a single axis (the N—C_α bond). In these calculations only intramolecular dipolar interactions between the carbamyl carbon and the three nitrogen-bound protons were considered, chemical shift anisotropy contributions being relatively small at 25.2 MHz. Surprisingly, the two dynamic models for which calculations were given corresponding to C_α and the carbamyl oxygen either *cis* or *trans*, gave very similar numerical results. This similarity is surprising given the strong geometric dependence of the interactions [equation (8)]. Led *et al.* suggest that as more and more protons contribute to the relaxation process, they will tend to occupy a greater variety of positions relative to the observed ^{13}C nucleus, giving an overall average effect which, in the case of the two models considered, led to very similar results. In the metal-free insulin, which exists in solution primarily as a dimer, an overall molecular correlation time of 6×10^{-8} s was found and internal correlation times of 10^{-9}, 6×10^{-10}, and 3×10^{-10} s for the carbamyl groups bonded to the phenylalanine and the glycine residues, respectively. Two values corresponding to the glycine-bound carbamyl group were determined since aggregation of the insulin monomers apparently results in both chemical shift inequivalence and mobility differences for the carbamylated glycine residues of the dimer. Internal mobility of the carbamyl groups was further related to conformational rearrangements occurring in the dimer.

The structure of the heme pocket of myoglobin and hemoglobin has been probed using the ^{13}C relaxation behavior of ethyl isocyanide coordinated directly to the heme iron (Gilman, 1979). The relaxation data for both protonated carbons was interpreted using a model equivalent to that utilized for methionine relaxation in dihydrofolate reductase (Figure 16), that is, restricted amplitude diffusion about the N—C bond and free internal diffusion of the methyl. The ranges and rates of motion deduced for ethyl isocyanide derivatives of sperm whale myoglobin, harbor seal myoglobin, and human hemoglobin in the relaxed state were found to be similar, suggesting close structural similarities of the heme pockets. An analogous conclusion was derived from the similar chemical shifts of the ethyl carbons in each derivative.

5. CONCLUSIONS

An effort has been made to summarize the NMR relaxation behavior resulting from different dynamic models describing the motion. The primary value of considering these dynamic models is to obtain a better physical characterization of the dynamic processes involved rather than a more accurate set of correlation times describing the motion. Although emphasis has been placed on the analysis of relaxation rates and NOE values which provide the most direct information on molecular motion, much dynamic information has been deduced indirectly through other types of NMR measurement. Exchange of amide protons provides indirect information on conformational equilibria of proteins (Hvidt and Nielsen, 1966; Englander et al., 1972; Tsuboi and Nakanishi, 1979). Chemical shift behavior in the presence of lanthanide ions indicates internal motion about the C_α—C_β bond of Val-109 of lysozyme (Dobson, 1977; Campbell et al., 1975a), analogous to the result for Phe proton shifts in the presence of lanthanide ions discussed in Section 4.3. Relaxation behavior of ligands either tightly bound to proteins or undergoing chemical exchange between bound and free states has also been used to probe the dynamics of macromolecules (Sykes and Scott, 1972).

The relation of structure to biochemical activity has advanced rapidly due to the availability of crystallographic data; however the implications of protein dynamics on biochemical function have yet to be fully explored. One of the more obvious implications of the conformational flexibility determined from NMR studies is in the area of ligand–macromolecule recognition. The possibility of mutual conformational adjustments of both ligand and macromolecule has implications for the kinetics of association which favor a "zipper" model of association rather than a "lock and key" model (Burgen et al., 1975). Conformational changes of proteins have often been proposed to have an important role in catalysis. NMR results indicate that ligands alter not only the distribution of accessible conformations but the rates of conformational interconversion as well (cf. Figure 19). The possibility of studying the dynamic response of specific amino acid residues to substrates or inhibitors should lead to greater understanding of the role of conformational flexibility in the catalytic process. Correlation of dynamic data derived from

NMR and other approaches (Lakowicz and Weber, 1973; Eftink and Ghiron, 1975; Saviotti and Galley, 1974; Careri *et al.*, 1975; Wagner and Wüthrich, 1978; Hantgan and Taniuchi, 1978) should provide a clearer picture of the complex motion characterizing these molecules.

ACKNOWLEDGMENTS

Critical evaluations of the manuscript by N. A. Matwiyoff, E. Fukushima, T. W. Whaley, L. O. Morgan, P. G. Schmidt, S. B. Roeder, R. D. Brown, and R. L. Blakley are gratefully acknowledged. The technical assistance of F. Ruth Capron, Dale E. Armstrong, Julie Grilly, and Phyllis London in the preparation of this manuscript is also greatly appreciated.

APPENDIX A

The dependence of several of the more frequently encountered relaxation parameters on the spectral densities given in Sections 3 and 4 of this paper are summarized below.

I: Relaxation due to dipolar interaction between a pair of equivalent spin-$\frac{1}{2}$ nuclei:

$$\frac{1}{T_1} = \frac{3}{10} \frac{\gamma_I^4 \hbar^2}{r_{IJ}^6} [J(\omega_I) + 4J(2\omega_I)] \tag{A1}$$

$$\frac{1}{T_2} = \frac{3}{20} \frac{\gamma_I^4 \hbar^2}{r_{IJ}^6} [3J(0) + 5J(\omega_I) + 2J(2\omega_I)] \tag{A2}$$

II: Relaxation due to dipolar interaction between two inequivalent nuclei, $I = \frac{1}{2}$ and S:

$$\left(\frac{1}{T_1}\right)_I = \frac{2}{15} \frac{\gamma_I^2 \gamma_S^2 S(S+1)\hbar^2}{r_{IS}^6} [J(\omega_I - \omega_S) + 3J(\omega_I) + 6J(\omega_I + \omega_S)] \tag{A3}$$

$$\left(\frac{1}{T_2}\right)_I = $$

$$\frac{1}{15} \frac{\gamma_I^2 \gamma_S^2 S(S+1)\hbar^2}{r_{IS}^6} [J(\omega_I - \omega_S) + 3J(\omega_I) + 6J(\omega_I + \omega_S) + 4J(0) + 6J(\omega_S)] \tag{A4}$$

$$\text{NOE} = 1 + \frac{\gamma_S}{\gamma_I} \cdot \left[\frac{6J(\omega_I + \omega_S) - J(\omega_I - \omega_S)}{J(\omega_I - \omega_S) + 3J(\omega_I) + 6J(\omega_I + \omega_S)}\right] \tag{A5}$$

In the above expressions, subscript I indicates the observed nucleus, r_{IJ} and r_{IS} are internuclear distances, γ_I is the magnetogyric ratio of species I, S is the spin of nucleus S, and the ω_I and ω_S values are the Larmor frequencies for nuclei I and S, that is, $\omega_I = \gamma_I H_0$, $\omega_S = \gamma_S H_0$, where H_0 is the applied magnetic field. As in the text, the spectral densities $J(\omega)$ correspond to half of the Fourier transform of the rotational autocorrelation functions so that in the extreme narrowing limit $(\omega\tau \ll 1)$ $J(\omega) \to \tau$ for isotropic motion.

A detailed analysis of the validity of expressions (A1)–(A5) is beyond the scope of the present article. An important assumption concerning the application of these relations to arbitrary dynamic models is discussed by Soda and Chihara (1974). As noted in the text, nonexponential recovery of the z component of magnetization can result from either cross-relaxation effects (Freeman et al., 1974 and references therein) or cross-correlation effects (Werbelow and Grant, 1977). Cross-relaxation effects on ^{13}C spin-lattice relaxation are eliminated as a result of proton decoupling. Further, the effects of cross correlation terms on the initial spin-lattice relaxation recovery are generally small; the low sensitivity of many protein and peptide experiments restricts measurements of T_1 to delays $\tau \lesssim T_1$ in a 180°–τ–90° experiment during which deviations from exponential recovery are minimal. It is also worth noting that calculations for a methylene group based on the free internal rotation model indicate that the NOE values including cross correlations effects deviate most significantly from the value neglecting cross correlations if the correlation time for overall motion is close to $1/\omega_c$ (London and Avitabile, 1976). These effects will be extremely difficult to measure since the spin lattice relaxation is most markedly nonexponential for the same parameters, making reliable intensity data difficult to obtain.

Another important consideration which has generally been neglected in treatments of nuclear relaxation is the symmetry of the interaction. This problem has recently been discussed in the elegant article by Spiess (1978), in which the various relaxation interactions are cast into the form of irreducible tensors. The (second rank) tensors can be decomposed into a sum of tensors of rank 0, 1, 2 containing the isotropic, antisymmetric, and traceless symmetric part of the interaction. If the antisymmetric elements are nonvanishing or if the asymmetry parameter (η_λ in the notation of Spiess) is nonzero, the relaxation rate expressions are considerably more complicated than is generally assumed. These complications do not affect the expressions for relaxation resulting from nuclear dipole–dipole relaxation due to the symmetry of the interaction, but can be significant for other types of interaction. For example, one case of practical importance is the relaxation due to chemical shift anisotropy of nonprotonated aromatic residues in proteins. Assuming that internal motion of the residues is slow relative to the rate of isotropic protein tumbling, the relaxation rates for nuclear species I are given by

$$\frac{1}{T_1} = \gamma_I^2 H_0^2 \left[\frac{2}{3} (\Delta_a\sigma)^2 \frac{\tau_1}{1 + \omega_I^2 \tau_1^2} + \frac{2}{15} (\Delta\sigma)^2 \left(1 + \frac{\eta_{CS}^2}{3}\right) \frac{\tau_2}{1 + \omega_I^2 \tau_2^2} \right] \quad (A6)$$

$$\frac{1}{T_2} = \quad (A7)$$

$$\gamma_I^2 H_0^2 \left[\frac{1}{3} (\Delta_a\sigma)^2 \frac{\tau_1}{1 + \omega_I^2 \tau_1^2} + \frac{1}{45} (\Delta\sigma)^2 \left(1 + \frac{\eta_{CS}^2}{3}\right)\left(4\tau_2 + \frac{3\tau_2}{1 + \omega_I^2 \tau_2^2}\right) \right]$$

where $\tau_1 = (2D_o)^{-1}$, $\tau_2 = (6D_o)^{-1}$, $\Delta\sigma = \sigma_{zz} - (\sigma_{xx} + \sigma_{yy})/2$, and $(\Delta_a\sigma)^2 = (\rho_{xy}^2 + \rho_{xz}^2 + \rho_{yz}^2)$, the latter reflecting the antisymmetric components of the chemical shift tensor, η_{CS}. An interesting characteristic of the above relations is that the coefficient of the $\Delta\sigma_a$ terms is considerably larger than the coefficient of the $\Delta\sigma$ terms, thereby enhancing the relative importance of the antisymmetric components. These potential complications make the quantitative evaluation of nondipolar relaxation mechanisms difficult unless the symmetry properties of the molecule are known or the elements of the tensor can be obtained from solid-state measurements.

APPENDIX B

Elements of the B matrix used in the text (London and Avitabile, 1976, 1977a, b):

$$B_{00} = \tfrac{1}{4}(3\cos^2\beta - 1)^2$$

$$B_{10} = 3\sin^2\beta\,\cos^2\beta$$

$$B_{20} = \tfrac{3}{4}\sin^4\beta$$

$$B_{01} = \tfrac{3}{2}\sin^2\beta\,\cos^2\beta$$

$$B_{11} = \tfrac{1}{2}(1 - 3\cos^2\beta + 4\cos^4\beta)$$

$$B_{21} = \tfrac{1}{2}(1 - \cos^4\beta)$$

$$B_{02} = \tfrac{3}{8}\sin^4\beta$$

$$B_{12} = \tfrac{1}{2}(1 - \cos^4\beta)$$

$$B_{22} = \tfrac{1}{8}(1 + 6\cos^2\beta + \cos^4\beta)$$

A useful property of the B matrix is given by the following sum rule:

$$\sum_{i=0}^{2} B_{ij} = 1$$

where $j = 0$, 1, or 2.

REFERENCES

Abragam, A. (1961), *The Principles of Nuclear Magnetism*, Clarendon Press, Oxford, England.

Abraham, R. J. and W. A. Thomas (1964), *J. Chem. Soc.*, 3739.

Allerhand, A. (1970), *J. Chem. Phys.* **52**, 3596.

Allerhand, A. (1978), *Acc. Chem. Res.* **11**, 469.

Allerhand, A. and R. A. Komoroski (1973), *J. Am. Chem. Soc.* **95**, 8228.

Allerhand, A. and E. Oldfield (1973), *Biochemistry* **12**, 3428.

Allerhand, A., D. W. Cochran, and D. Doddrell (1970), *Proc. Nat. Acad. Sci. U.S.A.* **67**, 1093.

Allerhand, A., D. Doddrell, and R. Komorski (1971b), *J. Chem. Phys.* **55**, 189.

Allerhand, A., D. Doddrell, V. Glushko, D. W. Cochran, E. Wenkert, P. J. Lawson, and F. R. N. Gurd (1971a), *J. Am. Chem. Soc.* **93**, 544.

Anderson, T. E. (1973), *J. Magn. Reson.* **11**, 398.

Andree, P. J. (1978), *J. Magn. Reson.* **29**, 419.

Bauer, D., S. J. Opella, D. J. Nelson, and R. Pecora (1975), *J. Am. Chem. Soc.* **97**, 2580.

Berger, S., F. R. Kreissl, D. M. Grant, and J. D. Roberts (1975), *J. Am. Chem. Soc.* **97**, 1805.

Berne, B. J. (1972), *J. Chem. Phys.* **56**, 2164.

Blakley, R. L., L. D. Cocco, R. E. London, T. E. Walker, and N. A. Matwiyoff (1978), *Biochemistry* **17**, 2284.

Bleich, H. E., J. D. Cutnell, A. R. Day, R. J. Freer, J. A. Glasel, and J. F. McKelvy (1976a), *Proc. Nat. Acad. Sci. U.S.A.* **73**, 2589.

Bleich, H. E., J. D. Cutnell, and J. A. Glasel (1976b), *Biochemistry* **15**, 2455.

Bleich, H. E., and J. A. Glasel (1978), *Biopolymers* **17**, 2445.

Bloembergen, N., E. M. Purcell, and R. V. Pound (1948), *Phys. Rev.* **73**, 679.

Blomberg, F., W. Maurer, and H. Ruterjans (1976), *Proc. Nat. Acad. Sci. U.S.A.* **73**, 1409.

Blunt, J. W., and J. B. Stothers (1977), *J. Magn. Reson.* **27**, 515.

Bothner-By, A. A., and P. E. Johner (1978), *Biophys. J.* **24**, 779.

Bovee, W. M. M. J., and J. Smidt (1974), *Mol. Phys.* **28**, 1617.

Brandts, J. F., H. R. Halvorson, and M. Brennan (1975), *Biochemistry* **14**, 4953.

Brandts, J. F., M. Brennan, and L. N. Lin (1977), *Proc. Nat. Acad. Sci. U.S.A.* **74**, 4178.

Brown, L. R., A. DeMarco, R. Richarz, G. Wagner, and K. Wüthrich (1978), *Eur. J. Biochem.* **88**, 87.

Brown III, R. D., C. F. Brewer, and S. H. Koenig (1977), *Biochemistry* **16**, 3883.

Browne, D. T., G. L. Kenyon, E. L. Packer, H. Sternlicht, and D. M. Wilson (1973), *J. Am. Chem. Soc.* **95**, 1316.

Bull, T. E. (1978), *J. Magn. Reson.* **31**, 453.

Bull, T. E., J. E. Norne, P. Reimarsson, and B. Lindman (1978), *J. Am. Chem. Soc.* **100**, 4643.

Burgen, A. S. V., G. C. K. Roberts, and J. Feeney (1975), *Nature* **253**, 753.

Burns, P. D. and G. N. La Mar (1979), *J. Am. Chem. Soc.* **101**, 5844.

Campbell, I. D., and R. Freeman (1973), *J. Magn. Reson.* **11**, 143.

Campbell, I. D., C. M. Dobson, and R. J. P. Williams (1974), *J. Chem. Soc., Chem. Comm.* 888.

Campbell, I. D., C. M. Dobson, and R. J. P. Williams (1975a), *Proc. Royal Soc. London* **A345**, 41.

Campbell, I. D., R. Freeman, and D. L. Turner (1975b), *J. Magn. Reson.* **20**, 172.

Campbell, I. D., C. M. Dobson, and R. J. P. Williams (1975c), *Proc. Royal Soc. London* **B189**, 503.

Campbell, I. D., C. M. Dobson, G. R. Moore, S. J. Perkins, and R. J. P. Williams (1976), *FEBS Letters* **70**, 96.

Careri, G., P. Fasella, and E. Grotton (1975), *CRC Crit. Rev. Biochem.* **3**, 141.

Cave, A., C. M. Dobson, J. Parello, and R. J. P. Williams (1976), *FEBS Letters* **65**, 190.

Cheng, H. N., and F. A. Bovey (1977), *Biopolymers* **16**, 1465.

Cocco, L., R. L. Blakley, T. E. Walker, R. E. London, and N. A. Matwiyoff (1978), *Biochemistry* **17**, 4285.

Combrisson, S., B. P. Roques, and R. Oberlin (1976), *Tetrahedron Letters* **38**, 3455.

Connor, T. M. (1964), *Trans. Faraday Soc.* **60**, 1574.

Conti, F., and M. Paci (1971), *FEBS Letters* **17**, 149.

Cooper, A. (1976), *Proc. Nat. Acad. Sci. U.S.A.* **73**, 2740.

Cutnell, J. D., J. A. Glasel, and V. J. Hruby (1975), *Org. Magn. Reson.* **7**, 256.

Cutnell, J. D., and J. A. Glasel (1976a), *J. Am. Chem. Soc.* **98**, 264.

Cutnell, J. D., and J. A. Glasel (1976b), *J. Am. Chem. Soc.* **98**, 5742.

Deber, C. M., M. A. Moscarello, and D. D. Wood (1978), *Biochemistry* **17**, 898.

Deslauriers, R., R. Walter, and I. C. P. Smith (1973), *FEBS Letters* **37**, 27.

Deslauriers, R., I. C. P. Smith, and R. Walter (1974), *J. Biol. Chem.* **249**, 7006.

Deslauriers, R., A. C. M. Paiva, K. Schaumburg, and I. C. P. Smith (1975), *Biochemistry* **14**, 878.

Deslauriers, R., and R. L. Somorjai (1976), *J. Am. Chem. Soc.* **98**, 1931.

Deslauriers, R., and I. C. P. Smith (1976) in *Topics in Carbon-13 NMR Spectroscopy* G. C. Levy, Ed., Vol. 2, Wiley, New York, NY, pp. 1–80.

Deslauriers, R., R. A. Komoroski, G. C. Levy, A. C. M. Paiva, and I. C. P. Smith (1976a), *FEBS Letters* **62**, 50.

Deslauriers, R., Z. Grzonka, and R. Walter (1976b), *Biopolymers* **15**, 1677.

Deslauriers, R., and I. C. P. Smith (1977), *Biopolymers* **16**, 1245.

Deslauriers, R., G. C. Levy, W. H. McGregor, D. Sarantakis, and I. C. P. Smith (1977a), *Eur. J. Biochem.* **75**, 343.

Deslauriers, R., E. Ralston, and R. L. Somorjai (1977b), *J. Mol. Biol.* **113**, 697.

Deslauriers, R., R. L. Somorjai, and E. Ralston (1977c), *Nature* **266**, 746.

Deslauriers, R., I. C. P. Smith, G. C. Levy, R. Orlowski, and R. Walter (1978), *J. Am. Chem. Soc.* **100**, 3912.

DeTar, D. F., and N. P. Luthra (1977), *J. Am. Chem. Soc.* **99**, 1232.

DeWitt, J. L., M. A. Hemminga, and T. J. Schaafsma (1978), *J. Magn. Reson.* **31**, 97.

Dill, K., and A. Allerhand (1979), *J. Am. Chem. Soc.* **101**, 4376.

Dobson, C. M., and R. J. P. Williams (1975), *FEBS Letters* **56**, 362.

Dobson, C. M. (1977) in *NMR in Biology*, Dwek, R. A., I. D. Campbell, R. E. Richards, and R. J. P. Williams, Eds., Academic Press, London, pp. 63–94.

Doddrell, D., V. Glushko, and A. Allerhand (1972), *J. Chem. Phys.* **56**, 3683.

Eakin, R. T., L. O. Morgan, and N. A. Matwiyoff (1975), *Biochemistry* **14**, 4538.

Edzes, H. T., and E. T. Samulski (1977), *Nature* **265**, 521.

Edzes, H. T., and E. T. Samulski (1978), *J. Magn. Reson.* **31**, 207.

Eftink, M. R., and C. A. Ghiron (1975), *Proc. Nat. Acad. Sci. U.S.A.* **72**, 3290.

Ellenberger, M., L. Pogliani, K. Hauser, and J. Valat (1974), *Chem. Phys. Letters* **27**, 419.

Englander, S. W., N. W. Downer, and H. Teitelbaum (1972), *Ann. Rev. Biochem.* **41**, 903.

Evans, C. A., and D. L. Rabenstein (1974), *J. Am. Chem. Soc.* **96**, 7312.

Feeney, J., G. C. K. Roberts, B. Birdsall, D. V. Griffiths, R. W. King, P. Scudder, and A. S. V. Burgen (1977), *Proc. Royal Soc. London* **B196**, 267.

Fermandjian, S., S. Tran-Dinh, J. Savrda, E. Sala, R. Mermet-Bouvier, E. Bricas, and P. Fromageot (1975), *Biochim. Biophys. Acta* **399**, 313.

Ferretti, J. A., and G. R. Marshall (1978), *Biophys. J.* **21**, 79a.

Fossel, E. T., W. R. Veatch, Y. A. Ovchinnikov, and E. R. Blout (1974), *Biochemistry* **13**, 5264.

Fossel, E. T., K. R. K. Easwaran, and E. R. Blout (1975), *Biopolymers* **14**, 927.

Freeman, R., H. D. W. Hill, and R. Kaptein (1972), *J. Magn. Reson.* **7**, 327.

Freeman, R., H. D. Hill, B. L. Tomlinson, and L. D. Hall (1974), *J. Chem. Phys.* **61**, 4466.

Gelin, B. R., and M. Karplus (1975), *Proc. Natl. Acad. Sci. U.S.A.* **72**, 2002.

Ghesquiere, D., B. Ban, and C. Chachaty (1977), *Macromolecules* **10**, 743.

Gilman, J. G. (1979), *Biochemistry* **18**, 2273.

Glushko, V., P. J. Lawson, and F. R. N. Gurd (1972), *J. Biol. Chem.* **247**, 3176.

Goodman, R. A., E. Oldfield, and A. Allerhand (1973), *J. Am. Chem. Soc.* **95**, 7553.

Gordon, S. L., and K. Wüthrich (1978), *J. Am. Chem. Soc.* **100**, 7094.

Grant, D. M., R. J. Pugmire, E. P. Black, and K. A. Christensen (1973), *J. Am. Chem. Soc.* **95**, 8465.

Grimaldi, J. J., and B. D. Sykes (1975), *J. Biol. Chem.* **250**, 1618.

Haar, W., S. Fermandjian, J. Vicar, K. Blaha, and P. Fromageot (1975), *Proc. Natl. Acad. Sci. U.S.A.* **72**, 4948.

Hall, L. D., and H. D. W. Hill (1976), *J. Am. Chem. Soc.* **98**, 1269.

Hantgan, R. R., and H. Taniuchi (1978), *J. Biol. Chem.* **253**, 5373.

Haslinger, E., and R. M. Lynden-Bell (1978), *J. Magn. Reson.* **31**, 33.

Hawkes, G. E., W. M. Litchman, and E. W. Randall (1975a), *J. Magn. Reson.* **19**, 255.

Hawkes, G. E., E. W. Randall, and C. H. Bradley (1975b), *Nature* **257**, 767.

Hawkes, G. E., E. W. Randall, and W. E. Hull (1977), *J. Chem. Soc., Chem. Commun.*, 546.

Higashijima, T., M. Tasumi, and T. Miyazawa (1977), *Biopolymers* **16**, 1259.

Hoffmann, R. A. and S. Forsen (1966), *Prog. NMR Spectr.* **1**, 15.

Howarth, O. W. (1978), *J. Chem. Soc., Faraday Trans. II* **74**, 1031.

Howarth, O. W. and D. M. Lilley (1978), *Prog. NMR Spectr.* **12**, 1.

Horwitz, A. F. (1972), in *Membrane Molecular Biology*, Sinauer Associates, Stamford, CT, pp. 164–191.

Hu, C. M., and R. Zwanzig (1974), *J. Chem. Phys.* **60**, 4354.

Hubbard, T. S., and C. S. Johnson, Jr. (1975), *J. Chem. Phys.* **63**, 4933.

Hull, W. E., and B. D. Sykes (1974), *Biochemistry* **13**, 3431.

Hull, W. E., and B. D. Sykes (1975a), *J. Chem. Phys.* **63**, 867.

Hull, W. E., and B. D. Sykes (1975b), *J. Mol. Biol.* **98**, 121.

Hull, W. E., H. E. Kricheldorf, and M. Fehrle (1978), *Biopolymers* **17**, 2427.

Hunkapiller, M. W., S. H. Smallcombe, D. R. Whitaker, and J. H. Richards (1973), *Biochemistry* **12**, 4732.

Huntress Jr., W. T. (1970), *Adv. Magn. Reson.* **4**, 1.

Hvidt, A., and S. O. Nielsen (1966), *Adv. Protein Chem.* **21**, 287.

Irving, C. S., and A. Lapidot (1975), *J. Am. Chem. Soc.* **97**, 5945.

James, T. L., G. B. Matson, and I. D. Kuntz (1978), *J. Am. Chem. Soc.* **100**, 3590.

Jankowski, K. (1977), *Org. Magn. Reson.* **10**, 50.

Jankowski, K., F. Soler, and M. Ellenberger (1978), *J. Mol. Struct.* **48**, 63.

Jardetzky, O., K. Akasaka, D. Vogel, S. Morris, and K. C. Holmes (1978), *Nature* **273**, 564.

Jeffers, P. K., W. McI. Sutherland, and R. G. Khalifah (1978), *Biochemistry* **17**, 1305.

Jones, A. A. (1977), *J. Polymer Sci., Polymer Phys. Ed.* **15**, 863.

Jones, A. A. and W. H. Stockmayer (1977), *J. Polymer Sci., Polymer Phys. Ed.* **15**, 847.

Jones, C. R., C. T. Sikakana, S. P. Hehir, and W. A. Gibbons (1978a), *Biochem. Biophys. Res. Commun.* **83**, 1380.

Jones, C. R., C. T. Sikakana, S. Hehir, M. C. Kuo, and W. A. Gibbons (1978b), *Biophys. J.* **24**, 815.

Jones Jr., W. T., T. M. Rothgeb, and F. R. N. Gurd (1976), *J. Biol. Chem.* **251**, 7452.

Kalk, A., and H. J. C. Berendsen (1976), *J. Magn. Reson.* **24**, 343.

Karplus, M., and J. A. McCammon (1979), *Nature* **277**, 578.

Karplus, S., G. H. Snyder, and B. D. Sykes (1973), *Biochemistry* **12**, 1323.

Keim, P., R. A. Vigna, J. S. Morrow, R. C. Marshall, and F. R. N. Gurd (1973a), *J. Biol. Chem.* **248**, 6104.

Keim, P., R. A. Vigna, J. S. Morrow, R. C. Marshall, and F. R. N. Gurd (1973b), *J. Biol. Chem.* **248**, 7811.

Keim, P., R. A. Vigna, A. M. Nigen, J. S. Morrow, and F. R. N. Gurd (1974), *J. Biol. Chem.* **249**, 4149.

Khaled, M. A., M. M. Long, W. D. Thompson, R. J. Bradley, G. B. Brown, and D. W. Urry (1977), *Biochem. Biophys. Res. Commun.* **76**, 224.

Kimber, B. J., D. V. Griffiths, B. Birdsall, R. W. King, P. Scudder, J. Feeney, G. C. K. Roberts, and A. S. V. Burgen (1977), *Biochemistry* **16**, 3492.

Kimmich, R. and F. Noack (1970), *Z. Naturforsch.* **25a**, 299; 1680.

King, R., and O. Jardetzky (1978), *Chem. Phys. Lett.* **55**, 15.

King, R., R. Maas, M. Gassner, R. K. Nanda, W. W. Conover, and O. Jardetzky (1978), *Biophys. J.* **24**, 103.

Kintzinger, J. P., and J. M. Lehn (1971), *Mol. Phys.* **22**, 273.

Kintzinger, J. P., and J. M. Lehn (1974), *Mol. Phys.* **27**, 491.

Koenig, S. H., C. F. Brewer, and R. D. Brown III (1978), *Biochemistry* **17**, 4251.

Kricheldorf, H. R., W. E. Hull, and V. Formacek (1977), *Biopolymers* **16**, 1609.

Krishna, N. R., D. G. Agresti, J. D. Glickson, and R. Walter (1978), *Biophys. J.* **24**, 791.

Kuhlmann, K. F., D. M. Grant, and R. K. Harris (1970), *J. Chem. Phys.* **52**, 3439.

Kuo, W. S., O. J. Jacobus, G. B. Savitsky, and A. L. Beyerlein (1979), *J. Chem. Phys.* **70**, 1193.

Ladner, K. H., D. K. Dalling, and D. M. Grant (1976), *J. Phys. Chem.* **80**, 1783.

Lakowicz, J. R., and G. Weber (1973), *Biochemistry* **12**, 4161.

Lapidot, A., and C. S. Irving (1975), in *Proc. 2nd Intl. Conf. Stable Isotopes*, U.S.E.R.D.A. Oakbrook, Ill. E. R. Klein and P. D. Klein, Eds., pp. 427–444.

Lapidot, A., and C. S. Irving (1977), *Proc. Nat. Acad. Sci. U.S.A.* **74**, 1988.

Led, J. J., D. M. Grant, W. J. Horton, F. Sundby, and K. Vilhelmsen (1975), *J. Am. Chem. Soc.* **97**, 5997.

Levine, Y. K., P. Partington, and G. C. K. Roberts (1973), *Mol. Phys.* **25**, 497.

Levine, Y. K., N. J. M. Birdsall, A. G. Lee, J. C. Metcalfe, P. Partington, and G. C. K. Roberts (1974), *J. Chem. Phys.* **60**, 2890.

Levy, G. C., D. E. Axelson, R. Schwartz, and J. Hochmann (1978), *J. Am. Chem. Soc.* **100**, 410.

Lin, L. N., and J. F. Brandts (1979), *Biochemistry* **18**, 43.

Llinas, M., W. Meier, and K. Wüthrich (1977), *Biochim. Biophys. Acta* **492**, 1.

London, R. E. (1978), *J. Am. Chem. Soc.* **100**, 2678.

London, R. E. (1979), *Int. J. Peptide Protein Res.* **14**, 377.

London, R. E., and J. Avitabile (1976), *J. Chem. Phys.* **65**, 2443.

London, R. E., and J. Avitabile (1977a), *J. Chem. Phys.* **66**, 4254.

London, R. E., and J. Avitabile (1977b), *J. Am. Chem. Soc.* **99**, 7765.

London, R. E., and J. Avitabile (1978), *J. Am. Chem. Soc.* **100**, 7159.

London, R. E., C. T. Gregg, and N. A. Matwiyoff (1975), *Science* **188**, 266.

London, R. E., M. P. Eastman, and N. A. Matwiyoff (1977a), *J. Phys. Chem.* **80**, 884.

London, R. E., T. E. Walker, V. H. Kollman, and N. A. Matwiyoff (1977b), *J. Magn. Reson.* **26**, 213.

London, R. E., J. M. Stewart, J. R. Cann, and N. A. Matwiyoff (1978), *Biochemistry* **17**, 2270.

London, R. E., J. M. Stewart, R. Williams, J. R. Cann, and N. A. Matwiyoff (1979b), *J. Am. Chem. Soc.* **101**, 2455.

London, R. E., J. P. Groff, and R. L. Blakley (1979a), *Biochem. Biophys. Res. Commun.* **86**, 779.

Lu, P., M. Jarema, K. Mosser, and W. E. Daniel, Jr. (1976), *Proc. Nat. Acad. Sci. U.S.A.* **73**, 3471.

Lyerla, Jr., J. R., and G. C. Levy (1974), in *Topics in Carbon-13 NMR Spectroscopy*, Vol. 1, p. 79.

Lyerla, Jr., J. R., H. M. McIntyre, and D. A. Torchia (1974), *Macromolecules* **7**, 11.

Lyerla, Jr., J. R., and D. A. Torchia (1975), *Biochemistry* **14**, 315.

Maia, H. L., K. G. Orrell, and H. N. Rydon (1976), *J. Chem. Soc., Perk. II*, 761.

Marshall, A. G., P. G. Schmidt, and B. D. Sykes (1972), *Biochemistry* **11**, 3875.

Matson, G. B. (1977), *J. Chem. Phys.* **67**, 5152.

Matthews, D. A., R. A. Alden, J. T. Bolin, S. T. Freer, R. Hamlin, N. Xuong, J. Kraut, M. Poe, M. Williams, and K. Hoogsteen (1977), *Science* **197**, 452.

McCammon, J. A., B. R. Gelin, and M. Karplus (1977), *Nature* **267**, 585.

McClung, R. E. D. (1969), *J. Chem. Phys.* **51**, 3842.

McClung, R. E. D., and H. Versmold (1972), *J. Chem. Phys.* **57**, 2569.

McConnell, H. M. (1958), *J. Chem. Phys.* **28**, 430.

Molday, R. S., S. W. Englander, and R. G. Kallen (1972), *Biochemistry* **11**, 150.

Monnerie, L., and F. Geny (1969), *J. Chim. Phys.* **66**, 1691.

Moore, G. R., and R. J. P. Williams (1975), *FEBS Letters* **53**, 334.

Nelson, D. J., S. J. Opella, and O. Jardetzky (1976), *Biochemistry* **15**, 5552.

Niccolai, N., M. P. deLeon deMiles, S. P. Hehir, and W. A. Gibbons (1978), *J. Am. Chem. Soc.* **100**, 6528.

Nigen, A. M., P. Keim, R. C. Marshall, J. S. Morrow, R. A. Vigna, and F. R. N. Gurd (1973), *J. Biol. Chem.* **248**, 3724.

Noack, F. (1971), in *NMR: Basic Principles and Progress*, Diehl, P., E. Fluck, and R. Kosfeld, Eds., Springer Verlag, New York. Vol. 3, pp. 83–144.

Noggle, J. H., and R. E. Schirmer (1971), *The Nuclear Overhauser Effect*, Academic Press, New York, NY.

Norton, R. S., A. O. Clouse, R. Addleman, and A. Allerhand (1977), *J. Am. Chem. Soc.* **99**, 79.

Oldfield, E., and A. Allerhand (1975), *J. Am. Chem. Soc.* **97**, 221.

Oldfield, E., R. S. Norton, and A. Allerhand (1975), *J. Biol. Chem.* **250**, 6368.

Opella, S. J., D. J. Nelson, and O. Jardetzky (1974), *J. Am. Chem. Soc.* **96**, 7157.

Pechhold, W. (1968), *Kolloid Z. Z. Polym.* **228**, 1.

Perrin, F. (1936), *J. Phys. Radium* **5**, 497.

Poe, M., and K. Hoogsteen (1978), *J. Biol. Chem.* **253**, 543.

Pogliani, L., M. Ellenberger, and J. Valat (1975), *Org. Magn. Reson.* **7**, 61.

Pople, J. A., W. G. Schneider, and J. J. Bernstein (1959), *High Resolution Nuclear Magnetic Resonance*, McGraw-Hill, New York, NY.

Prestegard, J. H., and D. M. Grant (1978), *J. Am. Chem. Soc.* **100**, 4664.

Reeke, Jr., G. N., J. W. Becker, and G. M. Edelman (1978), *Proc. Nat. Acad. Sci. U.S.A.* **75**, 2286.

Roberts, G. C. K., J. Feeney, B. Birdsall, B. J. Kimber, D. V. Griffiths, R. W. King, and A. S. V. Burgen (1977), in *NMR in Biology*, R. A. Dwek, I. D. Campbell, R. E. Richards, and R. J. P. Williams, Eds., Academic Press, Oxford, England, pp. 95–109.

Roques, B. P., C. Garbay-Jaurequiberry, S. Combrisson, and R. Oberlin (1977), *Biopolymers* **16**, 937.

Rowan III, R., J. A. McCammon, and B. D. Sykes (1974), *J. Am. Chem. Soc.* **96**, 4773.

Saito, H., H. H. Mantsch, and I. C. P. Smith (1973), *J. Am. Chem. Soc.* **95**, 8453.

Saito, H., T. Ohki, M. Kodama, and C. Nagata (1978), *Biopolymers* **17**, 2587.

Saviotti, M. L., and W. C. Galley (1974), *Proc. Nat. Acad. Sci. U.S.A.* **71**, 4154.

Schaefer, J., and D. F. S. Natusch (1972), *Macromolecules* **5**, 416.

Schaefer, J. (1973), *Macromolecules* **6**, 882.

Shimizu, H. (1962), *J. Chem. Phys.* **37**, 765.

Smith, I. C. P., R. Deslauriers, and K. Schaumburg (1975), in *Proc. 4th American Peptide Symp.*, Walter, R. and J. Meienhofer, Eds., Ann Arbor Science Press, Ann Arbor, Mich. p. 97.

Soda, G., and H. Chihara (1974), *J. Phys. Soc. Japan* **36**, 954.

Sogn, J. A., W. A. Gibbons, and E. W. Randall (1973), *Biochemistry* **12**, 2100.

Solomon, I. (1955), *Phys. Rev.* **99**, 559.

Somorjai, R. L., and R. Deslauriers (1976), *J. Am. Chem. Soc.* **98**, 6460.

Spiess, H. W. (1978), in *NMR: Basic Principles and Progress* Diehl, P., E. Fluck, and R. Kosfeld, Eds., Springer Verlag, New York, Vol. 15, p. 55.

Steele, W. A. (1976), *Adv. Chem. Phys.* **34**, 1.

Steigel, A. (1978), in *NMR: Basic Principles and Progress* Vol. 15, p. 1.

Sternlicht, H., D. T. Browne, G. L. Kenyon, E. L. Packer, and D. M. Wilson (1973), *Biochem. Biophys. Res. Commun.* **50**, 42.

Stilbs, P., and M. E. Moseley (1979), *J. Magn. Reson.* **33**, 209.

Sykes, B. D., and M. D. Scott (1972), *Ann. Rev. Biophys. Bioeng.* **1**, 27.

Sykes, B. D., W. E. Hull, and G. H. Snyder (1978), *Biophys. J.* **21**, 137.

Tancréde, P., R. Deslauriers, W. H. McGregor, E. Ralston, D. Sarantakis, R. L. Somorjai, and I. C. P. Smith (1978), *Biochemistry* **17**, 2905.

Thomas, W. A., and M. K. Williams (1972), *J. Chem. Soc., Chem. Commun.*, 994.

Torchia, D. A. (1971), *Macromolecules* **4**, 440.

Torchia, D. A., and J. R. Lyerla, Jr. (1974), *Biopolymers* **13**, 97.

Torchia, D. A., J. R. Lyerla Jr., and A. J. Quattrone (1975), *Biochemistry* **14**, 887.

Torchia, D. A., and D. L. VanderHart (1976), *J. Mol. Biol.* **104**, 315.

Traüble, H. (1971), *J. Membrane Biol.* **4**, 193.

Tsuboi, M., and M. Nakanishi (1979), *Adv. Biophys.* **12**, 101.

Tsutsumi, A. (1979), *Mol. Phys.* **37**, 111.

Uiterkamp, A. J. M. S., I. M. Armitage, J. H. Prestegard, J. Slomski, and J. E. Coleman (1978), *Biochemistry* **17**, 3730.

Urry, D. W., T. L. Trapane, and M. A. Khaled (1978), *J. Am. Chem. Soc.* **100**, 7744.

Valeur, B., J. P. Jarry, F. Geny, and L. Monnerie (1975), *J. Polym. Sci., Polymer Phys. Ed.* **13**, 667.

Van Putte, K. (1970), *J. Magn. Reson.* **2**, 23.

Venkatachalam, C. M., B. J. Price, and S. Krimm (1974), *Macromolecules* **7**, 212.

Venkatachalam, C. M., B. J. Price, and S. Krimm (1975), *Biopolymers* **14**, 1121.

Visscher, R. B., and F. R. N. Gurd (1975), *J. Biol. Chem.* **250**, 2238.

Wagner, G., and K. Wüthrich (1978), *Nature* **275**, 247.

Wallach, D. (1967), *J. Chem. Phys.* **47**, 5258.

Walter, R., I. C. P. Smith, and R. Deslauriers (1974), *Biochem. Biophys. Res. Commun.* **58,** 216.

Werbelow, L. G., and D. M. Grant (1975a), *J. Chem. Phys.* **63,** 544.

Werbelow, L. G., and D. M. Grant (1975b), *J. Chem. Phys.* **63,** 4742.

Werbelow, L. G., and D. M. Grant (1977), *Adv. Magn. Reson.* **9,** 189.

Wilbur, D. J., R. S. Norton, A. O. Clouse, R. Addleman, and A. Allerhand (1976), *J. Am. Chem. Soc.* **98,** 8250.

Wittebort, R. J., T. M. Rothgeb, A. Szabo, and F. R. N. Gurd (1979), *Proc. Natl. Acad. Sci. U.S.A.* **76,** 1059.

Wittebort, R. J., and A. Szabo (1978), *J. Chem. Phys.* **69,** 1722.

Woessner, D. E. (1962a), *J. Chem. Phys.* **36,** 1.

Woessner, D. E. (1962b), *J. Chem. Phys.* **37,** 647.

Woessner, D. E. (1965), *J. Chem. Phys.* **42,** 1855.

Woessner, D. E., B. S. Snowden, Jr., and G. H. Meyer (1969), *J. Chem. Phys.* **50,** 719.

Wokaun, A., and R. R. Ernst (1978), *Mol. Phys.* **36,** 317.

Wright, D. A., D. E. Axelson, and G. C. Levy (1979), in *Topics in Carbon-13 NMR Spectroscopy* Vol. 3., 103.

Wüthrich, K., and R. Baumann (1976), *Org. Magn. Reson.* **8,** 532.

Wüthrich, K., and G. Wagner (1975), *FEBS Letters* **50,** 265.

Zens, A. P., T. J. Williams, J. C. Wisowaty, R. R. Fisher, R. B. Dunlap, T. A. Bryson, and P. D. Ellis (1975), *J. Am. Chem. Soc.* **97,** 2850.

Two

Medical Imaging by NMR

David I. Hoult

**Biomedical Engineering and Instrumentation Branch
Division of Research Services
National Institutes of Health
Bethesda, MD 20205**

1. INTRODUCTION

A prerequisite for the usual high-resolution NMR experiment is the availability of a highly homogenous (typical variance 1 part in 10^8) magnetic field in which to place the sample of interest. If the field varies over the active sample volume, the Larmor frequency of the nuclei also varies, and so a particular nuclear species no longer has a unique frequency. Rather, the spectral line of that species is broadened and distorted in a manner which reflects the variations of field. Normally, of course, such broadening is annoying as it impairs the resolution of the experiment. However, it has long been recognized (Gabillard, 1951, 1952) that deliberate employment of inhomogenous magnetic fields can yield spatial information. Consider, for

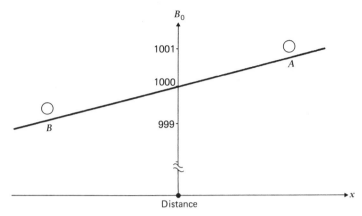

Figure 1. When NMR is conducted in a field which has a gradient, the Larmor frequency of the sample is a direct indication of its spatial position. Thus sample A has a higher frequency than sample B and the position of a sample can therefore be readily found.

example, the situation shown in Figure 1 where the magnetic field varies in a linear manner with distance x. The frequency of a small sample placed in such a field is a direct measure of the position x of the sample. Thus at point A, the sample will have a higher frequency than at point B. There is, in fact, nothing to stop our putting several samples in the field and finding their positions, and if the samples have distinguishing features (other than chemical shift) such as differing concentrations or spin-lattice relaxation times (T_1) then we can, by utilizing the usual store of NMR techniques, find unambiguously the x position of each sample. It is but a short step from the use of several discrete samples to the analysis of a continuous sample in order to determine, say, the variation of sample concentration with distance, and Figure 2 shows the proton *absorption* spectrum that one might expect from two beakers of water filled to different levels. (The dispersion spectrum reflects *changes* in sample concentration.)

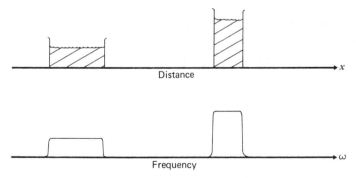

Figure 2. If we place two rectangular beakers of water in the field of Figure 1 and take the NMR spectrum, the absorption part of the spectrum is a good representation of the water distribution.

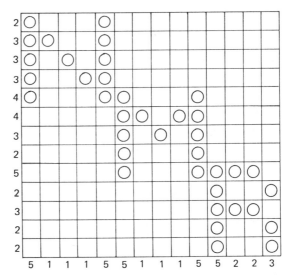

Figure 3. The letters NMR formed by a set of small samples, and their projections on the two axes. There are insufficient data to reconstruct the letters, as evidenced by the fact that if the letter N is reversed, the projections do not alter.

Now it would be useful to be able to produce images or locate our samples in two or even three dimensions instead of just one, and we might imagine, at first sight, that all that was necessary to obtain information about the second dimension was to change the direction of the field gradient of Figure 1 from the x axis to the y axis. That this is not so may be seen from Figure 3, where a set of small samples have been arranged to form the letters NMR. The numbers on the two axes are measures of the amplitudes of the signals received when the gradients are in the two directions. Each set of numbers represents effectively a one-dimensional projection of the letters. That these two projections are insufficient to define the matrix of Figure 3 may be clearly seen by reversing the letter N thus: ИMR. The projections remain the same even though the abbreviation is dyslexic, and more information is needed to define adequately the matrix. The ways in which more information can be obtained are now numerous, and we therefore examine them briefly before considering other aspects of the subject in more detail.

2. TWO-DIMENSIONAL IMAGING

Lauterbur (1973) in his now-classic paper in *Nature,* described a reconstructional method utilizing not just projections at 90°, but rather a set of projections at a variety of angles. Each projection contains some unique information, and by suitable mathematical manipulation this information can be utilized to form an image of the original object. The exact details of this

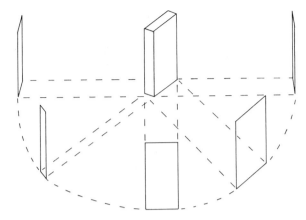

Figure 4. Projections of a rectangular block. Imagine a beam of light shining from behind the block at various angles. The shadows cast on a wall round the block are projections of the block, and, of course, their shapes are dependent upon the projection angle.

technique of "filtered back-projection reconstruction" are outside the scope of the present article (for references see Houndsfield and Ambrose, 1973, Lauterbur, 1973, and Brooks and diChiro, 1976). The method has gained wide acceptance through its use in computer-assisted X-ray tomography. Figure 4 illustrates how projections at various angles have differing information content. The back-projection is shown in Figure 5 and is an approximation to the original rectangular section. Many more projections and filtering are needed before the image is acceptable, but Figure 5 does demonstrate the

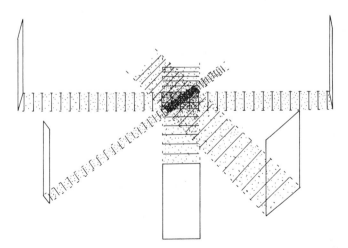

Figure 5. We now imagine that the projections of Figure 4 are solid entities which cast shadows as shown (back projection). The area of deepest shadow is where the block of Figure 4 was located, and if a sufficient number of projections is taken an accurate reconstruction of the rectangle may be obtained.

principles involved. Lauterbur, in esoteric fashion, named his technique "zeugmatography" from the Greek "ζευγμα" meaning a yoke, or "that which joins together." His idea was that the field, with its gradient, couples the spatial frame and the transmitter B_1 field together, permitting spatial resolution of an order very much less than the wavelength of the B_1 irradiation, and thereby circumventing the restrictions of the Uncertainty Principle.

Other techniques soon followed, for example, Fourier zeugmatography. With this method, Kumar *et al.* (1975a, b) removed the problem of the second dimension from the spatial (frequency) domain to the time domain. The principle behind the method is one common to many branches of spectroscopy where two-dimensional information has to be analyzed—2D Fourier transformation. Suppose we apply a 90° pulse to a small sample at position (x, y) followed by a gradient in the x direction for a time t_1. Let the Larmor frequency relative to some reference be given by $\omega_0 = -\gamma B_{01} x$ where γ is the gyromagnetic ratio and B_{01} is a constant. Then at the end of time t_1, the phase of the signal from the small sample at x is given by $\phi = -\gamma B_{01} x t_1$. In other words, the phase of the signal reflects its x location. Let us now switch the gradient rapidly into the y direction and collect the free induction decay. As usual, the frequency of the decay is governed by the y position ($\omega_0 = -\gamma B_{02} y$, say) and Fourier transformation yields y information, but the *phase* ϕ of the spectrum is dependent on x and t_1. Let us store the spectrum and repeat the experiment, but this time, we increase t_1. Clearly, the phase of the new spectrum changes and the rate of change of phase with increasing t_1 is a direct measure of the x position: $\partial\phi/\partial t_1 = -\gamma B_{01} x$. $\partial\phi/\partial t_1$ is, of course, frequency, and so by Fourier analyzing the rate of change of phase of each point in the spectrum as t_1 increases, we obtain information as to the x coordinate of the sample. This is shown in Figure 6. Mathematically, this procedure may be expressed as follows. Let the spin-density function be $M(x, y)$. Then, following a 90° pulse, the signal received from an elementary area $dx\ dy$ is given by

$$d\xi(t_1, t_2) \propto M(x, y) \exp\{-[j(\gamma B_{01} x t_1 + \gamma B_{02} y t_2) + t/T_2]\}\ dx\ dy \quad (1)$$

where $j = \sqrt{-1}$ and signal accumulation commences at time $t_2 = 0$. B_{02} is a constant and T_2 is the spin–spin relaxation time. The total received signal is therefore

$$\xi(t_1, t_2) \propto \int_x \int_y M(x, y) \exp\{-[j\gamma(B_{01} x t_1 + B_{02} y t_2)]\}\ dx\ dy\ \exp(-t/T_2) \quad (2)$$

Let $\xi(t_1, t_2) = \rho(t_1, t_2) \exp(-t/T_2)$.

Then from normal Fourier relationships, we see that

$$M(x, y) \propto \int_{t_1=0=t_2}^{\infty} \int^{\infty} \rho(t_1, t_2) \exp\{+j\gamma(B_{01} x t_1 + B_{02} y t_2)\}\ dt_1\ dt_2 \quad (3)$$

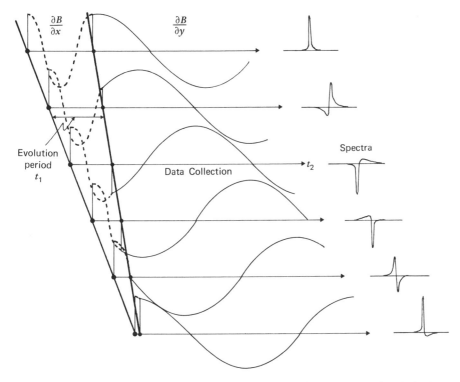

Figure 6. Fourier Zeugmatography. During the evolution period t_1, the gradient is in the x direction. The frequency of the FID (and therefore the phase at time t_1) is thus dependent on the x position. The gradient is now switched into the y direction and the signal collected. The *frequency* of the FID is now dependent upon the y position, but the *phase* is determined by t_1 and x. Variation of t_1 therefore reveals the x position of the sample.

It follows that the Fourier transform of ξ is the convolution of $M(x, y)$ with a Lorentzian line of width $1/\pi T_2$ (in hertz), and so long as $M(x, y)$ is sufficiently diffused over frequencies very much greater than the natural linewidth (i.e., the gradients are strong enough) this result is, to a good approximation, the desired spin-density function $M(x, y)$.

A similar technique, rotating frame zeugmatography proposed by Hoult (1979a), removes the need for rapid changes of gradient by applying an incline in the B_1 field; in other words, the flip angle associated with a pulse of length t_1 varies with distance y. Thus at one side of the sample, the flip angle might be 30° whereas at the other side the angle might be 150°. That this technique, in association with a static field gradient in the x direction, produces resolution in both directions may be understood as follows. Let us consider our small sample placed once again at position (x, y). Then following a pulse of length t_1, the magnetic moment M_o has been tipped through an angle $\theta = \gamma(B_{10} + B_{11}y)t_1$, where B_{10} and B_{11} are constants. Thus the free induction decay following the pulse is given by

$$d\xi \propto M(x, y) \sin \{\gamma(B_{10} + B_{11}y)t_1\} \exp \{j\gamma B_{01}xt - t/T_2\} \, dx \, dy \quad (4)$$

Following Fourier transformation, the frequency of the signal is dependent on position x, but the amplitude varies with position y as shown in Figure 7. The amplitude, however, also varies with t_1 sinusoidally, and so if we repeat the experiment but increment t_1, we obtain a matrix of spectra in which the amplitude of a peak oscillates at a frequency which is dependent on position y, as shown in Figure 8. A second Fourier transformation down the columns of the matrix therefore reveals the y position. The method once again employs a two-dimensional Fourier transformation.

The three methods described above have one facet in common: they collect information simultaneously from all points in the plane of interest and as we shall see later, they are therefore inherently sensitive. So far in our analyses, we have mainly considered for clarity the case of a small sample located at some position (x, y) but no restrictions have been made on x or y—what applies to a sample at one point applies equally to a sample at another and so we can gain information impartially and simultaneously from all parts of a distributed object. This is not so for the next class of methods to be discussed. These techniques, we shall see, collect information from lines across the sample. Consider first the method of echo planar imaging used by Mansfield and Pykett (1978). Once again we place a small sample at position (x, y) and apply a static gradient in the x direction so that the frequency of the free induction decay (FID) is dependent upon x as usual. However, let us

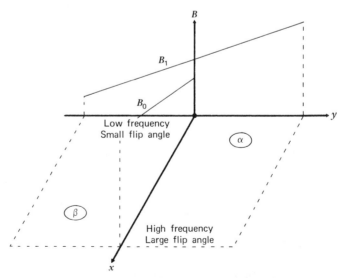

Figure 7. Rotating Frame Zeugmatography. Not only does the static magnetic field B_0 have a gradient, but also the transmitting field B_1. Thus the flip angle of the experiment depends upon the y position as well as the pulse length.

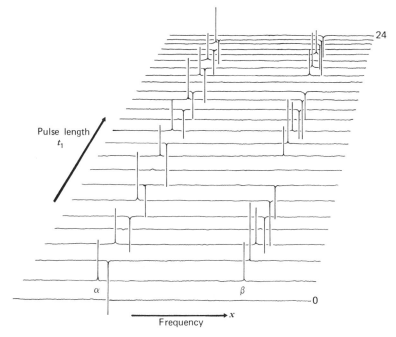

Figure 8. Referring to the two small samples α and β of figure 7, we see that as the pulse length t_1 increases, the amplitude of the α peak oscillates more rapidly than that of the β peak, because α is in a stronger B_1 field. Thus, while frequency indicates the x position, the rapidity of oscillation indicates the y position.

also assume, for simplicity, that we happen to be on resonance by a judicious choice of the spectrometer frequency. We therefore observe, following a pulse, an ordinary FID as shown in Figure 9a. Let us now apply, in the y direction, a strong gradient which is periodically reversed, as shown. In the middle of each cycle, the spins attain the same phase in the rotating xy frame that they had immediately after the pulse; it is as if time $t = 0$ recurred every cycle. Of course, when the y gradient is applied, the sample is no longer at resonance; rather its frequency is $\pm \gamma B_{02} y$. The positive frequency is shown in Figure 9c. Now let us consider Fourier transformation of this signal. The first step in transformation is multiplication by a sinusoid, followed by integration, and we expect a large result when the sinusoid has the same frequency *and phase* as the signal of interest. In general, this is not the case. Figure 9d shows a sinusoid of the same frequency as the sample, but the phase relationship in the echo is incorrect. It is a simple matter to show that the phase relationship is correct in the echoes only when the frequency of the sinusoid is a multiple n of the frequency of gradient alternation ω_m, and so we expect a large signal only at certain positions in the spectrum, corresponding to various values of y. Of course, if we change the x position the above analysis no longer holds, but in practice, once the signals have been

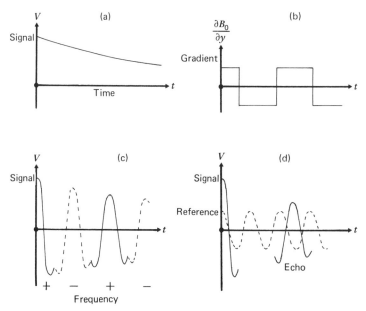

Figure 9. Echo Planar Imaging. A static gradient is applied in the x direction, and (a) shows an ordinary FID from a small sample which happens to be on resonance. If the sample is now subjected to an alternating gradient in the y direction (b), the sample frequency swings between positive and negative values (relative to the on-resonance condition) as the gradient alternates. (c) shows the cosine component of the signal, and it is clear that echoes are formed. However, even though the echo has the same *frequency* as the signal, we see from (d) that it does not in general have the same *phase*.

recorded in the computer, being "on resonance" is a matter only of mathematics and provided $\gamma B_{01} x$ does not exceed ω_m, the peak in the spectrum is shifted from $n\omega_m$ to $n\omega_m + \gamma B_{01} x$ and we have an unambiguous assignment of the x position for a particular y region. The signal at frequency $n\omega_m$ can be shown to come from a range of y positions about $y = n\omega_m / \gamma B_{02}$ and so the method has a sensitivity approaching the other methods described above. In addition it has the potential of being implemented rather rapidly. Figure 10 shows the spectrum one would obtain from an annulus. The image is constructed from the various zones of the spectrum.

Another method of obtaining information from a number of lines in a sample is Hinshaw's multiple sensitive point method (Andrew *et al.*, 1977). Once again a static gradient is applied in the x direction, but in the y direction an alternating gradient is applied in a manner similar to that described above. For times long in comparison to the period of oscillation, only sample at $y = 0$ (where the field does not alter) is capable of satisfying a resonant condition and so signal can only be obtained from a line at $y = 0$. This line is stimulated by a train of $\frac{1}{2}\pi$ pulses and a steady-state condition builds up in which signal from the line bunches about multiples of the train frequency ω_m. Thus, signal

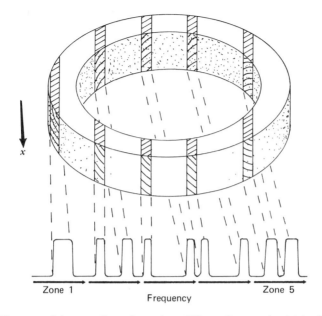

Figure 10. Because of the y gradient alternation of Figure 9, we only obtain signal from the hatched regions of the annulus. However, the static x gradient gives resolution in the x direction and so each sideband is spread out so that it describes concentration versus x; in effect, the spectrum is zoned, each zone being centered on a sideband.

from about $x = n\omega_m/\gamma B_{01}$ shows in a Fourier analysis as a spectral line at frequency $n\omega_m$. The "sensitive points" on the x axis may now be driven through the sample in the y direction as shown in Figure 11 by changing the y origin electronically. The changes in sample concentration are mirrored in the changes in the sideband amplitudes and so a map of concentration versus distance for each of the points can be produced, that is, an image may be formed. Another possibility (Hinshaw, 1976) is to stimulate only the sensitive line with a pulse, record the decay, and obtain the spin density along that line by Fourier transformation. There are, however, other ways of performing such an experiment by the use of selective excitation (Garroway *et al.*, 1974; Lauterbur *et al.*, 1975a, b; Sutherland and Hutchison, 1978). If a gradient is applied in the y direction and the sample is excited with a tailored pulse whose frequency spectrum covers only the y range of interest, the pulse might, for example, be shaped as sinc t, giving a rectangular frequency distribution. The response of the sample mimics the pulse shape for flip angles of 50° or less and as the receiver is turned off when the transmitter is on, negligible signal is observed. However, if the gradient is reversed at the end of the pulse, an echo is formed (Hoult, 1977, 1979b; Fernbach and Proctor, 1955; Sutherland and Hutchison, 1978) and the echo contains signal only from the y range of interest. If the y gradient is now terminated in the middle of the echo as shown in Figure 12, and replaced with a gradient in the

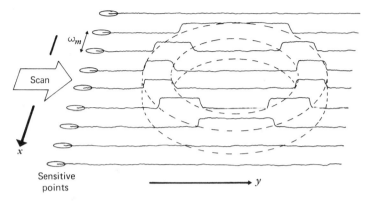

Figure 11. The Multiple Sensitive Point Method. By using a sequence of 90° pulses, signal is obtained from the many sideband frequencies. As a gradient is present in the x direction, this implies that signal is received from a number of x positions. The locality of these positions is restricted to $y = 0$ by application of an alternating gradient in the y direction, so creating multiple sensitive points. These points can then be scanned through the sample by electronically altering the y origin, and the figure shows the results one would expect for an annular sample.

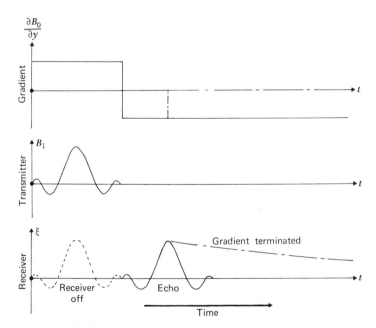

Figure 12. If a selective pulse is applied in the presence of a powerful field gradient so that only a fraction of the sample is excited, the received signal mimics the pulse and no signal is observed for the receiver is turned off. However, by reversing the gradient, an echo can be formed and if the gradient is terminated in the middle of the echo, evolution of the magnetization ceases, leaving a coherent signal to be processed as desired.

x direction, the free induction decay may be accumulated in the usual way and Fourier transformed to give a line image. Other possibilities include exciting several widely spread *y* portions simultaneously and stringing out in a line the *x* information from each of them in a manner similar to that used in echoplanar imaging. Whichever method is used, the underlying theme is the same: a line (or lines) is excited and viewed.

Finally, we turn to the least sensitive methods which aim to produce signal from a localized volume. These methods do not use the zeugmatographic principle, for they do not employ static linear field gradients. In the precursor of the multiple sensitive point method, Hinshaw (1974, 1976) applies alternating gradients in all three orthogonal directions *x*, *y*, and *z*. Thus there is a unique origin point in space which does not experience fluctuations in field strength and which therefore satisfies fully the conditions for resonance. Needless to say, the three gradients must alternate at unrelated frequencies. The signal from the sensitive point is a measure of the spin density at that point and, to form an image, the point is scanned through the sample in raster fashion and recorded. Damadian and colleagues (1976, 1977, 1978; Minkoff *et al.*, 1977; Goldsmith *et al.*, 1977) localize their volume of interest with the aid of an inhomogenous magnetic field which produces a saddle-shaped field profile—a technique they term "Field Focusing NMR" (FONAR). Following selective excitation, appreciable signal is obtained only from the region of the saddle point; signal from other regions is either off-resonance or is rapidly dephased. The principle is illustrated in Figure 13. Yamada *et al.* (1978) in a variation of the FONAR method, have produced a field profile which has a localized minimum. The magnet required to produce

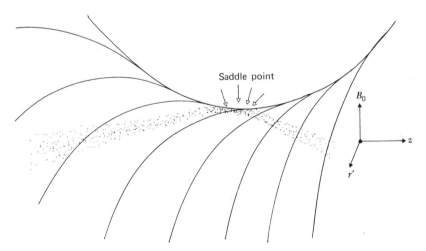

Figure 13. If a saddle-shaped magnetic field is applied to the sample, only the shaded region responds to a selective pulse. However, most of that region lies in a strong field gradient and so, following the pulse, negligible signal remains (See Figure 12). Only from the saddle-point (arrowed) where there is no gradient do we obtain appreciable signal.

such a profile is, however, considerably more complex than that needed to generate a saddle-shaped field, and it remains to be seen whether the technique will be easily applied.

Such then, briefly, are the methods of NMR imaging. Each, quite apart from considerations of speed and sensitivity, has its advantages and drawbacks. One always has to remember that not only is spin density a variable, but that T_1 and T_2 can also influence the results obtained, as can a whole host of technical factors. The image is a multiparametered variable and the choice of technique may well be determined by the system under study. Let us therefore consider some of the factors which influence the formation of an image.

3. SENSITIVITY, SPEED, AND RESOLUTION

Consider once again our small sample sitting somewhere within the region of interest. What are the factors which determine the signal-to-noise ratio that we may obtain from the sample in a given time? This question has, basically, nothing to do with imaging and has been considered in detail by several authors. The approach adopted in the present article is that of Hoult and Richards (1976). They first consider the signal-to-noise ratio ψ from the sample following Fourier transformation of the FID from a 90° pulse. Their conclusions are that

$$\psi = \frac{\alpha \omega_0^{7/4} \, V \, T_2^{1/2} \, C}{a \, T_s \, T_c^{3/4}} \tag{5}$$

where, for a particular nuclear species (usually ^{1}H) α is a constant which depends on the receiving coil geometry and is generally greatest for a solenoidal coil, ω_0 is the Larmor frequency, V is the sample volume, C is the sample concentration, a is the receiving coil radius, T_s is the sample temperature, and T_c is the coil temperature. Obviously, ψ is maximized in the largest magnetic field with a cooled sample and coil, the latter having the smallest dimensions possible. However, major difficulties arise in the pursuit of these objectives. The most serious concerns the "constant" α. For given coil dimensions, there is a limiting frequency above which α tends to decrease rapidly. This frequency is *approximately* given by, for a solenoidal receiving coil,

$$\omega_{\max} \sim 2\pi 10^6/a \tag{6}$$

where a is in meters. Thus a coil 30 cm in diameter designed to fit over a person's head will not operate satisfactorily above about 6 MHz. This phenomenon, caused by distributed capacitance in the coil, is discussed fully elsewhere (Hoult, 1978), but it is clear that it imposes a very severe limitation upon the sensitivity of an imaging experiment. We may not employ without limit the most powerful dependence of equation (5). Having reached a limit, we must therefore turn to the other dependencies. The minimum coil size is fixed by the total region of interest (e.g., a person's head), even though we may only be interested in a small volume therein. In most cases of biological interest too, the sample temperature is fixed. The coil temperature is the only remaining variable and we may therefore hope to cool the coil to reduce the Johnson noise it produces. However, this too can be a false hope when the sample is biological, for in such cases the sample is electrically conducting and dielectrically lossy, and the noise contribution from the sample itself may well dominate matters. This loss mechanism is discussed fully by Hoult and Lauterbur (1979) and Gadian and Robinson (1979), and their conclusion is that in many cases, particularly when close to the maximum operating frequency, it is not worth cooling the coils. Yet another factor in this conspiracy of Nature is that the electrical conductivity of the sample limits the depth within that sample from which signal may be received (see Appendix A). In most biological systems with a salt concentration in the region of 120 mM, the effective penetration depth δ is approximately given by

$$\delta \sim \frac{1.4 \times 10^3}{\sqrt{\omega_0}} \qquad (7)$$

Thus at 6 MHz, $\delta \sim 22$ cm. As penetration occurs in differing degrees from all sides of the sample, this "skin effect" is fortunately not a limiting factor. Nevertheless, it may influence results in some circumstances and we must be mindful of its existence and in particular, of the phase changes which accompany it. The conclusion is, as is usual in NMR spectroscopy, that we are severely limited in sensitivity. Hoult and Lauterbur (1979) have estimated that the *maximum* signal-to-noise ratio available from a volume V of water in milliliters, with $T_2 = 0.1$ s, within a solenoidal receiving coil of radius a is given by

$$\psi \sim V\omega_0^{7/4} \times 10^{-11}/a \qquad (8)$$

Thus we might expect, with luck, from a coil at 6 MHz around a person's head ($a = 0.15$ m) a signal-to-noise ratio of 1000 from 1 ml. In practice, because of the factors described above, a value of 500 might be more reasonable, and further, it is very easy to lose sensitivity through instrumental factors and interference. The sensitivity can, of course, be improved by averaging and then the value becomes dependent on T_1 and the number of accumulations in a given time in the usual way (Ernst and Anderson, 1966).

Having determined, for the pertinent experimental conditions, the sensitivity available from our small sample of volume V, we can now return to imaging and estimate the sensitivity and resolution available to us. (A discussion of this topic is given by Brunner and Ernst, 1979.) Remaining for the moment with two-dimensional imaging, we can consider a plane of sample to comprise $m \times n$ small samples, each of volume V, as shown in Figure 14.

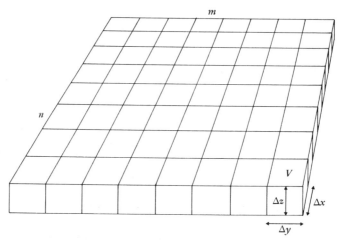

Figure 14. A plane of sample can be considered as an assembly of mn elements (voxels) each of volume V. Each element produces a picture element (pixel) with a certain signal/noise ratio, and the ideal imaging method collects information simultaneously from all elements.

With the first three imaging methods discussed (namely, back-projection reconstruction, Fourier zeugmatography, and rotating frame zeugmatography) signal is obtained simultaneously from all $m \times n$ volume elements. Thus if, in some way, all the necessary information for producing an image could be recorded in one FID, we would have an $m \times n$ image, where *each* picture element would have a sensitivity ψ given by equation (5). We must now ask what is an acceptable sensitivity, say ψ_a. Typically, for acceptable picture quality, this might be in the range of 10 to 50. From equation (5), or more specifically for water with $T_2 = 0.1$ s, we can then find V from equation (8). Knowing V and the sample dimensions (including depth), it is a trivial matter to compute m and n and the resolution available. For example, suppose we are looking at water content in a plane of depth Δz. Let $V = \Delta x \, \Delta y \, \Delta z \times 10^6$ where V is in milliliters and the dimensions are in meters. Then from equation (8)

$$\Delta x \, \Delta y \sim \frac{10^5 \, a \, \psi_a}{\omega_0^{7/4} \, \Delta z} \tag{9}$$

For a square element ($\Delta x = \Delta y$) we then have that

$$\Delta x = \Delta y = \left(\frac{10^5 \, a \, \psi_a}{\omega_0^{7/4} \, \Delta z} \right)^{1/2} \tag{10}$$

Thus at 6 MHz a solenoidal coil of diameter 30 cm about the head might be capable of giving a linear resolution of 2.2 mm in a plane Δz 5 mm deep, with a signal-to-noise ratio of 30 for water. If we were prepared to accept a sensitivity of only 10, then Δx would be 1.3 mm. The resolution would be better but the sensitivity would be worse and the picture would be noisy.

So far, we have related directly sensitivity and resolution for a two-dimensional situation and, ignoring the delicate question of how to perform the experiment, have roughly calculated the resolution available from a single FID. To obtain better resolution, we must have more sensitivity; then we can obtain ψ_a from a smaller volume. The only way left to improve the signal is to average, and of course as soon as accumulation enters the picture, T_1 becomes a factor determining sensitivity. If we wish to obtain an image which is independent of T_1, then we must repeat the experiment so slowly that full recovery of the longitudinal magnetization takes place between pulses. The increase of signal-to-noise ratio is correspondingly slow, however. It is possible to obtain water relaxation times as long as 2.5 s and so, for an improvement of a factor of 5 in linear resolution in a two-dimensional experiment, an increase of 25 in sensitivity is called for and hence an experiment lasting over 2 h. If we wished to decrease the thickness of the plane by a factor of 5 as well, the experiment would take 5^6 ($5T_1$) or 54 h. The price of high resolution is clearly exorbitant, having a sixth-power time dependency. We may reduce this time somewhat (typically by a factor of 2.5) by repeating the experiment more rapidly and using, if possible, an optimum flip angle given by $\theta = \arccos [\exp (-T_r/T_1)]$, where T_r is the repetition period, but the price paid is that the signal amplitude becomes T_1 dependent, which may or may not be useful. Clearly, the basic resolution is determined overridingly by the physical constants of the system and we are not at liberty to improve resolution at will. It is, however, worth noting from equation (10) that the linear resolution in a two-dimensional image is almost inversely proportional to frequency within the constraints already discussed, and clearly this factor should not be ignored lightly. It is also apparent that, allowing for the uncertainties in equation (10), it ought to be possible to achieve a resolution of 2 mm in a head scan (slice thickness 5 mm) in times considerably less than 1 min, provided an efficient imaging method is used.

We must now consider what happens when other imaging techniques are employed. Consider, for example, the situation where only one line in the plane is excited. Once again, we obtain a resolution given typically by equation (10), but we have only looked at a fraction of the plane of interest, as shown in Figure 15. We have to repeat the experiment n times to observe n lines, and so at first sight one might imagine that a line scanning technique might take n times as long as a two-dimensional method. As n might be 100 or more, an accumulation time of many minutes might be envisaged. However, in most biological systems, $T_2 < T_1$. Suppose, taking previous values, that $T_2 = 0.1$ s and $T_1 = 2.5$ s. In a time $5T_1$, we can stimulate and observe 50 lines allowing $2.5T_2$ for each accumulation and so we have recouped much of the method's disadvantage. An exact comparison with a two-dimensional method depends very much on the experimental conditions. If we repeat a two-dimensional experiment more rapidly than every $5T_1$ in order to gain T_1 information then the disparity between the two methods becomes much greater. A *tentative* conclusion therefore is the following: when biological

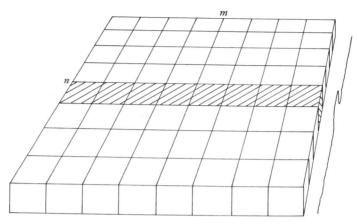

Figure 15. Line imaging techniques obtain signal from only a section of the plane of interest. However, this state of affairs is not as disadvantageous as might be supposed, for if $T_2 \ll T_1$, many lines can be imaged in time T_1.

images independent of T_1 values are required, there is little to choose between two-dimensional techniques and line- or multiple line-scanning methods. If T_1 information is also desired, then two-dimensional methods are faster. (Note that we have as yet introduced no constraints based on technical problems other than sensitivity. The practical implications of the above statements will be examined later.) An exception to the statement is echo-planar imaging (Mansfield and Pykett, 1978). In its dependencies it probably lies closer to two-dimensional techniques, but the method has not yet been sufficiently developed to permit a detailed evaluation.

To conclude this section, we consider the FONAR and sensitive-point techniques. Both are scanning techniques and obtain information only from one volume element at a time. They are therefore relatively inefficient and obtaining images with them can take times approaching an hour for the experimental conditions considered. Their chief advantage lies in their ability to obtain spectra from localized regions of the body. With both methods the focal point is scanned in raster fashion through the sample and the image is built up directly without the need for Fourier transformation. Such then are the interdependencies of sensitivity, resolution, and time. Let us now look at some of the technical factors involved in obtaining an image.

4. TECHNICAL CONSTRAINTS

Let us continue to consider the head as a sample as it is probable that one of the major uses of NMR imaging will be clinical. We have seen that from the point of view of sensitivity, an image with a resolution of about 2 mm may be obtained in a single FID. The image thus contains approximately 10,000 independent pieces of information if we consider that a slice through the head has an area of about 4×10^{-2} m^2. To collect this information in one FID lasting about 0.1 s implies that a datum point must be collected every 10 μs and that the bandwidth of the receiving system must be at least 50 kHz. Further there is the implication that the imaging technique used is 100% efficient and that the FID contains no redundant information. This is exceedingly unlikely. However, neglecting such details, there is a much more important constraint on the system concerning probe design. The calculations on sensitivity assumed that the best possible coil configuration for the experiment was employed—namely a solenoidal receiving coil. Such a coil, with correct design (Hoult, 1978), can have an unloaded quality factor (Q) approaching 1,000 in the frequency range 1 to 6 MHz. [With the head inside, this value may drop somewhat due to inductive coupling (Hoult and Lauterbur, 1979). Dielectric losses *must*, for safety's sake, be eliminated with the aid of a Faraday screen.] The bandwidth of the probe, as measured between the -3 dB points is therefore of the order of 1–10 kHz, depending on the experimental conditions. This bandwidth is clearly insufficient to meet our requirement of 50 kHz. Of course the coil can be resistively loaded to increase the bandwidth by an order of magnitude, but then the sensitivity decreases by a factor of 3 or 4 and the resolution is degraded. A partial solution to the problem is to cool the damping resistance (either literally or by electronic ingenuity) so that it generates less noise (Hoult, 1979c; Radeka, 1974) but the problem is really fundamental—we do not have sufficient bandwidth. This being the case, we are forced to spread the data accumulation over several FID's if we desire good resolution. As a compensation, we obtain good sensitivity, but it also means that the sample must remain stationary for at least several seconds. It must also be remembered that once we have met the constraints of bandwidth, as defined by the -3 dB points, the signal amplitude and phase change with frequency by up to 30% and 90°, respectively, and a computer correction may be needed to compensate for these variations. The situation is summarized in Figure 16.

Other major design problems associated with the bandwidth concern the recovery time of the receiver after the pulse, the rf power needed to generate a pulse and the power requirements necessary to produce field gradients. The recovery time problem can largely be solved by suitable damping techniques (Hoult, 1979c), but the power requirements can demand considerable engineering. Suppose we employ a single coil system and wish uniformly to irradiate the bandwidth of the coil, that is, $\pm\omega_0/2Q$. Then in the rotating frame:

$$\omega_1 = \gamma B_1 \gg \omega_0/2Q \tag{11}$$

Let us assume that the shortest pulse of interest is a 90° pulse. ω_1 must be established in a time short compared to the pulse width and so the rise time

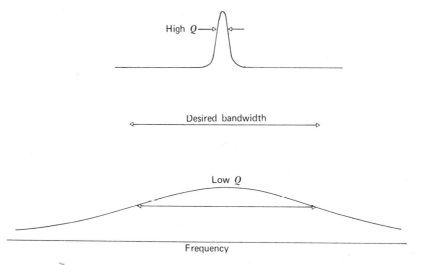

Figure 16. Good sensitivity is obtained from a high Q coil. However, the bandwidth of such a receiving coil is often insufficient for the implementation of rapid imaging, and so several FID's must be accumulated. Decreasing the Q to obtain the desired bandwidth results in reduced signal/noise ratios, and so, once again, several FID's must be stored.

constant τ of B_1 (caused by the Q of the probe) must be very much less than the pulse width, as shown in Figure 17. Thus

$$\tau = 2Q_{eff}/\omega_0 \ll \pi/2\omega_1 \tag{12}$$

where Q_{eff} is the effective Q of the probe and transmitter together. Thus from Equation (11)

$$Q_{eff} \ll \pi Q/2 \tag{13}$$

In other words, to irradiate the full bandwidth, the probe must be considerably damped when transmitting. We therefore need an estimate of the transmitter power under such circumstances. In Appendix B, it is shown that to produce a 90° pulse whose rise and fall time constant τ is one-fifth of the pulse width, the instantaneous power dissipation is given by

$$W \sim 1.5 \times 10^{-9} \, \omega_1^3 \, a^3 \tag{14}$$

and the effective Q of the probe and transmitter together is given by

$$Q_{eff} = \frac{\pi}{20}\left(\frac{\omega_0}{\omega_1}\right) \tag{15}$$

Suppose at the limits of receiver bandwidth $\pm\omega_0/2Q$ we require that ω_1 be three times $\omega_0/2Q$, that is, the effective B_1 field is only lifted out of the

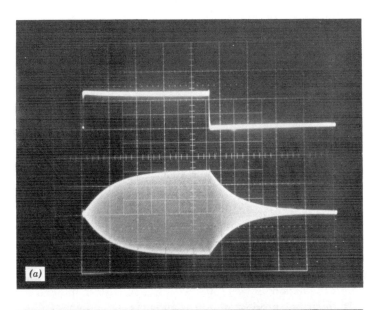

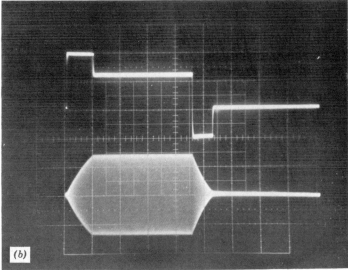

Figure 17. In (*a*), the rise and fall times of the rf pulse have a time constant which is about $\frac{1}{8}$ of the pulse width, resulting in some distortion of the pulse shape. The upper trace is the transmitter modulation function. If extra power is available, the waveform can be considerably improved as is shown in (*b*). A full discussion of the techniques involved is given by Hoult (1979c).

rotating xy plane by 18°. Then from equation (14) the power needed is

$$W \sim 5 \times 10^{-9} \left(\frac{\omega_0 a}{Q}\right)^3 \tag{16}$$

and

$$Q_{\text{eff}} = \pi Q/30 \tag{17}$$

So, for $\omega_0 = 2\pi\, 6 \times 10^6$, $a = 0.15$ m, and $Q = 500$, $W \sim 7.2$ kW. Note the cubic dependence on bandwidth and coil size in equation (16). Admittedly, some of this power may be dissipated in the internal resistance of the transmitter if the latter is used to reduce the effective Q of the probe, but a 2 kW transmitter is still needed if the full bandwidth of the receiver is to be irradiated satisfactorily without too much phase "glitch" (Ellett *et al.*, 1971).

Let us now turn to the question of field gradients. The gradient needed to produce a spread in proton frequency across the head of say 10 kHz is approximately 10^{-3} T m^{-1} (0.1 G cm^{-1}). The size of the gradient coils is determined by the linearity of field that they must produce, but the power they consume is dependent on the mass of conductor employed; for a given coil arrangement, the product of power and mass is a constant. Consider first a coil system, of the type shown in Figure 18, which produces a linear gradient along the main field axis z. The field about the origin is given by (Hoult, 1973):

$$B_z = \mu_0\, nI\, (0.64\, z/a^2) \tag{18}$$

and this formula is accurate to 1% provided $z/a \leqslant 0.4$. Thus to provide a linear gradient of 10^{-3} T m^{-1} to 1% accuracy across the head, $a \geqslant 0.33$ m, and it follows from equation (18) that $nI = 135$ A-turns. The power dissipation in each coil is approximately given by

$$W \sim 2\pi \rho a n^2 I^2/A \tag{19}$$

where ρ is the specific resistivity of the coil conductor and A is its total cross-sectional area. So for a copper coil, if we wish to limit the dissipated power to 10 W per coil, the conductor must be nearly a centimeter thick and its weight is about 1.2 kg per coil. These figures are perfectly acceptable, so the only decision left concerns n, the number of turns on the coil. It is preferable that n be very much greater than unity in order to minimize the stray fields from current return paths and to reduce the engineering demands of the power supply, but as n increases, so does the inductance of the coil system and if an experiment involving rapidly switched gradients is performed, then large, potentially destructive, transient voltages may be generated. An approximate formula for the inductance of the gradient coil system is

$$L \sim \left[5.8 \log_{10}\left(\frac{14}{\sqrt{A}}\right) - 4.63\right] \frac{an^2}{10^6} \tag{20}$$

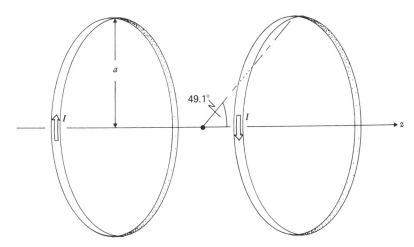

Figure 18. The coil system needed to generate a first-order field gradient in the axial direction. First-order transverse gradients can be produced by employing two such coil systems set at ±45° to the z axis. The currents in the second coil system should be reversed with respect to those in the first.

Thus for the system considered, $L \sim 4.6n^2 \mu H$. Suppose we wish to reverse the gradient in a time Δt. Then the induced voltage across the coil system is given by

$$V \sim 9.2 \times 10^{-6} \frac{n^2 I}{\Delta t} \tag{21}$$

So, for $n = 10$ and $\Delta t = 100 \ \mu s$, $V \sim 125$ V. The instantaneous power requirements of the supply are rather formidable—over 1 kW—and it is generally acknowledged that the production of rapidly switched gradients is no trivial matter. If anything, rapid gradient shutdown is more difficult than reversal, for capacitance discharge, such as is shown in Figure 19, may be used to reverse the current without dissipation of energy, but to eliminate the gradient, the stored energy must be given to a resistance. The use of silicon controlled rectifiers (RCA G5001 series) can aid considerably in the implementation of Figure 19.

The production of rapidly switched transverse gradients is an even more difficult matter, for the coils which produce such gradients tend to be rather inefficient if access to the working volume is not to be restricted, and so a greater current density and more coils are needed to produce similar gradients. This in turn implies higher instantaneous voltages if fast switching times are to be achieved. One must also not neglect the eddy current damping caused by conductors close to the coils. Such damping can badly perturb the local linearity of the gradient when switching and can cause rather disturbing anomalies in the final images. As a rough rule of thumb, all other conductors should be at least a coil radius (a) away from the gradient

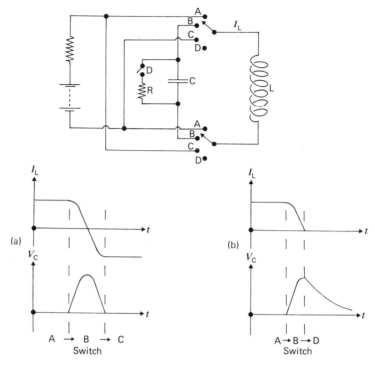

Figure 19. The most efficient way to reverse the current through gradient coils (L) is to utilize a capacitor (C) to resonate the coils. (*a*) shows the switching sequence needed, and the current and voltage associated respectively with L and C. If the gradient is to be terminated, the stored energy may be transferred to the capacitor and then dissipated in a resistor, as shown in Figure 6. The switches should be of the "make before break" variety.

coils. Thus the internal diameter of the magnet should be at least 1.2 m. Clearly, there are considerable technical problems to be overcome if an acceptable image is to be produced in a reasonable time, and at the time of writing, much research effort is being expended by several groups in the hope of overcoming some of the problems mentioned.

Before leaving the topic of technical constraints, mention should be made of the magnets which are employed to generate the main field B_0. The frequency requirements discussed earlier place the field values (~0.1 T) within the realm of conventional air-cored electromagnets dissipating up to 20 kW, and Figure 20 shows a commercially available design large enough to take an adult within its confines. A particularly attractive prospect is the imaging of babies, where higher sensitivity and lower power dissipations are possible, and a spherical magnet consuming 7 kW for a field of 0.13 T is being built in the author's laboratory. The magnet comprises two movable hemispherical windings which, when tilted towards one another, create a powerful transverse field gradient without consuming extra power. It is possible by

Figure 20. An electromagnet suitable for NMR imaging. The field produced is about 0.1 T (Courtesy of Oxford Instruments Ltd)

employing superconducting technology to generate much larger fields (Figure 21), but as we have seen, for imaging purposes, there is probably not much to be gained by such an exercise. The cryostat design of a magnet system where the bore is horizontal is not trivial and large amounts of helium tend to boil off. The majority of experimenters have therefore opted for conventional electromagnets, even though problems of thermal expansion and ''creep'' can have a detrimental effect on homogeneity. Normally however, the major threat to homogeneity with such magnets is the building in which the system is housed. Steel beam in floors and ceilings can cause serious field perturbations, as can passing vehicles. Lauterbur uses passive

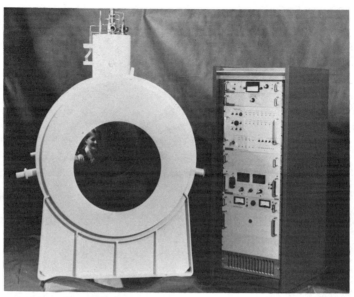

Figure 21. A superconducting magnet with a bore of 0.63 m and a maximum field of 2.5 T. (Courtesy of Intermagnetics General Corporation).

shims made of steel plate to annul the odd order components of such effects while Mansfield houses his equipment in a specially built iron-free laboratory. Other experimenters employ conventional shimming systems and magnetic shielding. If the latter is employed, it can also be used to advantage to screen the spectrometer from external rf interference which can sometimes be troublesome. A major source of such interference at the low frequencies used in imaging is the computer associated with the spectrometer, and it is as well to keep it far removed from the magnet and the probe.

5. THE THIRD DIMENSION

In our discussion so far we have studiously avoided the topic of the third dimension in order to examine clearly the other intricacies of the techniques of imaging. However, in the author's opinion, the third dimension presents a problem of considerable magnitude, and it therefore behooves us to spend some time examining its implications. We have seen that we might expect from 1 ml of water within a person's head a signal-to-noise ratio of 500 from an excellently designed system. Now

$$\psi \sim 500 = A \sqrt{\frac{T_2}{W_n}}$$

where A is the amplitude of the free induction decay and W_n is the mean square noise per unit bandwidth. Hence, for $T_2 = 0.1$ s, $A/\sqrt{W_n} \sim 1580$. For a large bandwidth of 10 kHz, the signal-to-noise ratio of the free induction decay is therefore 15.8. However, the head contains at least 2 l of matter and so if we wish to obtain information simultaneously from the entire head in order to perform a three dimensional imaging experiment, we must be able to handle a signal-to-noise ratio of *at least* 10^4 at the start of the FID. Such a ratio is at least an order of magnitude too large for the NMR receiver to handle without severe distortion, the limiting components being the phase sensitive detectors followed by the amplifiers and analog-to-digital converters. We must reduce the signal amplitude to reduce the distortion, and there are two choices available: the first is to spoil the sensitivity of the spectrometer and pay the price in terms of time and resolution; the second is to select a slice through the sample and obtain signal only from that slice. Let us therefore consider the problems of slice selection.

In selecting a slice, there are two alternatives open to us. The first is to irradiate only the slice of interest and the second is to saturate all but the slice of interest. Both methods are fraught with difficulties. Clearly, the easiest way of irradiating only a given region is to apply a field gradient across that region and then to irradiate with a waveform which contains only the appropriate frequency components—a so-called "selective pulse." Hoult (1979b) has considered this experiment in detail and has shown that when the on-resonance pulse flip-angle is less than 30°, negligible signal persists after the pulse. The signal occurs at the same time as the pulse and of course, when the transmitter is turned on, the receiver is turned off and no signal is received. This phenomenon, mentioned earlier, is shown in Figure 22, and there are several ways to circumvent the problem. The first is simply to increase the pulse flip angle to say 180°. A free induction decay then results. This somewhat unexpected phenomenon is too complex to be analyzed in detail here; suffice it to say that it is a manifestation of the nonlinearity of a driven magnetic resonance system and is well understood. The important point is that this signal is very unrepresentative of the magnetization in the region of interest, and therefore large selective pulses should be avoided. The second circumvention allows us to stay in the linear region. We may form an echo of the signal by reversal of the field gradient as shown in Figure 22. For flip angles of 30° or less, all the NMR signals from the slice of interest are then, to a good approximation, in phase in the center of the echo. At this point, the gradient may then be removed and applied in an orthogonal direction in order to allow two-dimensional resolution with an appropriate technique. A problem arises, however, if we wish to use this echo method with larger flip angles. Once again the nonlinearity of the NMR system spoils the technique. With larger flip angles, signals from different regions of the sample no longer become phase coherent at the same point in time. Suppose, for example, that the selective pulse were square (B_1 constant), of 90° flip angle and of duration τ. Then signals from frequencies

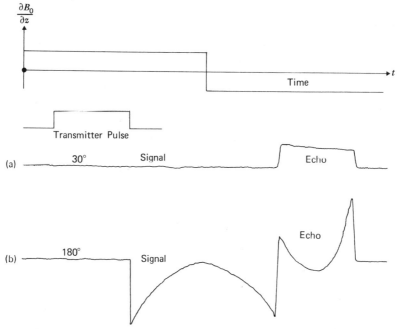

Figure 22. When a powerful field gradient is applied to a homogenous sample, the receiver output mimics the transmitted voltage for selective pulses with small flip angles. As the receiver is turned off when the transmitter is on, no signal is therefore observed, as shown in (a). However, by reversing the field gradient, an echo of the pulse may readily be formed. When the NMR system is driven into the "nonlinear" region, for example, by a 180° selective pulse (b), signal is now observed, and a very distorted echo is obtained. The signal does not arise simply from those frequencies contained in a Fourier analysis of the pulse; the nonlinearity of the system drastically changes the frequency spectrum.

$|\Delta w| \gg 1/\tau$ come into phase in the middle of the echo, while it may be shown (Hoult, 1973) that signals from frequencies $|\Delta w| \ll 1/\tau$ come into phase 64% of the way through the echo. Thus when we terminate the gradient, the phase of the remaining free induction decay depends somewhat upon the distribution of magnetization across the slice. This phenomenon is also observed when the third method for circumventing the selective pulse problem is employed. With this technique, the selective pulse and the gradient are simultaneously curtailed, freezing the evolution of the magnetization and leaving a large signal to be gainfully employed. However, a Fourier analysis of a suddenly curtailed pulse reveals that the imaginary component is of antisymmetric form extending out a considerable frequency distance. Now the integral of an antisymmetric function is of course zero and so with a homogenous sample, this component would not be troublesome—it would contribute no signal. However, with an *inhomogenous* sample, as shown in Figure 23, the phase of the signal is clearly influenced and so with this technique, phase is sample dependent even if the flip angle is kept to

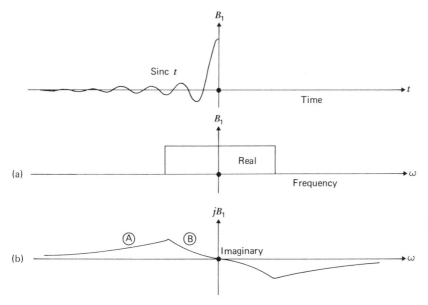

Figure 23. One way of obtaining a signal is to curtail suddenly a selective pulse and the field gradient. The magnetization, given no chance to dephase, generates a large signal. However, it is clear from a Fourier analysis of such a selective pulse (*a*) and (*b*) that the phases of the signals from different points (for example, A and B) vary drastically, and so, with an inhomogenous sample, the phase of the signal can change with sample composition.

below 30°. The echo method, combined with small flip angles, would therefore seem to give the most agreeable signal, and the latter can then be manipulated as desired with suitable variations of field gradients using, for example, Mansfield's method of echo planar imaging or Fourier zeugmatography. Note that methods which require further rf pulses are not to be recommended as the rest of the sample would then be excited.

Turning now to selective saturation, we wish to saturate all but a slice of the sample. Once again, we may apply a field gradient across the sample, but now we irradiate the specimen with broadband noise with a hole in the center, as shown in Figure 24. Hopefully, saturation will not occur in the hole, but once again we must beware of the nonlinearity of the system causing a line in the hole to pick up radiation from beyond the hole. To illustrate this danger, consider the situation shown in Figure 25 where we irradiate one line with a view to saturation, while wishing to preserve the other, both lines having the same relaxation times. Abragam (1961) shows that the steady-state solution of the Bloch equations gives

$$M_z = \frac{1 + (\Delta\omega\, T_2)^2}{1 + (\Delta\omega T_2)^2 + \gamma^2 B_1^2 T_1 T_2} M_0 \tag{22}$$

On resonance, satisfactory saturation may be accomplished if $\gamma^2 B_1^2 T_1 T_2 \geqslant$

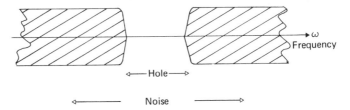

Figure 24. In the presence of a field gradient, broadband noise can saturate a large volume. If a "hole" is left in the noise, then a plane of sample remains unsaturated. However, the width of that plane depends on T_1 and T_2.

100. Thus off resonance, to obtain less than 10% saturation $\Delta\omega T_2 \geqslant 30$. So for $T_2 = 0.1$ s, we must be off resonance by more than 50 Hz. It follows that the width of the unsaturated region is considerably less than that of the noise hole, and further, the exact width is dependent on T_1 and T_2. Thus, quite apart from considerations of pulse repetition rate, we obtain, if we employ selective saturation, an additional dependence on T_1 and T_2 in the image: as T_1 increases the signal becomes smaller. It also becomes smaller as T_2 decreases, the exact details of the dependencies being determined by the nature of the saturating noise. It should also be pointed out that if saturation is to be effective, it must be powerful saturation. The signal from a thin slice could easily be dominated by the integrated remnant signal from the rest of the sample. It is also of interest to note how long we have after saturation ceases before the remnant magnetization becomes dominant due to relaxation. If the slice thickness is a fraction ϵ of the total thickness, then the remnant dominates for times greater than ϵT_1. (Note that we have not considered M_x and M_y. Contributions from these components following saturation can be cancelled by appropriate phasing techniques, Hoult, 1976.)

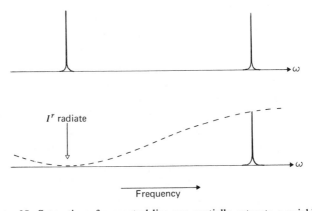

Figure 25. Saturation of a spectral line can partially saturate a neighbor.

As T_1 may have values as short as 50 ms under some circumstances, a 5% slice thickness allows us less than 2.5 ms to change the gradient direction and then to apply the appropriate pulse.

A possible, if rather unsatisfactory, way around this inconvenience is to apply a B_1 noise saturating field which has a cubic dependency upon distance. The magnetization after saturation then varies as

$$M_z = \frac{M_0}{1 + \alpha^2 z^6 \gamma^2 T_1 T_2} \qquad (23)$$

where $B_1 = \alpha z^3$. A large range of $T_1 T_2$ values is of course encountered, which may cause the effective width of the unsaturated region to vary up to twofold, but there is with this method no need to change the field-gradient direction. The spread of frequencies caused by a gradient in the x direction can be covered by the noise; the problems arise in designing suitable coils.

Such then are the problems associated with the third dimension. We must now turn our attention to possible applications and the safety of the method.

6. HAZARDS

In considering the safety of NMR imaging, three factors are of obvious relevance: the irradiating field B_1, the static field B_0 and changes in the static field dB_0/dt. It is important to realize that we are *not* considering here the biological effects of electromagnetic radiation. The vast majority of literature on the subject of radiation safety deals with the incidence of an electromagnetic wave (from, for example, a radar station or a microwave oven) on animals and man. In an NMR spectrometer, our aim is to expose the sample to an alternating B_1 magnetic field while minimizing exposure to an alternating electric field (for example, by inserting a Faraday shield between the probe and the sample). We are in the so-called "near field" region and this fact must alter our approach to any consideration of safety and cause us to examine carefully the foundations of published data.

With this caution in mind, let us consider first the effects of the B_1 field. (Introductory reading material is Neuder, 1978; Durney *et al.*, 1978; Johnson and Shore, 1976.) The interaction of the field with a biological system is characterized by classical electrodynamics rather than by quantum mechanics. We are dealing with Maxwell's equations as applied to a poor and dielectrically lossy conductor which has internal structure. Thus the application of an alternating magnetic field causes power to be dissipated in the sample, currents to flow in conducting loops, and electromotive forces (EMFs) to be generated between points. A wide divergence of safety standards exists due to different criteria being used for specifying the hazard. In Russia and Eastern European countries, behavioral responses of animals and the influence on the central nervous system form a basis for specifying an acceptable *electromagnetic* power density, whereas in the Western countries, the hazard is based on an acceptable thermal dissipation for humans. An exact calculation of any of the effects mentioned is impossible for the body; models using simple shapes such as spheres or ellipsoids may be employed to calculate approximate results, but in general the results depend upon the frequencies involved, and the orientation of the sample with respect to the direction of the field, which can influence results profoundly. (In this context, it is worth noting that even with a spherical sample of homogenous conductor, the attenuation of B_1 field with sample radius is not independent of the angle θ between the radius and B_1. See Appendix A.) It is possible, however, in the case of NMR imaging to obtain reasonable estimates of the thermal dissipation involved in an experiment, and further, it is possible to obtain safely an experimentally accurate measurement of this figure. What can *not* be obtained easily is an estimate of localized heating effects, but as the frequencies used in imaging are well below the resonant frequencies of organs and cavities in the body, it is unlikely that strong localized heating will take place. It has already been mentioned that for an imaging experiment to be successful, the penetration depth of the magnetic field must be considerably greater than the dimensions of the sample. This being the case, we may consider the body as a conductor of conductivity σ of approximately $1 \ \Omega^{-1} \ m^{-1}$ subjected to a homogenous B_1 field. Hoult and Lauterbur (1979) have shown that the power dissipated in a conducting sphere of saline solution of radius b, subjected to an alternating field of amplitude B_1 and frequency ω_0, is given by

$$W = \pi\omega_0^2 B_1^2 \sigma b^5/15 \tag{24}$$

and so an estimate of W may be obtained for, say, the head. For example, suppose a B_1 field of 8.6×10^{-4}T (corresponding, in the rotating frame to ω_1 = 18 kHz) is applied for 5 ms once a second, at a frequency of 5 MHz. Such a situation might be encountered with Hoult's method of rotating frame zeugmatography. Then W is about 20 W. The maximum short-term rate at which excess heat can be removed from the body by perspiration is about 1.3 kW so the power is well below this limit. On the other hand, this power is dissipated in the head alone and exceeds the recommended absorption rate of 3.1 W kg^{-1}. But, such high power is only, presumably, applied for a short length of time and the temperature rise in a minute would be less than 0.3°C, which is not excessive. Some carefully controlled animal experiments are clearly needed in an attempt to define maximum short-term levels of B_1 field, and we must not forget the Eastern safety standards which cause us to ponder whether or not damage may be caused by the induced currents and EMFs. The B_1 field considered induces an EMF of 850 V at say 5 MHz in a ring of conductor with a radius of 0.12 m. Now of course this EMF is only manifest if the ring has a break in it, and it is very unlikely that such a situation is possible at radio frequencies in a biological system. Nevertheless, caution is indicated. Returning to experimental methods of measuring the power dissipation safely, the simplest method is, with the correct experimental setup, to measure the change in the Q of the receiving coil when the sample is placed therein. Details are discussed by Hoult and Lauterbur (1979) and Gadian and Robinson (1979). The effective resistance that the sample adds to that of the coil is given by

$$R_m = \omega_0 L \left(\frac{1}{Q_s} - \frac{1}{Q_0} \right) \qquad (25)$$

where L is the inductance of the coil, Q_0 is the free-standing Q and Q_s is the Q value with the sample in place. Thus the fraction of applied power which is dissipated in the sample is

$$\frac{R_m}{R_m + R_0} = 1 - \frac{Q_s}{Q_0} \qquad (26)$$

where R_0 is the resistance of the coil. $R_0 = \omega_0 L / Q_0$. This formula applies whether a solenoid or a saddle-shaped coil is used, so it is possible by measuring Q to determine power absorption in the body, and in general it would appear that the heat generated by imaging experiments is of little danger. However, considerable tests with animals need to be performed to check for impairment of nervous function.

Turning now to switched field gradients which are employed in some experiments, we are faced with the possibility of inducing direct EMFs and currents in the body for short periods of time on a repetitive basis. Suppose, for example, that the field at one part of the head changes by 4.8×10^{-4}T in 100 μs (corresponding to a change in proton frequency of 20 kHz). Then the EMF induced in a ring of conductor of radius 0.12 m is about 200 mV. It is possible that rapid repetition of such an EMF (say at 10 Hz) could cause seizure. Budinger (1978, 1979), however, has pointed out a more serious danger when the torso is subjected to change of field. Many experiments (Walter, 1970) have shown that remarkably small values of current in the frequency range 10 to 200 Hz can cause ventricular fibrillation. For example, in experiments reported by Weinberg (1970), one dog consistently went into ventricular fibrillation when a current of only 20 μA was applied between its left ventricle and chest. Now it is very difficult to know what rate of change of magnetic field could induce such currents as the resistance of the body is highly nonlinear at very low frequencies and low voltages. However, if we postulate a toroid of conductor of diameter 0.2 m, cross-sectional area 1.2 $\times$ 10^{-3} m^2, and conductivity 1 Ω^{-1}m^{-1}, we can see that the rate of change of field mentioned earlier induces a current of 290 μA for a period of 100 μs. Whether such a current for such a short period of time would be dangerous is unknown. To move into charted areas, a field of amplitude 2.4×10^{-4}T alternating sinusoidally at 200 Hz would induce an EMF of about 1.5 mV and a current of about 3 μA. This figure is certainly safe but further investigation is definitely called for as the postulate above is very crude, and at the time of writing, it seems likely that changes of magnetic field constitute the greatest hazard for a patient—especially a sick person. In this context, the possibility of a power supply failure to the magnet must also be considered. It is imperative that the magnet be shunted with a diode (under normal circumstances, reverse biased) so that in the event of a mishap, the current decreases slowly. Even so, the time constant of magnetic field decay is likely to be only a fraction of a second, and a field change of 0.5 T s^{-1} can induce, in the toroidal model considered, an initial direct current of 30 μA lasting easily long enough to cause concern about possible fibrillation. Without diode protection, the possibility of a fatality must be seriously considered.

Finally, we must examine the possible hazard associated with the static field B_0. Evidence for such a hazard at the low fields associated with imaging is negligible, but there is a dearth of good data on the subject. Sheppard and Eisenbud (1977) have reviewed the available literature and offer little that is substantive; known effects include an increase in the steroid levels of monkeys exposed to 2×10^{-2}T (Friedman and Carey, 1972) and an anomalously large change in the dielectric constant of lysozyme (Ahmed *et al.*, 1975). Many years of experience with cyclotron magnets have produced no firm evidence of any hazard, and the general conclusion is that static homogenous fields are probably safe, despite the prevalent rumor that NMR spectroscopists have more female offspring than male!

7. CONCLUSION

To conclude this review, we take a brief look at possible future applications of imaging. It is quite clear from the considerations of sensitivity presented earlier that the prospects for performing medical imaging on any nucleus other than protons are exceedingly poor. The sensitivity of protons with a concentration in water of 110 M is barely good enough to produce acceptable images; the sensitivity of, for example, phosphorus with a typical biological concentration of 50 mM total renders any attempt at imaging useless. The best that may be expected in reasonable times is a signal from some gross region, for example, the heart or the head. Such a signal would, of course, be very interesting and attempts are underway in several laboratories to obtain a localized ^{31}P spectrum from perfused organs, but these experiments lie outside the scope of the present article. If protons are used, a wealth of potential medical applications present themselves, a few being the detections of cancerous tissue, hydrocephalus, intercranial bleeding, cystic malformation, infarcts, bone cavities, various abnormal growths, the long-term monitoring of paramagnetic ion concentration in various organs, bladder function, and so on. However, at the time of writing, the technique is too new to make definite conclusions as to its applicability; these must follow as the method gains clinical acceptance (Figure 26). If NMR imaging is safe, as seems likely, its major advantage over radiative methods is the freedom to repeat results and follow the time course of changes in the sample. The technology of imaging is advancing rapidly—there are now more than 50

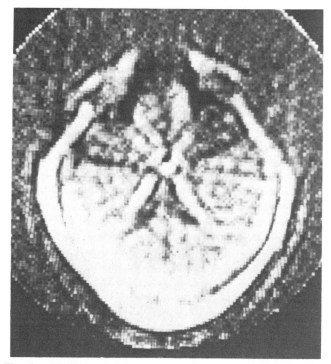

Figure 26. The first NMR head-scan, produced in 1978, showing eye orbits and brain ventricles. (Reproduced with permission, courtesy of EMI Central Research Labs., Hayes, Middlesex, England).

papers in the field—and it is to be hoped that the list of applications above is but a small fraction of the uses to which the techniques described will be applied as more researchers are able to use the method as an investigative clinical tool.

APPENDIX A.   THE SOLUTION OF MAXWELL'S EQUATIONS FOR AN ALTERNATING MAGNET FIELD IN A CONDUCTING SPHERE

Maxwell's equations for the medium may be written in the form

$$\text{div } \mathbf{D} = 0 \tag{A1}$$

$$\text{div } \mathbf{B} = 0 \tag{A2}$$

$$\text{curl } \mathbf{E} = -\frac{\partial \mathbf{B}}{\partial t} \tag{A3}$$

$$\text{curl } \mathbf{H} = \frac{\partial \mathbf{D}}{\partial t} + \sigma \mathbf{E} \tag{A4}$$

where $\mathbf{D} = \epsilon\epsilon_0\mathbf{E}$ is the electric displacement and $\mathbf{E}$ is the electric field in the same, $\mathbf{B} = \mu\mu_0\mathbf{H}$ is the induction field and $\mathbf{H}$ is the magnetic field in the sample. ϵ is the dielectric constant of the sample, while ϵ_0 is that of free space; μ is the permeability of the sample, and μ_0 that of free space, and finally, σ is the sample conductivity. Now

$$\text{curl curl } \mathbf{H} = \text{grad div } \mathbf{H} - \nabla^2\mathbf{H} - - \nabla^2\mathbf{H}$$

from equation (A2).

Thus

$$\nabla^2\mathbf{H} = \epsilon\epsilon_0\mu\mu_0\frac{\partial^2\mathbf{H}}{\partial t^2} + \sigma\mu\mu_0\frac{\partial\mathbf{H}}{\partial t}$$

from equations (A4) and (A3).

Let us assume that $\mathbf{H}$ varies throughout the sample as $\mathbf{H}_1\,e^{j\omega t}$ where $j = \sqrt{-1}$ and ω is frequency. Then substituting,

$$\nabla^2\mathbf{H}_1 = -\,\omega^2\epsilon\epsilon_0\mu\mu_0\mathbf{H}_1 + j\omega\sigma\mu\mu_0\mathbf{H}_1$$

or

$$\nabla^2\mathbf{H}_1 + k^2\mathbf{H}_1 = 0 \tag{A5}$$

where

$$k^2 = \omega^2\epsilon\epsilon_0\mu\mu_0 - j\omega\sigma\mu\mu_0 \tag{A6}$$

Now

$$\nabla^2\mathbf{H}_1 = \mathbf{i}\,\nabla^2 H_{1x} + \mathbf{j}\,\nabla^2 H_{1y} + \mathbf{k}\,\nabla^2 H_{1z}$$

where $\mathbf{i}$, $\mathbf{j}$, and $\mathbf{k}$ are unit vectors in the x, y, and z directions, respectively. Substitution in equation (A5) gives three equivalent equations. As a representative, let us consider the solution of

$$\nabla^2 H_1 + k^2 H_1 = 0 \tag{A7}$$

The simplest way to solve this equation for the sample considered is to make a transformation to spherical polar coordinates (r, θ, ϕ) which are shown in Figure 27. It may then be shown (Sneddon, 1961) that solutions of equation (A7) are of the form

$$H_1 = A_l^m r^{-1/2} J_{\pm(l+1/2)}(kr)\, P_l^m\,(\cos\theta)\, e^{jm\phi} \tag{A8}$$

where $J_{\pm(l+1/2)}\,(kr)$ are the spherical Bessel functions, $P_l^m\,(\cos\theta)$ are associated Legendre polynomials, and A_l^m are constants. To simplify the interpretation of these equations, we assume that the direction of the applied field is the z direction. (Of course, in the laboratory frame, this direction would be x or y, as z is the direction of the main field.) Then, by considerations of symmetry, we may conclude for our sample that $m = 0$. Equations (A8) then reduce to

$$H_1 = A_l r^{-1/2} J_{\pm(l+1/2)}(kr)\, P_l\,(\cos\theta) \tag{A9}$$

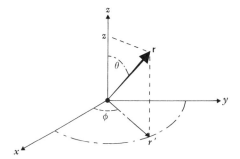

Figure 27. Spherical polar coordinates.

where P_l (cos θ) are the Legendre functions. Further, we know that at the center of the sample, H_1 is finite; it may then be shown that we eliminate those terms which contain a negative J index. Let us assume that penetration of the field into the sample is considerable, and the field within the sample is predominantly described by low values of the index l. Knowing that the field within the sample must be symmetrical, we may then write

$$H_1 = A_0 \frac{\sin kr}{kr} + A_2 \frac{(3 \cos^2 \theta - 1)}{kr}$$

$$\left[\left(\frac{3}{k^2 r^2} - 1\right) \sin kr - \frac{3}{kr} \cos kr\right] + \cdots \tag{A10}$$

Now outside the sample, $k \sim 0$ and so equation (A7) reduces to Laplace's equation $\nabla^2 H_1 = 0$. Let us try as an acceptable solution the sum of the spherical harmonics:

$$H_1 = C_0 + C_2 \frac{(3 \cos^2 \theta - 1)}{r^3} + \cdots \tag{A11}$$

where the following boundary conditions pertain:

$H_1 \rightarrow C_0$ as $r \rightarrow \infty$

H_1 constant for $r = 0$

H_1 continuous at $r = a$, the surface of the sphere

$\partial H_1 / \partial z$ continuous for $\theta = 0$

Thus, from equations (A10) and (A11), equating fields at the surface,

$$C_0 = A_0 \frac{\sin ka}{ka} \tag{A12}$$

$$\frac{C_2}{a^3} = \frac{A_2}{ka} \left[\left(\frac{3}{k^2 a^2} - 1\right) \sin ka - \frac{3}{ka} \cos ka\right] \tag{A13}$$

and equating derivatives along the z axis at the surface:

$$- \frac{6C_2}{a^4} = A_0 \left[- \frac{\sin ka}{ka^2} + \frac{\cos ka}{a} \right] + \frac{2A_2}{k}$$

$$\left[\left(\frac{-9}{k^2 a^4} + \frac{4}{a^2} \right) \sin ka + \left(\frac{9}{ka^3} - \frac{k}{a} \right) \cos ka \right]$$

Substituting from equation (A13), we therefore find that

$$A_2 = \tfrac{1}{2} A_0 \tag{A14}$$

Hence, inside the sample

$$H_1 = A_0 \left\{ \frac{\sin kr}{kr} + \frac{(3 \cos^2 \theta - 1)}{2 kr} \right.$$

$$\left. \left[\left(\frac{3}{k^2 r^2} - 1 \right) \sin kr - \frac{3}{kr} \cos kr \right] + \cdots \right\} \tag{A15}$$

Now $k^2 \sim -j\omega\sigma\mu\mu_0$ if the sphere has a conductivity comparable to that of 100 mM NaCl and the frequency is in the range 1–30 MHz. Thus writing the "skin depth" δ as $(\omega_0 \sigma \mu \mu_0/2)^{-1/2}$, we have that $k = (1 - j)/\delta$ and substituting and expanding in equation (A15) we find that:

$$H_1 \sim A_0 \left\{ 1 + \frac{jr^2}{\delta^2} \left[\frac{1}{3} - 0.0667(3 \cos^2 \theta - 1) \right] - \frac{r^4}{\delta^4} \right.$$

$$\left. \left[\frac{1}{30} - 9.525 \times 10^{-3} (3 \cos^2 \theta - 1) \right] + \cdots \right\} \tag{A16}$$

We may draw several conclusions from this equation. (1) The attenuation of H_1 as the field penetrates the sample is *not* the same in the axial and radial directions. The variation along the z axis (the direction of the field) is considerably less than in the transverse direction. (2) The phase of the field varies in all directions, to a good approximation, as the square of the radius (though of course the coefficient depends on direction), while the amplitude of the field varies as the fourth power of the radius. (3) If we substitute $\delta \sim$ 25 cm at 5 MHz, a reasonable value for the human body, we find that in a radius a of 10 cm, the phase varies by only about 3.7° in the transverse direction and by only about 1.8° in the axial direction. The amplitude varies by about 0.1% in the transverse direction and by about 0.01% in the axial direction. We conclude therefore that penetration effects are negligible at the frequencies usually encountered in NMR body imaging.

APPENDIX B. THE POWER REQUIRED TO IRRADIATE THE PROBE BANDWIDTH

Let the required B_1 field in the rotating frame be described by the relationship $\gamma \tilde{B}_1 = \omega_1$. Then in the laboratory frame, the required B_1 field has an amplitude $2\tilde{B}_1 = 2\omega_1/\gamma$. The B_1 field produced by a current of rms value I is, for a solenoidal coil whose length is approximately equal to its diameter:

$$B_1 = \frac{2\omega_1}{\gamma} \sim \frac{In\mu_0}{2a} \tag{B1}$$

where n is the number of turns on the coil. For a given coil we therefore know the required current. A 90° pulse, neglecting rise and fall time effects, is of duration

$$t_{90} = \pi/2\omega_1 \tag{B2}$$

If these effects are to be negligible, then their time constant τ must be very much less than t_{90}. For example

$$\tau \leqslant \pi/10\omega_1 \tag{B3}$$

Now

$$Q_{eff} = \tau\omega_0/2 \tag{B4}$$

where Q_{eff} is the effective Q of the tuned circuit, damped as is necessary, and ω_0 is the resonant frequency. Thus, from equations (B3) and (B4),

$$Q_{eff} \leqslant \frac{\pi}{20}\left(\frac{\omega_0}{\omega_1}\right) \tag{B5}$$

Now the inductance of a solenoidal coil whose length is a little less than its diameter (the optimum ratio for maximum Q) is approximately given by

$$L \sim 1.58\, n^2 a \times 10^{-6} \tag{B6}$$

Thus the effective resistance associated with the probe must be

$$r = \frac{\omega_0 L}{Q} \geqslant 20 \times 1.58 \times 10^{-6}\,\frac{\omega_0\omega_1 n^2 a}{\pi\omega_0} \tag{B7}$$

It follows from equations (B1) and (B7) that the power dissipation is given by

$$W \geqslant 1.49 \times 10^{-9}\,\omega_1^3 a^3 \tag{B8}$$

Now the bandwidth of the receiving system is $\Delta\omega = \pm\omega_0/2Q$. Let us assume that for satisfactory irradiation $\omega_1 \geqslant 5\Delta\omega$. Then

$$W \geqslant 2.33 \times 10^{-8}\left(\frac{\omega_0 a}{Q}\right)^3 \tag{B9}$$

Thus for $\omega_0 = 2\pi 6 \times 10^6$, $a = 0.15$ m, $Q = 500$, and $W \geqslant 33$ kW

REFERENCES

The list of references below is not exhaustive. A complete bibliography may be found in
Lauterbur, P. C., *IEEE Trans. Nucl. Sci.* **26**, 2808 (1979).

Abragam, A., *The Principles of Nuclear Magnetism* (1961), Clarendon Press, Oxford, England, p. 46.

Ahmed, N. A. G., J. H. Calderwood, H. Frohlich, and C. W. Smith (1975), *Phys. Letters* **53A**, 129.

Andrew, E. R., P. A. Bottomley, W. S. Hinshaw, G. N. Holland, W. S. Moore, and C. Simaroj (1977), *Phys. Med. Biol.* **22**, 971.

Brooks, R. A., and G. diChiro *Phys. Med. Biol.* (1976), **21**, 689.

Brunner, P., and R. R. Ernst (1979), *J. Magn. Res.* **33**, 83.

Budinger, T., (1978), Lawrence Berkeley Laboratory Report No. 5694, University of California.

Budinger, T. (1979), *IEEE Trans. Nucl. Sci.* **26**, 2821.

Damadian, R., L. Minkoff, M. Goldsmith, M. Stanford, and K. Koutcher (1976), *Science* **194**, 1430.

Damadian, R., M. Goldsmith, and L. Minkoff (1977), *Physiol. Chem. Phys.* **9**, 97.

Damadian, R., L. Minkoff, M. Goldsmith, and J. Koutcher (1978), *Naturwiss.* **65**, 250.

Durney, C. H., C. C. Johnson, P. W. Barber, H. Massoudi, M. F. Iskander, J. L. Lords, D. K. Ryser, S. J. Allen, and J. C. Mitchell (1978), *Radiofrequency Radiation Dosimetry Handbook*, 2nd ed., Report SAM-TR-78-22, USAF School of Aerospace Medicine, Brooks Air Force Base, TX.

Ellett, Jr., J. D., M. G. Gibby, U. Haeberlen, L. M. Huber, H. Mehring, A. Pines, and J. S. Waugh (1971), in *Advances in Magnetic Resonance*, J. S. Waugh, ed., Vol. 5, Academic Press, New York, NY, p. 142.

Ernst, R. R., an W. A. Anderson (1966), *Rev. Sci. Instr.* **37**, 93.

Fernback, S., and W. G. Proctor (1955), *J. Appl. Phys.,* **26**, 170 (1955).

Friedman, H., and R. J. Carey (1972), *Physiol. Behav.* **9**, 171.

Gabillard, R. (1951), *Compt. Rend.* **232**, 1551.

Gabillard, R. (1952), *Phys. Rev.* **85**, 694.

Gadian, D., and F. N. H. Robinson (1979), *J. Magn. Reson.* **34**, 449.

Garroway, A. N., P. K. Grannell, and P. Mansfield (1974), *J. Phys. Chem.* **7**, L457.

Goldsmith, M., R. Damadian, M. Stanford, and M. Lipkowitz (1977), *Physiol. Chem. Phys.* **9**, 105.

Hinshaw, W. S. (1974), *Phys. Letters* **48A**, 87.

Hinshaw, W. S. (1976), *J. Appl. Phys.* **47**, 3709.

Hoult, D. I. (1973), D. Phil Thesis, Oxford University.

Hoult, D. I. (1976), *J. Magn. Reson.* **21**, 337.

Hoult, D. I., and R. E. Richards (1976), *J. Magn. Reson.* **24**, 71.

Hoult, D. I. (1977), *J. Magn. Reson.* **26**, 165.

Hoult, D. I. (1978), *Prog. NMR Spectry.* **12**, 41.

Hoult, D. I., and P. C. Lauterbur (1979), *J. Magn. Reson.* **34**, 425 (1979).

Hoult, D. I. (1979a), *J. Magn. Reson.* **33**, 183.

Hoult, D. I. (1979b), *J. Magn. Reson.* **35**, 69.

Hoult, D. I. (1979c), *Rev. Sci. Instr.* **50**, 193.

Hounsfield, G. N. (Part I) and Ambrose, J. (Part 2) (1973), *Brit. J. Radiol.* **46**, 1016.

Johnson, C. C., and M. L. Shore, eds. (1976), *Biological Effects of Electromagnetic Waves*, U.S. Dept. of Health, Education, and Welfare. HEW Publication (FDA) 77-8011.

Kumar, A., D. Welti, and R. R. Ernst (1975a), *Naturwiss.* **62**, 34.

Kumar, A., D. Welti, and R. R. Ernst (1975b), *J. Magn. Reson.* **18**, 69.

Lauterbur, P. C. (1973), *Nature* **242**, 190.

Lauterbur, P. C., W. V. House, Jr., D. M. Kramer, C. N. Chen, F. W. Porretto, and C. S. Dulcey, Jr. (1975a) in *Image Processing for 2-D and 3-D Reconstruction from Projections;*

Theory and Practice in Medicine and the Physical Sciences, Optical Society of America, p. MA10-1.

Lauterbur, P. C., D. M. Kramer, W. V. House, Jr., and C. N. Chen (1975b), *J. Amer. Chem. Soc.* **97**, 6866.

Mansfield, P., and I. L. Pykett (1978), *J. Magn. Res.* **29**, 355.

Minkoff, L., R. Damadian, T. E. Thomas, N. Hu, M. Goldsmith, J. Koutcher, and M. Stanford (1977), *Physiol. Chem. Phys.* **9**, 101.

Neuder, S. M. (1978), *Electromagnetic Fields in Biological Media*, U.S. Department of Health, Education, and Welfare, HEW Publication (FDA) 78-8068.

Radeka, V. (1974), *IEEE Trans. Nucl. Sci.* **21**, 51.

Sheppard, A. R., and M. Eisenbud (1977), *Biological Effects of Electric and Magnetic Fields of Extremely Low Frequency*, New York University Press, New York, NY.

Sneddon, I. N. (1961). *Special Functions of Mathematical Physics and Chemistry*, Oliver and Boyd, Edinburgh, U.K.

Sutherland, R. J., and J. M. S. Hutchison (1978), *J. Phys. E.* **11**, 79.

Walter, C. W., ed., (1970), *Electrical Hazards in Hospitals*, National Academy of Sciences, Washington, DC.

Weinberg, D. I. (1970), in *Electrical Hazards in Hospitals*, C. W. Walter, ed., National Academy of Sciences, Washington, DC, p. 11.

Yamada, Y., K. Tanaka, E. Yamamoto, and Z. Abe (1978), *8th Int. Conf. Magnetic Resonance in Biological Systems*, Tokyo, Japan (1978).

Three

NMR Studies of Drug Metabolism and Mechanism of Action

Gerald Zon

Department of Chemistry
The Catholic University of America
Washington, D.C. 20064

1. INTRODUCTION

The purpose of this article is to provide a brief overview of how NMR spectroscopy can be used to study various aspects of drug metabolism and mechanisms of drug action at the molecular level. Prior reviews of NMR applications within the general areas of drug metabolism, pharmacology, and medicine have dealt primarily with structural characterization of either isolated drug metabolites or drug conjugates and quantitative analysis of pharmaceutical preparations (Case, 1973; Case, 1976; Rackham, 1976; O'Neill and Pringuer, 1975). In the majority of this earlier work, which predated routine accessibility to multinuclear Fourier transform (FT) NMR spectrometers, continuous-wave (CW) ^{1}H NMR was utilized and spectral interpretations were, for the most part, limited to fundamental types of chemical shift/coupling constant *versus* structure correlations and the use of integrated signal intensities for direct measurement of component concentration. Quantitative NMR analysis of drugs and related pharmaceutical materials by NMR continues to be an active area of research, as may be seen from the following list of representative studies:

Determination of amyl nitrite in its inhalant form (Turczan and Medwick, 1976.
Rapid determination of dimethyl polysiloxane (Anhowry *et al.*, 1976).
Analysis of a four-component mixture of phenylglycine derivatives (Warren *et al.*, 1976).
Assay of pilocarpine ophthalmic formulations by ^{13}C NMR (Neville *et al.*, 1977).

Isomeric impurity identification in the anti-inflammatory agent benoxa-profen (Browner *et al.*, 1976).

Analysis of corticosteroids as trichloroacetyl isocyanate derivatives (Lanouette *et al.*, 1976).

Determination of trimethoprim and sulfamethoazole in tablets and powders (Rodriguez *et al.*, 1977).

Assay of clofibrate capsules (El-Fatatry and Aboul-Enein, 1978).

Quantitative analysis of aspirin tablets (Vinson and Kozak, 1978).

Diastereomer assay of the semisynthetic penicillin phenethicillin (Wilson *et al.*, 1977).

Stereochemical identification of β-chloromorphide, a clandestine opiate constituent (Yeh *et al.*, 1976).

Identification of a toxic impurity in illicit amphetamines ("mini-bennies") (Kram, 1977).

The spectroscopic principles which underlie such applications are the subject of various monographs (Wehrli and Wirthlin, 1978; Abraham and Loftus, 1978); consequently, the present review attempts to highlight relatively novel NMR studies dealing with, for example, dynamical intra- or intermolecular processes in either nonequilibrium or equilibrated systems. The information obtained by NMR measurements of this type may often be either impossible or very difficult to obtain by other analytical methods and, moreover, it is hoped that basic concepts presented in the following sections will stimulate further extensions of NMR spectroscopic methods into the challenging areas of drug metabolism, mechanism of drug action, and drug design.

As a final introductory note, it should be mentioned that computer-assisted searching of the 1972 through mid-1979 literature indexed by *Chemical Abstracts Service* and *Exerpta Medica* provided approximately 1,500 references to NMR studies of drugs, drug metabolism, pharmacology, pharmacokinetics, medicine, and related subjects. For obvious practical reasons the material selected for presentation has been chosen for pedagogical reasons and is meant to be an illustrative rather than comprehensive review of the available literature.

2. ALKYLATING AGENTS

Alkylating agents may be defined as compounds which are capable of undergoing displacement reactions with nucleophiles to form alkylated products under typical physiological conditions:

$$R—X + :Nucl \rightarrow R—\overset{+}{N}ucl + X^-.$$

Many alkylating agents are known to function as potent mutagens and carcinogens while others, somewhat ironically, serve as effective chemo-

Scheme I

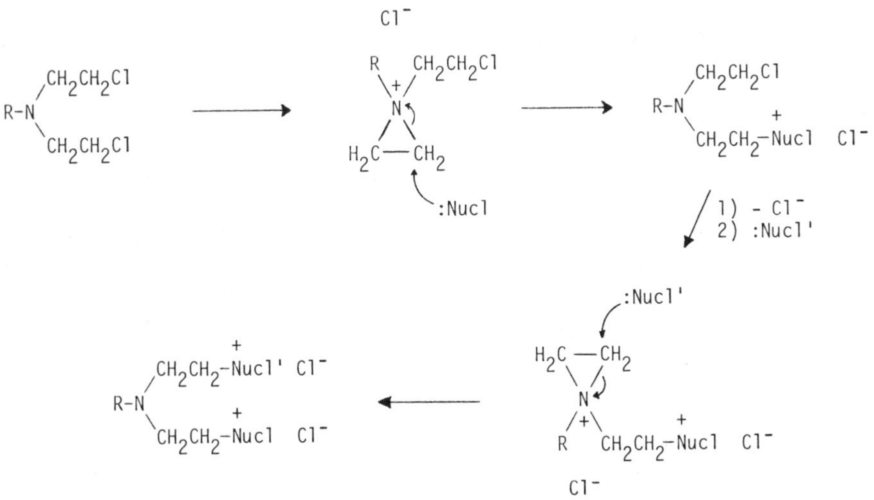

$$
\begin{array}{c}
\text{R-N}\begin{array}{c}\nearrow CH_2CH_2Cl\\ \searrow CH_2CH_2Cl\end{array}
\longrightarrow
\quad Cl^-\\
\end{array}
$$

R-N(CH₂CH₂Cl)₂ → [aziridinium ion intermediate with Cl⁻, R, CH₂CH₂Cl, N⁺, H₂C—CH₂, :Nucl] → R-N(CH₂CH₂Cl)(CH₂CH₂-Nucl)⁺ Cl⁻

1) - Cl⁻
2) :Nucl'

[second aziridinium ion: H₂C—CH₂, N⁺, R, CH₂CH₂-Nucl⁺ Cl⁻, Cl⁻, :Nucl']

← R-N(CH₂CH₂-Nucl'⁺ Cl⁻)(CH₂CH₂-Nucl⁺ Cl⁻)

therapeutic drugs against human cancers. Virtually all of the latter compounds have the capacity for bis-alkylation and are thought to exert their cytotoxicity, at least in part, by cross-linking DNA. A substantial number of bis-alkylating anticancer drugs are related to nor-nitrogen mustard (nor-HN2), $HN(CH_2CH_2Cl)_2$, and while details regarding either the sites of DNA alkylation or the mechanisms for *selective* cytotoxicity toward cancerous versus normal cells (oncostatic specificity) are not known at this time, the collective circumstantial evidence supports the multistep sequence shown in Scheme I for a generalized nor-HN2 derivative. According to this mechanism, initial intramolecular displacement of Cl^- to form a "strained" 3-membered ring aziridinium ion is followed by electrophilic attack (alkylation) of a nucleophilic site on DNA. Repetition of the aziridinium ion formation and subsequent alkylation of a proximate nucleophilic site on the opposite DNA strand thus afford cross-linked DNA. Recognition that the rate of intramolecular cyclization could be moderated by conjugation of the nitrogen lone-pair electron density led to the synthesis of hundreds of "latentiated" nor-HN2 derivatives represented by $RN(CH_2CH_2Cl)_2$, with R being a π-delocalizing substituent such as aryl or alkanoyl, and attempts to obtain oncostatic selectivity have presupposed preferential enzymatic cleavage of the R—N bond and/or facilitated transport via a "carrier" moiety (Workman and Double, 1978).

2.1. Nitrogen Mustards

^{19}F NMR data for $HN(CH_2CH_2F)_2$ was published over 20 years ago (Muller *et al.*, 1957); however, 8 years passed before publication of the first reports concerning the use of NMR to investigate the kinetics and reaction mechanism of such bis-(2-haloethyl)amine alkylating agents (Pettit and Smith, 1964; Pettit et al., 1965). In these ^{1}H NMR studies of symmetrically and unsymmetrically substituted N-methyl- and N-bis(2-haloethyl)amines, tentative evidence was presented for the detection of previously postulated aziridinyl intermediates. More detailed ^{1}H NMR studies (Levins and Papanastassiou, 1965) of similar compounds were published at about the same time and provided first-order rate data for intramolecular cyclization to "activated" aziridinium ion intermediates. Related ^{1}H NMR investigations (Levins and Rogers, 1965) dealt with the aqueous stability of phosphoramide mustard, $HO(H_2N)P(O)N(CH_2CH_2Cl)_2$, which was then undergoing preclinical toxicology studies as a possible anticancer drug. Monitoring the decrease in relative absorption intensity of CH_2Cl protons in the starting material as a function of time led to clinically useful half-life data for solutions of this unstable experimental drug, which would later be the subject of much closer scrutiny by NMR (see below).

2.2. Cyclophosphamide (CP)

Pioneering work (Friedman and Seligman, 1954; Friedman et al., 1963; Arnold et al., 1968) on the design and synthesis of phosphoramide mustard and related bis-alkylating agents of general structure $XYP(O)N(CH_2CH_2Cl)_2$ was based on the assumption that intracellular phosphoramidases might preferentially cleave P—N bonds and release nor-HN2 within cancerous cells and therefore affect a higher level of toxicity relative to normal cells. This very early prodrug strategy led to the synthesis of cyclophosphamide (CP, Scheme II), which subsequently became one of the most widely used clinical anticancer agents and, interestingly, was eventually shown to undergo oxidative "activation" in vivo rather than the originally planned hydrolytic process (Hill, 1975). Key metabolic transformations that have been established for CP are depicted in Scheme II (Hill, 1975; Friedman et al., 1979) and involve liver enzyme oxidation to 4-hydroxycyclophosphamide (4-HO-CP) followed by spontaneous tautomerization with aldophosphamide (AP), fragmentation of AP to yield phosphoramide mustard (PM), and intramolecular cyclization of PM (presumably as its conjugate base) to afford an aziridinium ion prior to the aforementioned sequence of DNA cross-linking events. The identification and study of these labile metabolites has been relatively difficult; however, significant advances were realized by application of newer mass spectroscopic techniques (Jarman et al., 1978). NMR methods have likewise proved to be useful in studies of CP and are briefly outlined in the following paragraphs.

2.2.1. Kinetic Analyses of 4-HO-CP and AP Metabolites. The complexity of the ^{1}H NMR spectra for CP and its metabolites relative to proton-decoupled singlet ^{31}P resonances for these molecules suggests distinct ad-

Scheme II

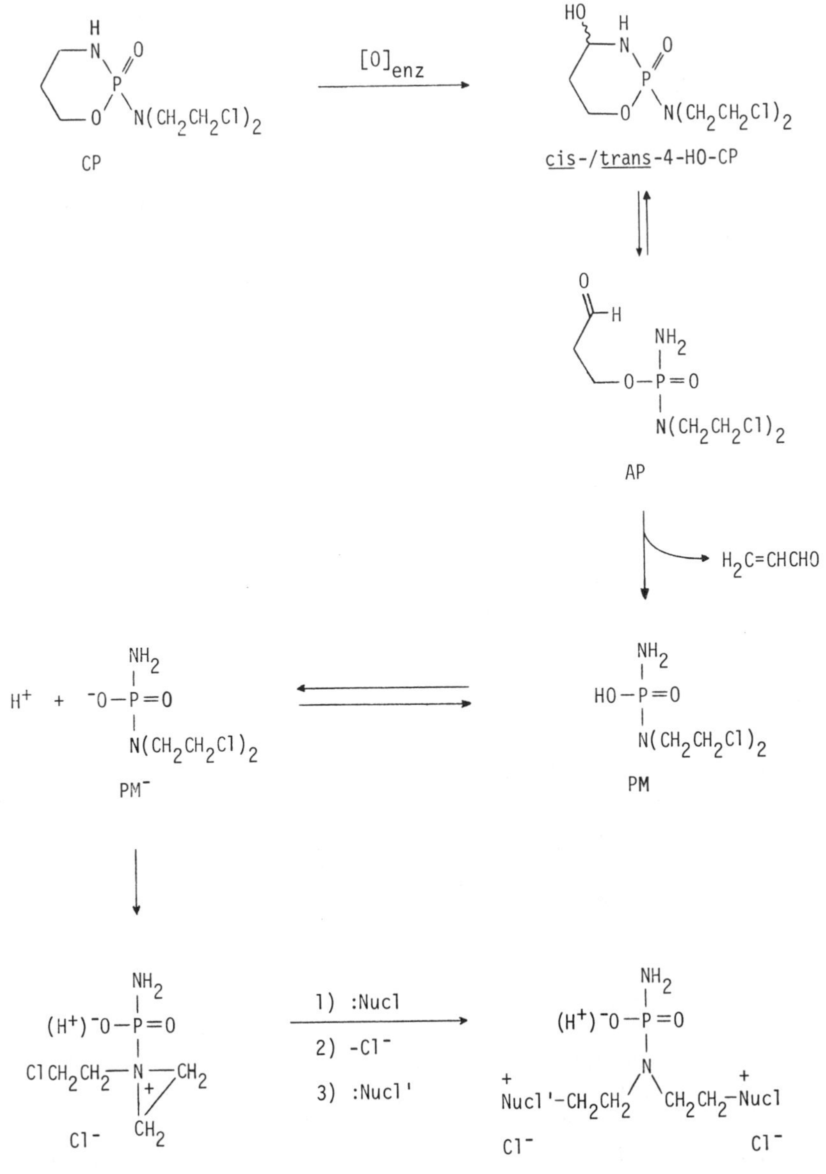

$[O]_{enz}$

CP

cis-/trans-4-HO-CP

AP

$H_2C=CHCHO$

NH_2
H^+ + $^-O-P=O$
$N(CH_2CH_2Cl)_2$
PM$^-$

NH_2
$HO-P=O$
$N(CH_2CH_2Cl)_2$
PM

NH_2
$(H^+)^-O-P=O$
$ClCH_2CH_2-N\overset{+}{\underset{CH_2}{|}}CH_2$
Cl^-

1) :Nucl
2) $-Cl^-$
3) :Nucl'

NH_2
$(H^+)^-O-P=O$
$Nucl'^--CH_2CH_2\overset{N}{\diagup}\diagdown CH_2CH_2-Nucl$
Cl^- Cl^-

vantages for utilizing the latter nucleus in NMR kinetic studies. Preliminary
^{31}P NMR studies (Zon *et al.*, unpublished work) at 40.25 MHz using "physiological solutions" (pH 7.4, 37°) of synthetic *cis*-4-HO-CP have provided
spectra showing the decrease in signal strength of starting material at 12.89
ppm (external H_3PO_4 reference) with concomitant growth of three signals at
12.68, 12.78, and 12.04 ppm tentatively ascribed to *trans*-4-HO-CP, AP, and

either AP hydrate or 3,4-dehydro-CP. During the apparent equilibration of these metabolites there is a gradual decrease in their combined signal intensity that obeys a first-order rate law and corresponds to irreversible fragmentation of AP into PM, which then undergoes relatively rapid hydrolysis (see below) to provide $H_2NPO_3H_2$ and, ultimately, H_3PO_4. The 75 min half-life derived from this data for disappearance of the 4-HO-CP/AP metabolite mixture compares reasonably well with the 107 min value previously determined (Hohorst *et al.*, 1976) for 4-HO-CP under somewhat different conditions (pH 7.0, 37°C).

2.2.2. Kinetic Analysis of Phosphoramide Mustard (PM) and Its Metabolites.

The importance of PM within the mechanism of action of CP has prompted a detailed evaluation of its alkylative chemistry (Engle *et al.*, 1979).[31] P NMR spectra of buffered solutions of synthetic PM afford reliable first-order rate constants (Figure 1) for disappearance of this metabolite, and variation of pH provided a pH-rate profile (Figure 2) that accurately quantified the decrease in rate with lowered pH originally predicted (Friedman, 1967) from consideration of electronic factors in the PM/PM⁻ (acid/conjugate base) equilibrium shown in Scheme II. Moreover, the fact that the half-life of PM decreased by a factor of about 10 between pH 6.0 and 7.4 led to the suggestion (Engle *et al.*, 1979) that localized "domains" of varying pH may, in part, be related to theories of CP onstatic specificity.

The distribution of hydrolysis products obtained from PM has also been found to be markedly pH dependent and, significantly, both the rate of formation and rate of disappearance of transient aziridinium ions could be studied over the pH range of 7–9 by monitoring the relative intensity changes for the aziridinium ion intermediate and product signals (Engle *et al.*, 1979). In the presence of excess 2-mercaptoethanol, $HSCH_2CH_2OH$, which serves as a nucleophilic "trapping agent," these aziridinium ions were not long-lived enough to allow for detection and, instead, the formation of a

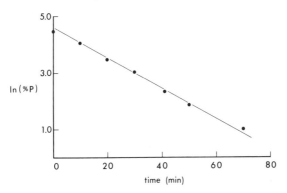

Figure 1. First-order kinetic plot for the decrease in the relative concentration of PM/PM⁻ expressed as ln (%P), where %P = 100 [(^{31}P NMR peak height of PM/PM⁻)/(sum of all ^{31}P NMR peak heights)]; $T = 37°C$, pH = 6.5.

monoalkylated intermediate can be seen (Figure 3). As the reaction proceeds, this signal decreases in intensity with ultimate formation of the expected bis-alkylated product, $HO(H_2N)P(O)N(CH_2CH_2SCH_2CH_2OH)_2$.

In contradistinction to these observations with 2-mercaptoethanol, excess guanosine 5′-monophosphate (GMP) had no detectable influence on the lifetime of aziridinium ions derived from PM at pH 7.4 (Zon, unpublished work). Thus, despite the highly reactive nature of such ions and the generally presumed nucleophilicity of the guanine base, alkylation of a ring

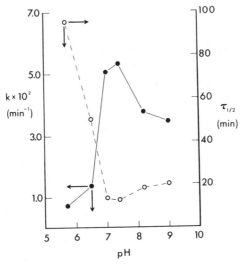

Figure 2. Combination plot showing the ^{31}P NMR-derived pH dependency of the rate of disappearance of PM/PM⁻ expressed as either a first-order rate constant (k, solid curve) or half-life ($\tau_{1/2}$, dashed curve); $T = 37°C$.

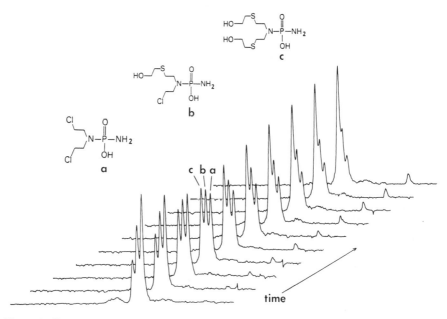

Figure 3. Stack-plot of ^{31}P NMR (109.3 MHz) spectra of PM, as a function of time, at pH 7.4 and 37°C in the presence of excess 2-mercaptoethanol, HSCH$_2$CH$_2$OH. Structures of PM starting material, the monoalkylated intermediate, and the final bis-alkylation product are shown in their protonated forms; however, at pH 7.4 these acids (pK$_a$ ~ 5) are extensively ionized.

N-position or the negatively charged phosphate group is appreciably slower than the alkylation of a mercapto group.

2.2.3. CP Hydrolysis and Rearrangement. The stability of CP in aqueous media used as administration vehicles is clinically important, and early hydrolytic studies (Friedman *et al.*, 1965) led to the proposal that the hydrolytic conversion of CP shown in Scheme III was initiated by intramolecular cyclization to a bicyclic intermediate rather than, for example, formation and subsequent trapping of nor-HN2. More recent "negative-labeling" experiments (Zon *et al.*, 1977) with specifically ^{2}H-labeled 3-methylcyclophosphamide (3-Me-CP), namely, 3-Me-CP-d_4 (Scheme III), gave ^{13}C NMR spectra of the corresponding N-methylated hydrolysis product which showed (Figure 4) relatively low intensity signals for those carbons bearing ^{2}H.[1] The final position of labeled carbons was consistent with

---

[1]For the CH$_2$Cl and CH$_2\overset{+}{N}$ positions of interest, replacement of the protons by deuterium causes each of these ^{13}C resonances to be split into a quintet due to spin-spin coupling with the two deuterons ($I = 1$). Additionally, the spin-lattice relaxation time of each carbon will be significantly lengthened via the greatly decreased magnetic moment of a deuteron relative to a proton ($\gamma_H/\gamma_D \sim 6.5$). These two factors, in concert, make the observation of these ^{13}C

Scheme III

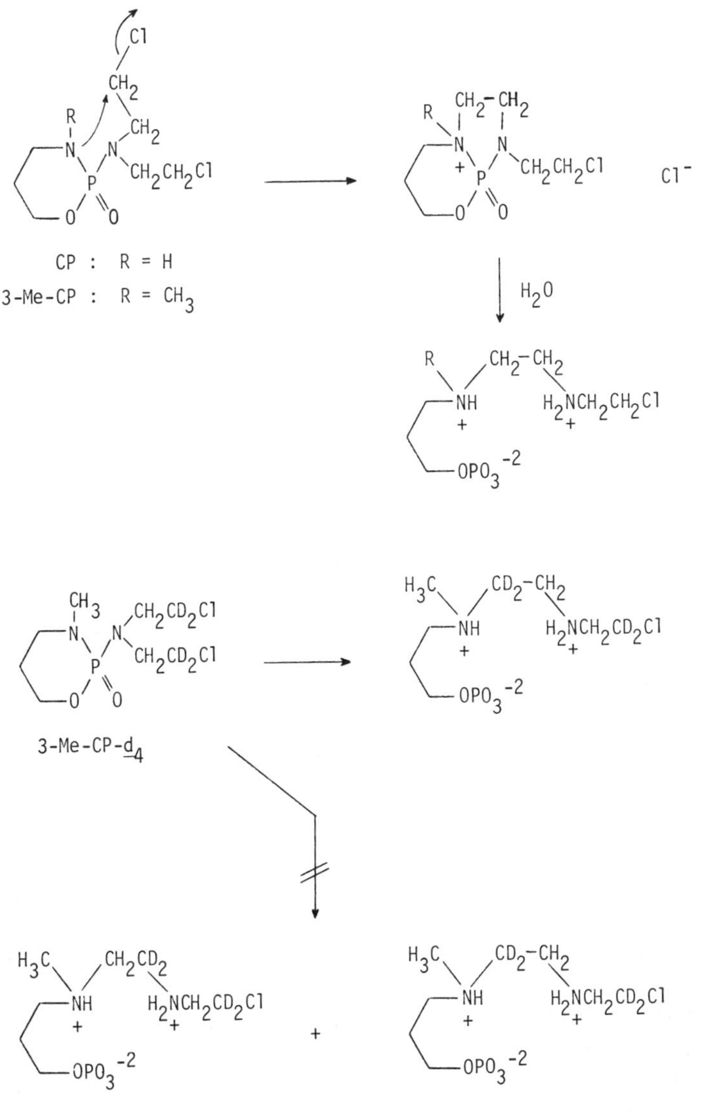

CP :   R = H
3-Me-CP :   R = CH_3

3-Me-CP-d_4

the proposed intramolecular displacement process and ruled out the nor-HN2 route (or any other mechanism) that would have "scrambled" the labeled carbons between the two nonequivalent (anisochronous) ^{13}C positions in the ethylenediammonium moiety.

---

resonances difficult relative to the remaining carbon atoms bearing spin-decoupled hydrogens. Hence, one may refer to this diminished ^{13}C signal intensity for ^2H-labeled carbons as "negative labeling" in contrast to the usual signal enhancements obtained from stable isotope incorporation.

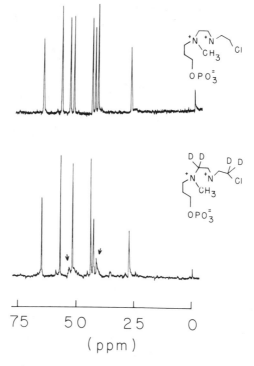

Figure 4. Aliphatic portion of the ^{13}C NMR (67.8 MHz) spectra of the products, isolated as oxalate salts, from the separate hydrolysis (30 min, 100°C, pH ~ 7) of 3-Me-CP (top) and 3-Me-CP-d_4 (bottom).

The rate-controlling nature of the initial intramolecular cyclization of CP to give the bicyclic intermediate was also supported by NMR kinetic measurements of the first-order rate constant (k) for disappearance of CP versus ^{2}H-labeled CP bearing an N(CH$_2$CD$_2$Cl)$_2$ functionality, as the value of $k_H/k_D = 1.14 \pm 0.07$ per deuterium implies a limiting dissociative mechanism (Zon *et al.*, 1977).

In connection with these mechanistic findings, it is worthwhile to recognize the possibility for biological metal ion-assisted interconversion of CP with its constitutional isomer, isophosphamide (IP, Scheme IV) which is also an anticancer agent and is currently under investigation as an experimental drug. To test for this rearrangement, ^{1}H NMR was used to monitor solutions of CP in the presence of Ag$^+$; however, the value of k was virtually unchanged and signals from IP were not detected, which militate against the likelihood of such "drug crossover" in the absence of enzymatic intervention.

Scheme IV

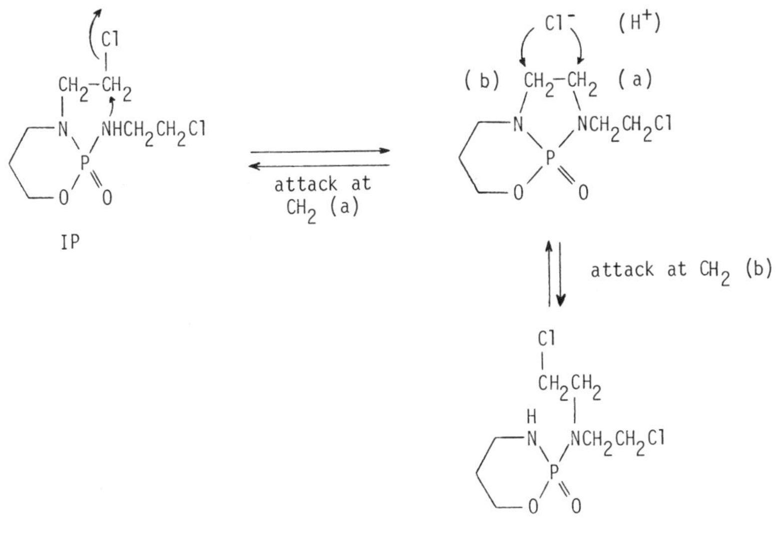

IP

attack at CH$_2$ (a)

(b) CH$_2$–CH$_2$ (a)

Cl$^-$ (H$^+$)

attack at CH$_2$ (b)

CP

2.2.4. Stereochemical Aspects of CP Metabolism.

The presence of an asymmetric phosphorus center in CP leads to the existence of nonequivalent mirror image forms of this drug, namely, the (R)- and (S)-CP enantiomers.

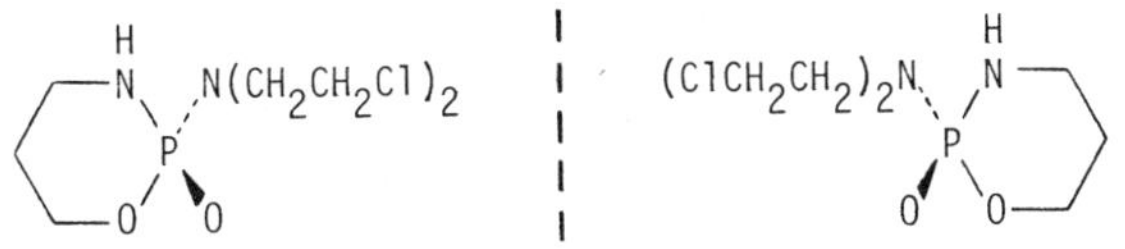

(R)-CP   (S)-CP

Enzymes and other chiral components in biological systems must, in principle, selectively interact with these enantiomers and their corresponding asymmetric metabolites, such as (R)- and (S)-4-HO-CP/AP. "Enantioselectivity" may therefore be a significant factor during metabolism of racemic CP, which is the form of the drug currently administered to patients, and, moreover, there exists the possibility that treatment of enantiomerically pure CP might afford improved cancer therapy (Cox et al., 1976; Tsui et al., 1979; Jarman et al., 1979). The enantiomeric purity of synthetic samples of (+)- and (−)-CP has thus been *unambiguously* established (Zon et al., 1977) by ^{1}H (Figure 5) as well as ^{31}P (Figure 6) NMR using either of the following chiral shift reagents: Eu(tfc)$_3$ and Eu-(hfc)$_3$.[2] Extension of this methodology

[2]Eu(tfc)$_3$ = tris[3-(trifluoromethylhydroxymethylene)-d-camphorato]europium (III); Eu(hfc)$_3$ = heptafluoropropyl analog of Eu(tfc)$_3$. Additional studies of CP enantiomers with Eu(hfc)$_3$ have

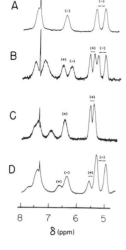

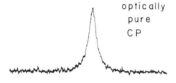

Figure 5. Partial ¹H NMR spectra (220 MHz, CDCl₃) recorded for various cyclophosphamide (CP) samples containing one equivalent of Eu(*tfc*)₃. The plus and minus signs refer to mustard proton absorptions assigned to (+)- and (−)-enantiomers of CP, respectively. The sharp peak at δ-7.3 is due to residual CHCl₃. Spectrum A; 0.06 *M* (−)-CP, pulse Fourier transform (FT) mode. Spectrum B; 0.2 *M* (±)-CP, continuous wave (CW) mode. Spectrum C; 0.2 *M* (±)-CP, CW mode. Spectrum D; 0.2 *M* 1:1 mixture of (−)-CP: (±)-CP, FT mode. Relatively small chemical-shift changes between spectra are ascribed to small concentration differences among the samples.

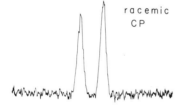

Figure 6. ³¹P NMR spectra (100 MHz, CDCl₃) of 0.2 *M* racemic (±)-cyclophosphamide (CP) in the presence of 1 equivalent Eu(tfc)₃ δ-76.74 and −8.86, external H₃PO₄ reference) and of optically pure (−)-CP under essentially identical conditions (δ −78.00). Mixing of these two samples yielded the expected 1:3 ratio of peaks (δ −75.88: −78.30).

to enantiomerically enriched samples of 4-HO₂-CP, 4-ketocyclophosphamide, and IP has indicated the superiority of Eu(hfc)₃, which causes less line broadening (Zon, Brandt and Egan, unpublished work; Ludeman *et al.,* 1979).

Application of the chiral shift reagent method to, for example, the analysis of isolated metabolites, is not obviated by the presence of impurities as is the traditional optical rotation method. Other attractive features of the NMR

been published (Kawashima *et al.,* 1978), and cobaltous adenosine 5′-triphosphate has been reported to serve as a water-soluble chiral shift reagent (Granot and Reuben, 1978). Enantiomerically pure 1-aryl-2,2,2-trifluoroethanols as solvents may also be used to determine optical purity by NMR (Pirkle and Hoekstra, 1976).

technique with chiral shift reagents include the ability to bypass prior determination of, or reliance upon, an absolute rotation, and the applicability to microgram quantities of sample via pulse FT-NMR. For those cases involving overlap of two resonance peaks from the same nucleus (or nuclei) in the enantiomers being examined with a chiral shift reagent, a so-called "Base Line Technique" has been recently developed in connection with the determination of optical purity for drug samples such as ibuprofen (Parfitt *et al.*, 1978). The technique utilizes the fact that the distance between the "true base line" (T) and junction between the two peaks, that is, the "false base line" (F). is proportional to $1 -$ percent optical purity (OP)/100 at a constant reagent to substrate molar ratio. A plot of $F - T$ differences versus $1 -$ percent OP/100 gave a straight line and the $<1\%$ standard deviation was reported to be superior to electronic integration, peak height measurement, counting of squares, and the cut-and-weigh methods.

2.3. N-Nitroso Compounds

N-(2-chloroethyl)-N-nitrosoureas such as BCNU and CCNU (Scheme V) are notable members of a successful class of anticancer drugs which are believed to undergo nonenzymatic fragmentation *in vivo* to form an isocyanate and an alkylating agent capable of cross-linking DNA (Lown *et al.*, 1978). A likely pathway for this decomposition is pictured in Scheme V, and evidence supporting the existence of the symmetrical chloronium ion intermediate (or its chemical equivalent) has been obtained by ^{2}H labeling in conjunction with mass spectroscopic product analysis to detect positional "scrambling" of the label (Brundrett *et al.*, 1976). Analogous labeling experiments with BCNU or CCNU which employ either ^{1}H, ^{13}C, or ^{15}N NMR may provide additional mechanistic information with regard to the proposed bis-alkylation of DNA by $[C_2H_4Cl]^+$ (Lown *et al.*. 1978) as well as providing an informative method for studying the initial fragmentation kinetics.

Scheme V

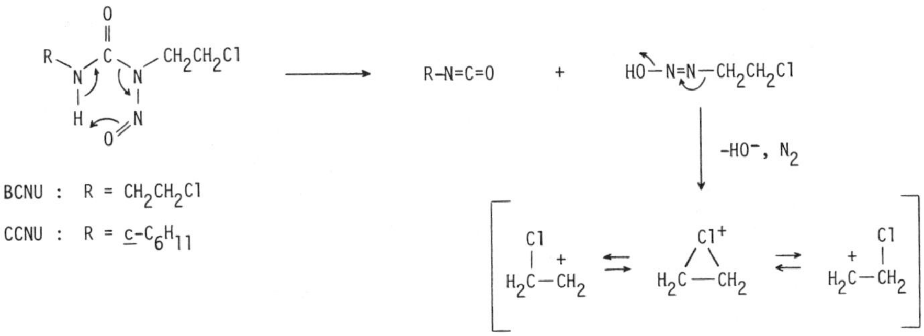

Scheme VI

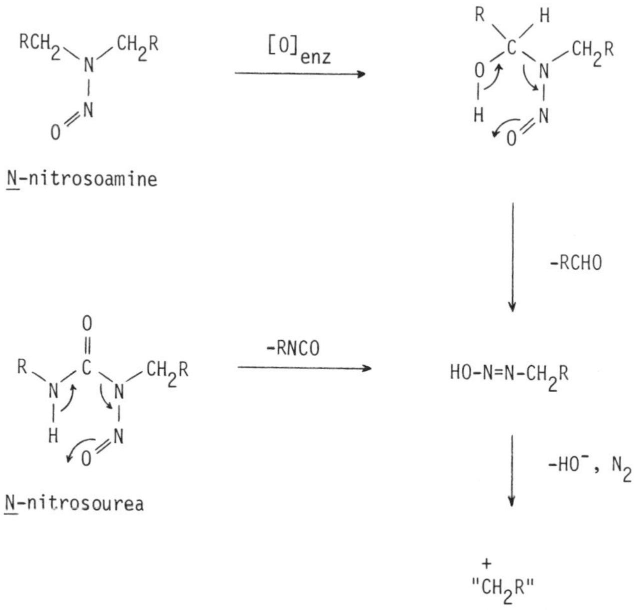
RCH2 N CH2R
N
O
N-nitrosoamine
[O]enz
R H
C
O N CH2R
H N
O
-RCHO
O
C
R N N CH2R
H N
O
N-nitrosourea
-RNCO
HO-N=N-CH2R
-HO⁻, N2
+
"CH2R"

The generally accepted metabolic pathway for N-nitrosoamines is outlined in Scheme VI and, as shown, is somewhat analogous to the aforementioned N-nitrosoureas; however, N-nitrosoamines function as monoalkylating agents and are potent mutagens and carcinogens (Singer, 1975). In view of this mechanistic similarity and the possibility for "crossover" between the metabolic pathways for these two classes of compounds, it is worthwhile to consider alternative nonenzymatic degradation mechanisms for N-nitrosoamines that have been investigated by ^{1}H NMR (Olah *et al.,* 1975) and have not as yet been tested in the N-nitrosourea systems.

A series of dialkylnitrosoamines [R-N(NO)-R′] ranging from dimethyl to *n*-butyl-*t*-butyl were examined in $ClSO_2F$ solutions of "magic acid" ($FSO_3H:SbF_5$) at $-60°$ and showed, in addition to restricted rotation about the N—N bond (Forlani *et al.,* 1979), protonated nitrosoamines which gave rise to a one-proton sharp singlet at 13–14 ppm that was not detectable at higher temperatures and reappeared upon cooling. This temperature-dependent resonance was assigned to the hydroxy proton of the O-protonated tautomer, as protonation at either nitrogen would be expected to produce a quadrupolar broadened NH resonance. The presence of such *N*-protonated ions in undetectably low concentration was, however, indirectly deduced from the identity of subsequently formed degradation products (see below).

Scheme VII

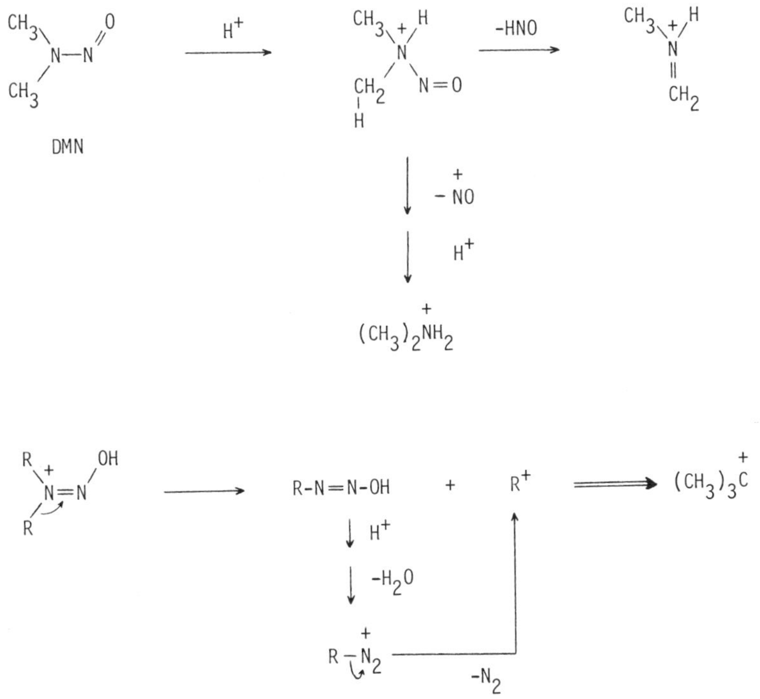

The first of three distinct protolytic fragmentation pathways was detected by the irreversible conversion of O-protonated dimethylnitrosoamine (DMN) into N-methylmethyleneimmonium ion, $H_2C=\overset{+}{N}HCH_3$ ($^1J_{NH} = 66$ Hz). However, the second type of reaction, denitrosation of dialkylnitrosoamines, was only detected for DMN in 100% H_2SO_4 solvent, where an upfield triplet for $(CH_3)_2\overset{+}{N}H_2$ was seen as a minor constituent (30%) together with $H_2C=\overset{+}{N}HCH_3$. Proposed mechanisms for the formation of these two types of products are shown in Scheme VII together with a suggested sequence for the third degradation pathway which was observed for the various propyl- and butyl-substituted nitrosoamines in "magic acid." In these cases the major NMR-detectable product was t-butyl cation, which was identified by noting its increased signal strength upon addition of t-butyl alcohol:

$$(CH_3)_3COH + H^+ \rightarrow (CH_3)_3C^+ + H_2O$$

The formation of t-butyl cation can result from S_N1 ionization of the O-protonated nitrosoamine, as shown, with further protolysis of the resulting alkyldiazonium hydroxide ($R-N=N-OH$) leading to an unstable $R\overset{+}{N}_2$

intermediate[3] which loses N_2 to form the second alkyl cation. These cations can, under "stable ion conditions," condense and rearrange to ultimately afford the more thermodynamically stable t-butyl cation as the major product.

In view of the fact that the aforementioned studies have established the chemical feasibility of three novel pathways for "activation" of dialkylnitrosoamines (and, by extension, N-nitrosoureas) under drastic protolytic conditions in "superacid" media, the investigators (Olah *et al.*, 1975) addressed the question of whether such reactions could occur in a living cell. It was concluded that these transformations are realistic if one considers the following factors. First, biological systems are nonhomogeneous and hydrogen ion concentration differences across membranes may be as large as 10^6. Second, pH is a *thermodynamic* concept and application of such data to *kinetic* phenomena involving the extent and even direction of a chemical reaction is unwarranted. Finally, the cleavage reactions under consideration are not restricted to activation by hydrogen ion and could, alternatively, be triggered by other electrophilic species such as metal ions which are present at enzyme active sites.

2.4. Miscellaneous Alkylators

2.4.1. Aziridinyl Compounds. Early studies (Pettit *et al.*, 1965; Levins and Papanastassiou, 1965; Levins and Rogers, 1965) of nitrogen mustards by proton magnetic resonance indicated that NMR was a preferred alternative to classical methods of analysis for alkylating functionalities, such as 4-(p-nitrobenzyl)pyridine (NBP) colorimetric assays (Friedman and Boger, 1961), chloride and hydrogen ion determinations (Bartlett *et al.*, 1949), and thiosulfate titrations (Golumbic *et al.*, 1946). ^{1}H NMR has been used to study the reactions of aziridinyl chemosterilants (Beroza and Borkovec, 1964), and a more recent application to aziridine-containing systems concerns the hydrolytically labile anticancer drug thiotepa. On the basis of colorimetric tests for SH and alkylating groups, it had been reported that thiotepa underwent acid-catalyzed (Benckhuijsen, 1968) and saline-induced (Maxwell *et al.*, 1974) reactions by strikingly different pathways to form the products shown. The significance of such transformations to metabolic and clinical aspects of thiotepa prompted a reinvestigation by ^{1}H NMR (Zon *et al.*, 1976), and both of the prior claims were disproven, as no evidence for characteristically upfield aziridinyl proton resonances was observed under the originally reported reaction conditions. Findings such as these represent a caveat regarding the validity of indirect functional group tests and provide

---

[3]Protonation of diazomethane ($H_2\bar{C}-\overset{+}{N}{\equiv}N \leftrightarrow H_2C{=}\overset{+}{N}{=}\bar{N}$) in FSO_3H: FSO_2Cl at $-120°$ has been recently shown (Berner and McGarrity, 1979) by ^{1}H and ^{13}C NMR to afford an 80:20 mixture of C- and N-protonated products ($CH_3-\overset{+}{N}{\equiv}N$ and $CH_2{=}\overset{+}{N}{=}NH$, respectively) which decompose to primarily methyl fluorosulfate (CH_3OSO_2F) upon warming to temperatures above $-85°C$.

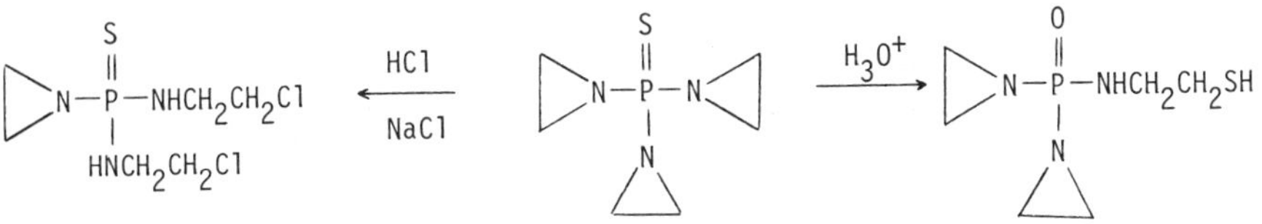

thiotepa

support for the use of direct NMR spectroscopic techniques whenever possible.

2.4.2. Prodrug Systems. The design of new classes of potential anticancer prodrugs capable of nonenzymatic "activation" *in vivo* has led to the synthesis and anticancer screening of various pro-phosphoramide mustards, pro-PM, and pro-nitrogen mustards, pro-(nor-HN2), which are shown in Scheme VIII together with their proposed fragmentation pathways (Chiu *et*

Scheme VIII

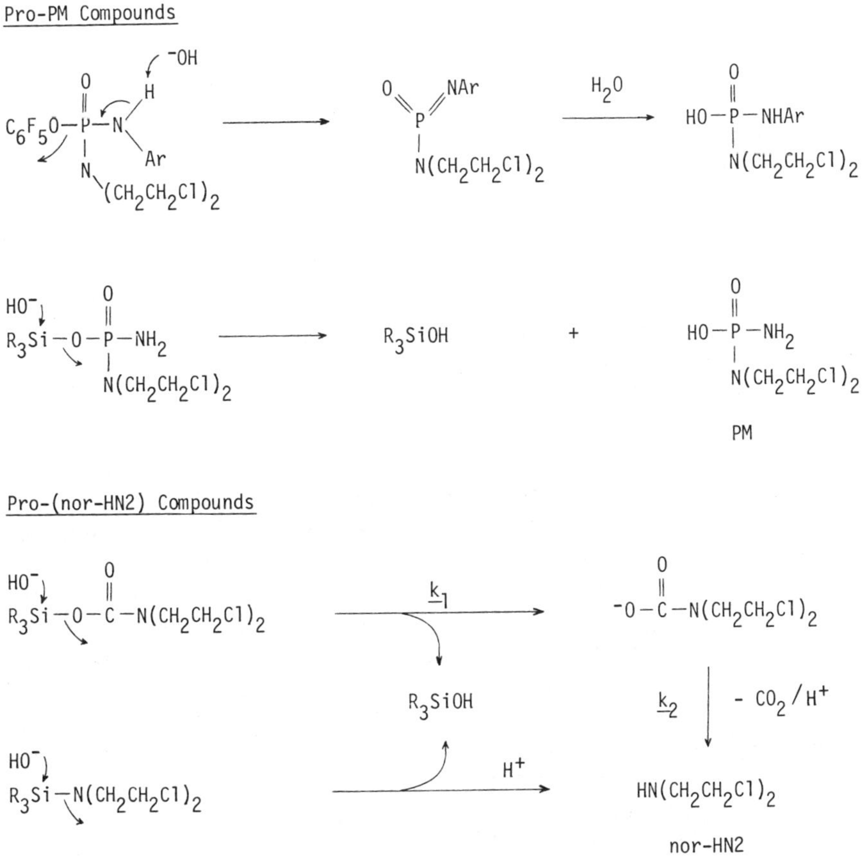

Pro-PM Compounds

Pro-(nor-HN2) Compounds

nor-HN2

al., 1979; Zon and Chiu, unpublished work). In all cases the rate of hydrolytic "activation" is pH dependent and may be controlled by incorporation of substituents having appropriate steric and/or electronic properties. Kinetic evaluation of these structure–reactivity concepts has been achieved by ^{1}H NMR monitoring of the relative concentrations of starting material versus products. For example, initial rates of disappearance of the C_6F_5O-substituted pro-PM were measured by aromatic proton peak area changes, which in turn allowed for the determination of a Hammett reaction constant (ρ value) of -0.9.

The triorganosilylated pro-(nor-HN2) compounds having $SiCH_3$ and/or $SiC(CH_3)_3$ groups were studied (Zon and Chiu, unpublished work) by utilizing chemical shift differences between singlet ^{1}H absorptions for these "sensor" groups in the carbamate starting materials versus triorganosilanol hydrolysis byproducts. The latter peaks generally appear at higher field, relative to the carbamate, which is presumably due to the removal of carbonyl anisotropy effects. These ^{1}H NMR kinetic measurements provided pseudo first-order rate constants for initial desilylation (k_1), while companion UV kinetics afforded the apparent first-order rate constants for the subsequent decarboxylation step (k_2). Together, the two techniques provided a novel means of evaluating the half-life of the unstable carbamic acid intermediate and also ruled out alternative "concerted" carbamate fragmentation mechanisms which bypass this metastable species.

2.5. Sites of Alkylation

An increasing effort is being directed toward determining the nature and effects of DNA and RNA alkylations by anticancer drugs (Lown *et al.*, 1978) as well as carcinogens and mutagens (Singer, 1975). For example, the covalent bonding of a benzo[*a*]pyrenetetrahydrodiol epoxide diastereomer to poly(G) has been shown, after hydrolysis, to involve substitution of the N-2 of guanine at the C-10 epoxide carbon position (Jeffrey *et al.*, 1976). However, such chemical degradation methods for locating the sites of nucleic acid reaction or cross-linking are subject to possible errors introduced by decomposition and/or rearrangement, and consequently there is a need to explore the utility of NMR as a tool in the development of new and improved analytical procedures in this area. Initial studies (Chang and Lee, 1976) of RNA alkylation by methyl methanesulfonate ($CH_3SO_3CH_3$) have confirmed the feasibility of this direct spectroscopic approach by demonstrating that the various alkylation products (sites of alkylation) could be monitored by ^{13}C resonances: phosphonomethyl, 53.4 ppm; 1-methyladenosine, 38.1 ppm; 7-methylguanosine, 36.2 ppm; 3-methylcytidine, 30.8 ppm; 1-methylguanosine, 28.7 ppm; and 3-methyluridine, 28.0 ppm. In addition, N-7 methylation of guanosine in RNA could be assessed from the relative integrated signal intensities of the C-5 position in guanosine (116.2 ppm) versus 7-methylguanosine (108.3 ppm). The extent of methylation may be obtained by concurrent radioactivity measurements with ^{14}C-labeled alkylating agent. Drawbacks associated with the high molecular weight and low solubility of RNA were minimized by use of 90% ^{13}C-enriched methyl methanesulfonate. An important advantage of this approach is that lower concentrations of agent can be used to study initial product distributions, which more closely simulate *in vivo* exposure levels and allow for assessment of kinetically controlled selectivity factors (Chang and Lee, 1978).

The first ^{13}C NMR comparison of methylated RNAs obtained from alkylations with either methyl methanesulfonate, dimethyl sulfate, 1-methyl-3-nitro-1-nitrosoguanidine, or 1-methyl-1-nitrosourea has revealed distinctly different product distributions (Figure 7) for the former and latter pair of agents, which are regarded as "weakly" and "strongly" mutagenic/carcinogenic, respectively (Chang and Lee, 1978). As noted by these investigators, it is premature to conclude whether or not this type of direct

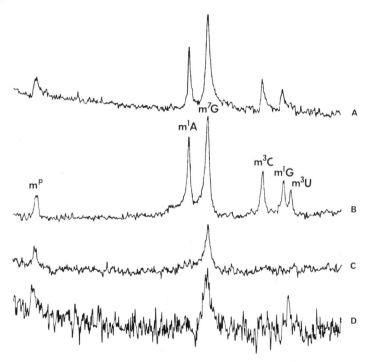

Figure 7. Proton-decoupled natural abundance ^{13}C NMR (100 MHz) spectra of methyl carbon resonances of methylated *Torula* yeast RNA (400 mg/1.5 ml) from the reactions with: A, methyl methanesulfonate (162,500 scans); B, dimethyl sulfate (148,700 scans); C, 1-methyl-1-nitrosourea (140,000 scans); D, 1-methyl-3-nitro-1-nitrosoguanidine (205,784 scans). m^{3}U = 3-methyluridine; m^{1}G = 1-methylguanosine; m^{3}C = 3-methylcytidine; m^{7}G = 7-methylguanosine; m^{1}A = 1-methyladenosine; m^P = phosphonomethyl. From integration of the m^{7}G signal, the relative percentages of N-7 methylation determined from spectra A, B, C, and D were 58, 38, 82, and 79%, respectively, with the estimated uncertainty range being 5–15%.

spectroscopic characterization may be used as a reliable criterion for *a priori* biological classification of various types of mono- (and bis-) alkylating agents; however, it is clear that the general method represented by their studies is feasible and thus represents a significant area for future research.

Another approach for determining the site of N-alkylation of nitrogen bases involves measurement of ^{15}N–^{13}C spin–spin coupling by ^{13}C NMR using ^{15}N-enriched materials. A series of ^{15}N-labeled benzyladenine models gave the ^{15}N–^{13}C spin coupling constants indicated (Wiemer *et al.*, 1976).

$^1J_{^{15}N-^{13}C} = 9.3$ Hz $^1J_{^{15}N-^{13}C} = 7.2$ Hz $^5J_{^{15}N-^{13}C}$ not obsvd

3. PLATINUM COMPOUNDS

Platinum coordination complexes constitute a relatively new class of anti-cancer agents which have been of considerable interest since the discovery that *cis*-diaminedichloroplatinum(II), *cis*-$PtCl_2(NH_3)_2$, exhibits activity against a broad spectrum of experimental tumors (Roberts and Thomson, 1979; Rosenberg, 1978). There is ample evidence (Roberts and Thompson, 1979; Zwelling *et al.*, 1979) which implicates DNA as the vital target for such platinum compounds and, more specifically, it appears that G + C content is a controlling factor in the reactivity of DNA (Stone *et al.*, 1976; Munchausen and Rahn, 1975). NMR and allied spectroscopic methods have been used to elucidate the interaction of such platinum complexes with DNA by examination of model nucleosides and nucleotides, as outlined in the following paragraphs.

cis-$PtCl_2(NH_3)_2$

^1H NMR analyses of preformed platinum–nucleoside complexes were among the earliest studies in this area, and spectra A and B in Figure 8 show free inosine (In) and its $[Pt(en)In_2]Cl_2$ complex (Kong and Theophanides, 1974a). Exchangeable protons (NH, OH) are "washed out" by D_2O and the

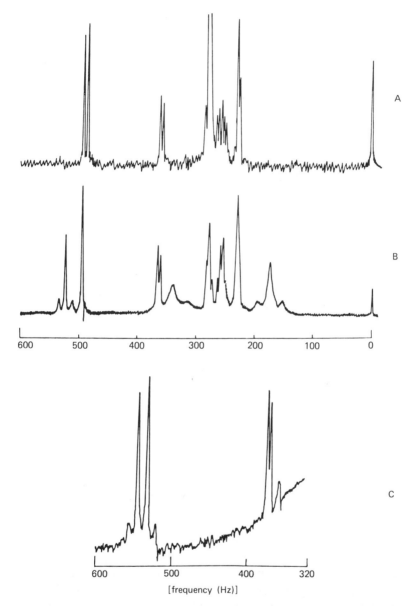

Figure 8. ^{1}H NMR (60 MHz) spectra in D_2O with DDS as the integral chemical shift reference. A: free inosine (In); B: $[Pt(en)In_2]Cl_2$; C: $[Pt(dien)Cl]Cl$ in large excess over adenosine $(Pt/A \geqslant 3)$.

sugar proton resonances (200–400 Hz) show little variation between spectra, as they do not directly participate in the ligand-to-metal bonding. The natural abundance of ^{195}Pt ($I = \frac{1}{2}$) is 33.7%, and the broad "triplet"[4] centered at 174 Hz in spectrum B is due to ^{195}Pt–^{1}H spin–spin coupling of the methylene groups of ethylenediamine (en) in [Pt(en)In$_2$]Cl$_2$, with $J = 44$ Hz. It can be seen in spectrum B that H-8 undergoes a greater downfield shift than H-2, which is consistent with coordination to Pt at N-7, and that the furthest downfield "triplet" assigned to H-8 in each case results from coupling to ^{195}Pt. The same type of downfield-shifted "triplet" signal was observed for H-8 in [Pt(en)L$_2$]Cl$_2$ complexes where L = guanosine (G) and xanthosine (X), which likewise supports N-7 binding.

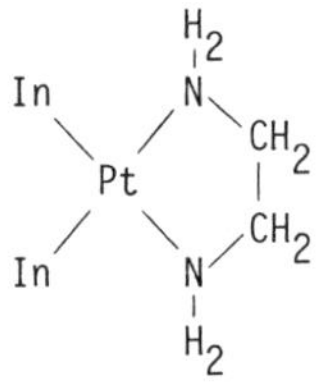

Pt(en)In$_2$

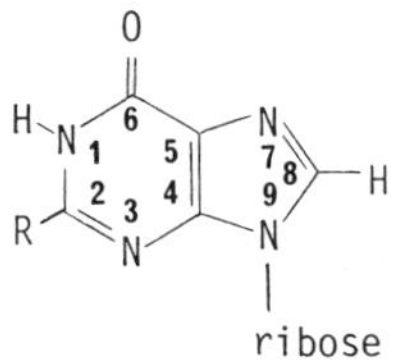

Inosine (In) : R = H

Guanosine (G) : R = NH$_2$

Xanthosine (X) : R = OH

By way of contrast, spectrum C in Figure 8 shows the downfield H-8 and H-2 absorptions for adenosine in the presence of excess [Pt(dien)Cl]Cl, where dien = diethylenetriamine (H$_2$NCH$_2$CH$_2$NHCH$_2$CH$_2$NH$_2$) and only one chlorine may be displaced (Kong and Theophanides, 1974b). There are two main peaks with two ^{195}Pt satellites, which suggested that H-8 and H-2 are each coupled with an adjacent ^{195}Pt and that mutual overlapping of the resultant "triplets" gave the observed four-line pattern. Assignment of the lower field "triplet" to H-8 was based on the absence of this multiplet when 8-^{2}H-adenosine was used.

---

[4]"Triplet" here refers to the superimposed uncoupled singlet absorption and ^{195}Pt-coupled doublet pattern.

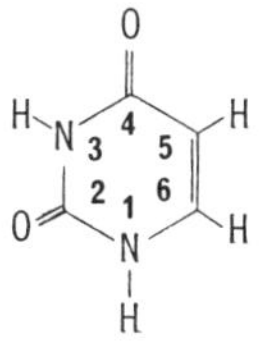

Uracil

cis-[Pt(NH$_3$)$_2$(H$_2$O)$_2$]$^{2+}$ reacts with uracil to produce "platinum-pyrimidine blue" which has proven to be a good antineoplastic agent that does not cause kidney damage (Davidson *et al.*, 1975). To examine the binding site of platinum to uracil, the reaction of uracil with triamineplatinum has been studied (Inagaki and Kidami, 1978), and it was found that two complexes are formed. ^{1}H NMR data indicated that one was a complex in which triamine-platinum displaced a proton and bound at the N-3 position while the other resulted from proton displacement and binding to N-1. The proton NMR pattern observed for the H-6 resonance in the latter complex (Figure 9) further illustrates the utility of ^{195}Pt coupling in spectral analysis, as coordination of platinum to N-1 affords the six-line doubled "triplet" with $^3J_{Pt-H} \sim$ 40 Hz and $^3J_{H-H} \sim$ 10 Hz.

^{1}H NMR has been used to study the reaction of [Pt(dien)Cl]Cl with cytidine 5'-monophosphate (CMP) and guanosine 5'-monophosphate (GMP), and it was concluded that coordination to platinum occurs through N-1 and N-7, respectively (Kong and Theophanides, 1975). These studies also addressed the question of possible interaction between the phosphate group and platinum; however, the observation of base proton shifts similar in magnitude to those found for cytidine–platinum complexes was taken as evidence against such bonding. Additional experiments along these lines could utilize the sensitivity of ^{31}P chemical shifts of monophosphate groups to electronic perturbations as a probe for this bonding mode or a rapid exchange process involving free and metal-complexed phosphate. In cases where phosphate–platinum coordination persists long enough for ^{31}P–^{195}Pt spin–spin coupling to be observed, information concerning angular relationships (geometry) might be obtained.

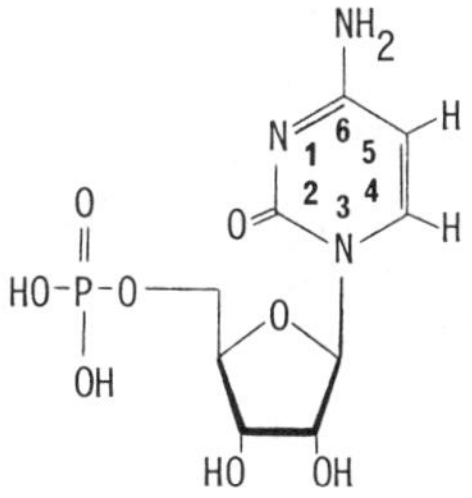

CMP

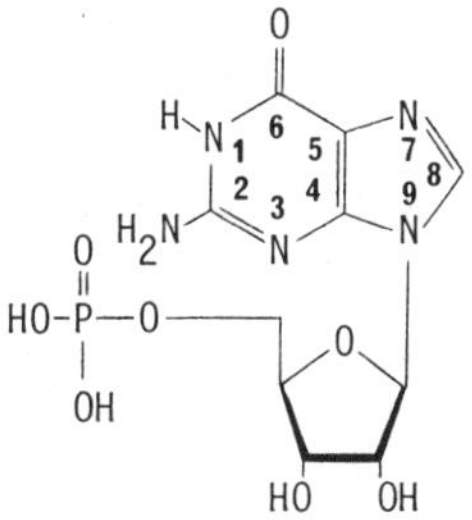

GMP

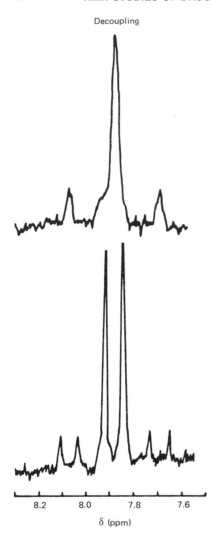

Figure 9. ^{1}H NMR (100 MHz) spectra of a complex of triamineplatinum and uracil in D_2O with DDS as the internal chemical shift reference.

It has been noted (Nelson *et al.*, 1976) that nucleic acid bases bear at most only two nonexchangeable protons in contradistinction to the four or five carbon atoms in pyrimidine and purine bases, respectively, and that this inherently greater "information content" of ^{13}C NMR spectra may prove to be useful. Carbon chemical shifts for nucleoside complexes with Pd(en)Cl$_2$ in water were therefore measured, and it was concluded that binding occurs at N-3 in cytidine, thymidine, and uridine, and at N-1 in guanosine (Nelson *et al.*, 1976). For the ^{13}C atoms adjacent to N-3 in the pyrimidine nucleosides, the "complexation shifts" ($\Delta\delta = \delta_{complex} - \delta_{free\ nucleoside}$) of the basic ligand are ~30% of corresponding "protonation shifts." It was further suggested that future ^{13}C NMR studies of Pt(II) complexes could possibly be assisted by ^{13}C–^{195}Pt spin–spin coupling constant data of the type which has been previously measured (Chow and Martin, 1974) in Pt(II) complexes with pyridine (pyr) ligands. However, only the three-bond $^{13}C-^{195}Pt$ couplings of ~40 Hz were resolved in $[Pt(pyr)_4]^{2+}$ and, furthermore, the two- and three-bond couplings that were seen in the bipyridyl complex $[Pt(en)(bipy)]^{2+}$ were similar in magnitude, namely 26–34 Hz. It is thus apparent that a lack of reliable discrimination between two- and three-bond $^{13}C-^{195}Pt$ spin interactions may diminish their diagnostic value in identifying key carbon positions in DNA bases of interest.

$$[Pt(pyr)_4]^{+2} \qquad\qquad [Pt(en)bipy]^{+2}$$

More recent ^1H NMR work with Pt(II) compounds and the 5'-monophosphate nucleotides GMP, CMP, AMP, and UMP has been reported (Mansy *et al.*, 1978) and has also employed Raman difference spectroscopy (RADS) as an adjunct analytical tool. It was concluded on the basis of RADS data that cis-$PtCl_2(NH_3)._2$, cis-$[(H_3N)_2Pt(OH_2)_2]^{2+}$, and $trans$-$[(H_3N)_2Pt\text{-}(OH_2)_2]^{2+}$ exhibit considerable specificity during competition experiments with mixtures of excess nucleotide: both purines reacted at $r = 0.05$ [$r =$ Pt(II): nucleotide], whereas there was no detectable reaction with either of the pyrimidine nucleotides even at $r = 0.20$. ^1H NMR data in analogous studies with enPt(II) at $r = 0.1$ and 0.2 were qualitatively in agreement, as downfield-shifted H-8 signals for $[enPt(GMP)_2]^{2-}$, $[enPt(AMP)_2]^{2-}$, and the mixed complex $[enPt(GMP)(AMP)]^{2-}$ were observed. The ~0.5 ppm downfield complexation shift in the case of N-7 binding of GMP may be compared with the ~1 ppm downfield shift caused by protonation at N-7.[5]

Diagnostic ^1H and ^{13}C chemical shift changes have also been measured in complexes of cis-$MCl_2(mit)_2$, where M = Pt or Pd and mit = 1-methyl-imidazole-2-thiol (Dehand and Jordanov, 1977). For example, binding of two guanosines to Pt via N-7 produced $\Delta\delta$ values which were largest for the proximate C-8 and C-5 positions:

	C-2	C-4	C-5	C-6	C-8
guanosine	154.7	152.0	117.7	157.7	136.5
$[Pt(mit)_2$ (guanosine)]Cl_2	155.1	153.2	120.1	159.3	142.7
$\Delta\delta$(ppm)	0.4	1.2	2.4	1.6	6.2

[5]Related studies (Chu *et al.*, 1978) with values of $r > 0.5$ afforded marked spectral broadening, which was ascribed to, *inter alia*, polymerization of the type represented by cis-$[(H_3N)_2$ $Pt(GMPH_{-1})]^- \leftrightarrow (1/n)$ cis-$[(H_3N)_2Pt(GMPH_{-1})^-]_n$. An additional complication is non-stoichiometry such as that found by ^1H and ^{13}C NMR in the case of antineoplastic Pt(II) complexes with inosine 5'-monophosphate (Marzilli, 1978).

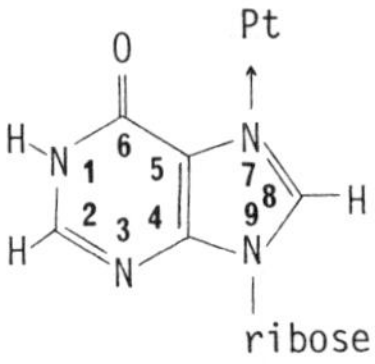

Secondary stereochemical information concerning the nature of Pt–nucleotide complexes has been obtained by consideration of diamagnetic anisotropy effects (Chu *et al.*, 1978). Thus, while perturbations in RADS can reflect changes in the purine base and are relatively little affected by the environment of the ligand, ¹H chemical shifts of the two GMP ligands in a Pt(GMP)₂ complex can be influenced, in a predictable way, by their mutual anisotropy effects, which in turn relate to their relative positioning about the metal bonding center. The Raman vibrational spectra of *cis*- and *trans*-[(H₃N)₂Pt(GMP)₂]²⁻ are virtually identical; however, H-8 for the *cis* isomer is 0.26 ppm *upfield* relative to the *trans* isomer. This is consistent with mutual GMP ring-current shieldings in the *cis* complex and serves to differentiate the two structures shown in Figure 10. The alternative GMP array for the *cis* isomer having a local mirror plane versus the indicated C_2 axis was ruled out by consideration of X-ray data. The fact that only one sharp H-8 resonance was observed suggested that only one of these structures occurs in solution.

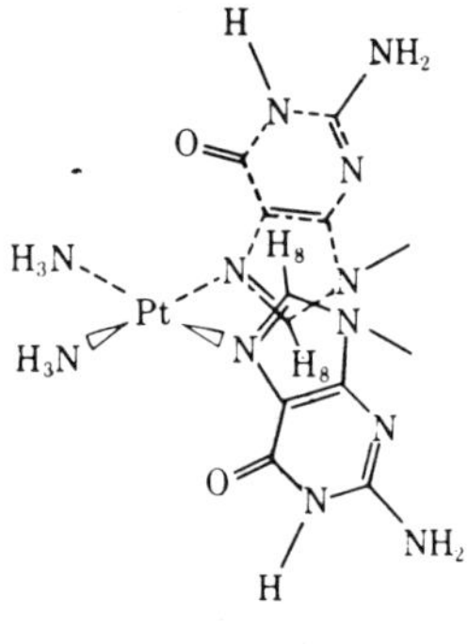

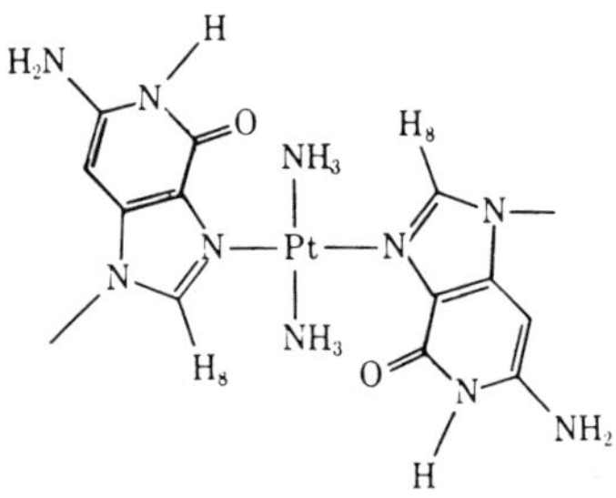

Figure 10. Isometric structures for [(H₃N)₂Pt(GMP)₂]²⁻; *cis* on top, *trans* at the bottom. Open valences at nitrogen indicate bonding to ribosylphosphate.

In concluding this section it should be mentioned that *dinucleotides* have been investigated from the viewpoint of possible selective complexation effects which result from "stacking" of the bases (Jordanov and Williams, 1978). The interaction of $Pt(en)Cl_2$ with cytidilyl-3' $\rightarrow$ 5'-guanosine phosphate (C3'p5''G) produced ^{1}H NMR (270 MHz) spectral changes that were significantly more complicated than those of the individual mononucleotides. From the time-dependent appearance of two sets of downfield-shifted C and G proton signals it was concluded that there are several binding modes of the dinucleotide with platinum, namely, reaction of the single base C before G and then intra- or intermolecular crosslinking. Chromatography led to isolation of two complexes which were identified as the expected "internally" and "externally" crosslinked products, Pt—C3'p5'G and (Pt—C3'p5'G)$_2$, respectively.

4. INTERCALATORS

In contrast to covalent bonding of a drug or drug metabolite to polynucleic acids, a number of antibiotics and anticancer agents are known to "associate" with DNA by attractive electrostatic interactions along the phosphate backbone and/or intercalation, wherein complexes are stabilized primarily by "stacking interactions" between aromatic rings of the drug and adjacent base pairs. NMR spectroscopy has proven to be a very powerful technique for studying these variants of drug–molecule association; however, in view of the fact that recently published reviews (Krugh, 1978; Patel, 1979a) of this subject are available, only very brief mention will be made here concerning some general features of the NMR approach.

Watson–Crick hydrogen-bonded ring NH resonances of stable nucleic acid base pairs occur in an unencumbered "window" region between 12 and 13 ppm in aqueous solutions (Kearns *et al.*, 1971; Patel and Tonelli, 1974). The linewidth of such signals can provide a measure of the lifetime of the hydrogen-bonded state and, for example, the thymidine H-3 proton of poly-(dAdT) exhibits a linewidth of ~90 Hz between 30 and 60° but broadens between 60 and 70°C, which is somewhat below the melting transition (72°C), and thus reveals that proton exchange with solvent occurs by transient opening of the duplex (Patel, 1979a).

^{31}P NMR can provide detailed structural information about the phosphate groups in nucleic acids, especially with the increased resolution that is possible at very high (superconducting) magnetic field strengths. Theoretical conformational calculations suggest that the transition from stacked to unstacked states results predominantly from a *gauche* to *trans* change in the phosphate backbone torsional angle and may account for downfield shifts of ^{31}P resonances that are observed during the previously mentioned poly(dA–dT) melting transition. More recently, the duplex-to-strand transitions ("unwinding") of the self-complementary sequence dG—dC—

dG—dC has been probed at the backbone phosphates by high-resolution ^{31}P NMR (Patel, 1979b). The observed variation of ^{31}P chemical shifts with temperature appeared to monitor changes in rotational angles about the O—P bonds in the postmelting transition temperature region, while the structure of the intercalation complex formed between this DNA duplex and the antitumor anthracycline antibiotic daunomycin was probed at both the nucleic acid and drug proton resonances as a function of temperature.

Dimethyl actinocynilbis(sarcosyl-L-valinate), with sarcosyl = CH_3NCH_2 CO, has been prepared as a model compound for conformational studies related to the anticancer intercalating drug actinomycin (Cavalieri *et al.*, 1978). Based on NMR, CD, and model-building data it was concluded that steric hindrance due to methylation of the phenoxazinone-bridging peptides does not allow for the coplanarity between peptide and heterocyclic moieties that exists in other actinomycin analogs and may, therefore, account for the lack of intercalation by this compound.

ACTINOMYCIN D

5. MISCELLANEOUS DRUG INTERACTIONS

In general, the reversible association of drugs or drug metabolites with molecules or paramagnetic ions can have a profound influence upon the linewidths and/or chemical shifts of NMR-active "sensor" nuclei within the drug or drug metabolite molecular structure. Resultant applications of NMR spectroscopy to the study of biological macromolecules and more complex molecular systems have been reviewed (Jardetzky and Jardetzky, 1971;

Campbell and Dobson, 1979). The following sections, by way of contrast, outline representative NMR investigations which deal with a wide range of *drug*-related "interactions" that in various ways contribute to understanding of drug action.

5.1. Polyene Antibiotics and Phospholipid Vesicles

Polyene antibiotics are a class of macrocyclic lactones having a lipophilic portion and a segment of four to seven conjugated C—C double bonds. The polyene antibiotics nystatin and amphotericin B have been developed for

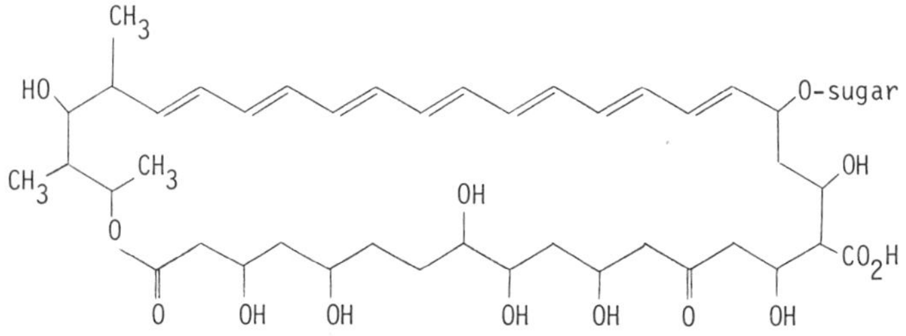

Amphotericin B

control of noncontagious, airborne, systemic infections, and their activity against fungal, algae, protozoan, and metazoan cells is generally ascribed (Weinberg, 1974) to a "complexing with sterols in the membranes with subsequent alteration in permeability and loss of essential and inorganic cell constituents." Details regarding this cellular disruption may be indirectly obtained by study of phospholipid vesicles, which are small (~300 Å) spherical shells composed of a single bilayer. Alterations in bilayer structure can be monitored by NMR spectroscopy, as the collapse to a multilayer structure can produce increased linewidths and, additionally, localized paramagnetic shift reagents can be employed to differentiate effectors which are either "inside" or "outside" of a vesicle. ^{1}H NMR investigations (Gent and Prestegard, 1976) which nicely illustrate these points are concerned with the influence of nystatin, amphotericin B, and filipin on cholesterol-containing vesicles. For example, adding polyene antibiotics to vesicles prepared from phosphatidylcholine (PC) and cholesterol (2:1) led to exten-

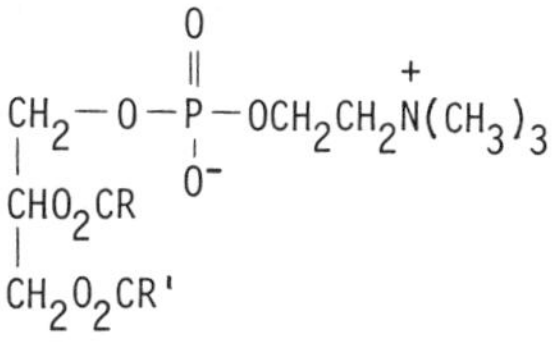

phosphatidylcholine

sive broadening and, in effect, an apparent loss of intensity of the choline methyl resonance (3.2 ppm, $\omega_{1/2}$ = 150Hz). While the exact origin of this broadening is debatable, an increase in vesicle size was supported by chromatographic behavior on a sepharose gel.

Changes in vesicle properties which occur at low antibiotic concentrations and which are not associated with vesicle destruction have been monitored by measuring the efflux rate of relatively easily monitored tetramethylammonium ion, $(CH_3)_4\overset{+}{N}$ (2.9 ppm), from within the vesicle (Gent and Prestegard, 1976). Thus, deionization of the outer vesicle solution followed by addition of $K_3Fe(CN)_6$ to upfield-shift outside ions was used to observe the decrease in external $(CH_3)_4\overset{+}{N}$ ion signal intensity upon treatment with polyene antibiotic. In this manner it was possible to quantitatively compare nystatin- and filipin-treated vesicles.

The relatively slow generation of pores which form upon treatment with amphotericin was suggested by experiments with added Mn^{2+} to broaden the phosphocholine (PC) methyl signals (Gent and Prestegard, 1976). Figure 11 shows the choline region for a vesicle to which Mn^{2+} is added 1 day after amphotericin treatment. An immediate loss of all the resolvable choline intensity occurs compared with the two-thirds intensity loss for the external cholines when Mn^{2+} is added to an untreated vesicle preparation. The

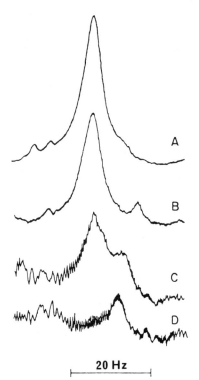

Figure 11. ^1H NMR (100 MHz) spectra of the choline and tetramethylammonium resonances for samples containing $2 \times 10^{-3}\,M$ cholesterol in $4:1$ phosphatidylcholine: cholesterol vesicles, with $0.5\,M$ tetramethylammonium ion originally trapped inside the vesicles. A: no antibiotic, $0.01\,M\,K_3Fe(CN)_6$; B: $1 \times 10^{-3}\,M$ amphotericin B, $0.10\,M\,K_3Fe(CN)_6$; C: no antibiotic, $0.005\,M\,MnCl_2$; D: $1 \times 10^{-3}\,M$ amphotericin B, $0.005\,M\,MnCl_2$.

20 Hz

authors suggested that there are sufficient pores after 1 day of antibiotic treatment to make all the choline accessible to Mn^{2+} in only a few minutes.

As a final note, polyene antibiotics with δ-hydroxylated keto groups, such as pimaricin, lucensomycin, and tetrins A and B, have the potential for keto-hemiketal tautomerism of the general type that is shown, which has been studied (Dornberger *et al.*, 1976) by ^{13}C NMR using off-resonance decoupling to identify key carbon positions. This structural equilibrium may be markedly pH and media dependent (aqueous compared with lipid-rich) and should be recognized as a possible complicating factor in polyene antibiotic–membrane investigations or conformational analyses of polyene antibiotics (Omura *et al.*, 1975).

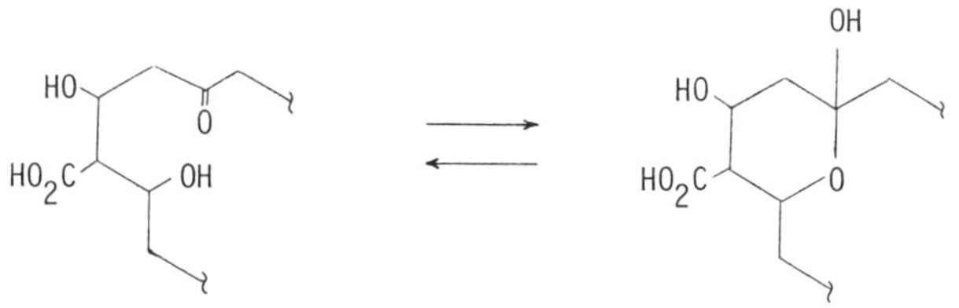

5.2. Amphiphilic Drugs and Phospholipids

Certain amphiphilic drugs show an unusually high affinity for lipid-rich tissues and thereby give rise to an impairment of lipid metabolism during chronic treatment. The reason for this effect might be either an interaction between drugs (metabolites) and lipid-metabolizing enzymes or between drugs (metabolites) and phospholipids as substrates. The latter possibility has been supported by NMR measurements which showed no interaction between phospholipase A_2 and chlorphentermine but did, on the other hand, reveal strong interaction between this amphiphilic drug and phospholipids (Seydel and Wassermann, 1973). More recent measurements of T_2 ($1/T_2^{obs} = \pi\omega_{1/2}$) using a series of drugs and various lipid materials has led to the conclusion that lipophilicity is correlateable with increased binding, as illustrated, for example, by the significantly greater degree of signal-broadening observed for chlorphentermine (log $P_{octanol} = 3.43$) relative to

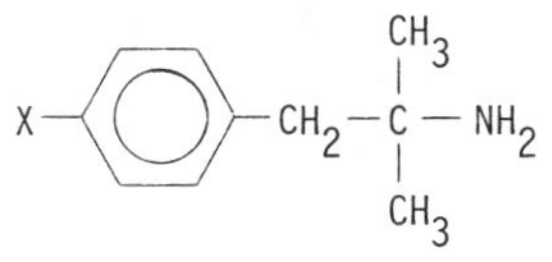

phentermine : X = H

chlorphentermine : X = Cl

phentermine (log $P_{octanol}$ = 2.45) (Seydel and Wassermann, 1976).[6] Strong interactions were seen with PC and phosphatidylethanolamine, whereas less polar lipids like diacylglycerol and digalactosyldiglyceride showed no evidence of drug interaction. The effect of added cholesterol, which is an essential constituent of natural lipid mixtures, was also studied however, for preformed chlorphentermine–PC "complex," added cholesterol had no observable effect on the extent of line broadening.[7]

5.3. Anesthetics and Membranes

5.3.1. Halothane.

The mechanisms of action for anesthetic agents have been extensively studied, and a comprehensive review of current biophysical concepts has been published (Kaufman, 1977). Drug-induced phase changes in membrane phospholipids with attendant consequences for membrane-associated proteins is generally believed to be a primary factor, and NMR spectroscopy has therefore been used to study relatively simple model systems such as phospholipid vesicles.

Portions of the fatty acid chains of the lipids in sarcoplasmic reticulum are sufficiently mobile for resolution by ^{1}H NMR and conditions which increase the amount of mobile (melted) lipid should lead to increased absorption intensity of these protons (Vanderkooi et al., 1977). Such is the case for addition of either halothane ($CF_3CHBrCl$) or chloroform. From Figure 12 it can be seen that the relative ratio of fatty acid peaks to protein peaks increases, which supports the conclusion (Vanderkooi et al., 1977) that in a biological membrane the general anesthetic shifts the phase transition of the phospholipid and thereby increases the pool size of melted phospholipids.[8] Significantly, the surgical concentration of anesthetic is estimated to be 8 nmol/mg membrane protein of 25,000 sites/μm^2 and this is approximately the concentration at which the NMR spectra of sarcoplasmic reticulum is affected.

The influence of halothane on membrane structure has also been studied by ^{19}F NMR (Koehler et al., 1977a). When halothane was added to dipalmitoylphosphatidylcholine vesicles, a "broadened base" appears on the initial sharp doublet (6.6 ppm, $^3J_{F-H}$ = 5 Hz) and, with further addition, an upfield broadened singlet grows in intensity relative to the initial doublet.

---

[6]Based on the definition of log $P_{octanol}$, which relates to the partitioning of a compound between octanol and an aqueous buffer, it is generally assumed that larger log $P_{octanol}$ values correspond to molecules with greater lipophilic character.

[7]The interaction of plant growth hormones such as gibberellic acid and 3-indoleacetic acid with phospholipids has been studied by ^{1}H NMR (Wood et al., 1974), and more recent ^{31}P chemical shift measurements of phospholipid head group nuclei have been used for determination of 3-indoleacetic acid dissociation constants (Marker et al., 1978).

[8]Halothane is a chiral molecule and exists as two enantiomers. The resulting possibility for enantioselective interactions between (R)- and (S)-halothane and a chiral domain of a given membrane can, in principle, be directly detected and quantitatively studied by either ^{13}C or, more likely, ^{19}F NMR.

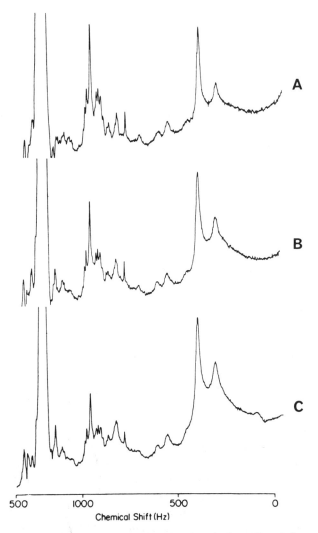

Figure 12. ^{1}H NMR (220 MHz) spectra of sarcoplasmic reticulum (40 mg/ml) suspended in 10 mM phosphate/D$_2$0 buffer at pH 7.0 and 17°C. The two major peaks between 250 and 500 Hz are assigned to phospholipid methylene and terminal methyl protons. A : control spectrum; B : 38 mM halothane added; C : 76 mM halothane added.

Heating a 1:1 halothane:vesicle mixture from 25 to 35°C likewise affected this doublet-to-singlet spectral change, which was reversible, and a plot of the temperature and halothane concentration at the appearance of the singlet absorption when extrapolated to zero halothane concentration intersects at 41°C, which is the transition temperature of the phospholipid. The ^{19}F NMR spectra of the drug–vesicle mixture (1:1) obtained at a variety of delay times suggested the presence of at least two distinct populations of halothane

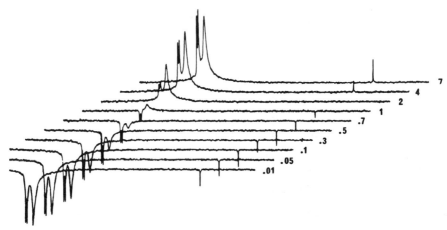

Figure 13. Stack-plot of partially relaxed FT-^{19}F NMR spectra of $2.1 \times 10^{-1} M$ halothane in an aqueous suspension of $5.4 \times 10^{-2} M$ phosphatidylcholine ($T = 25°C$). The upfield singlet is due to $(CF_3)_2CO/D_2O$ external reference, and the delay times in seconds are as indicated to right of each spectrum.

molecules with chemical shifts close to the halothane doublet seen in buffer solutions. On the other hand, partially relaxed spectra (Figure 13) of a 2:1 halothane:vesicle preparation revealed different T_1 values and, consequently, different mobilities for the doublet and singlet species.

To rationalize these NMR observations, it was proposed (Koehler *et al.*, 1977a) that at temperatures below the phospholipid gel to liquid-crystalline phase transition, halothane exists in three distinct environments. As depicted in Figure 14 halothane in the bulk aqueous buffer phase (A) is in equilibrium with those molecules located at the polar-head-group–bulk-phase interface (A′), and these in turn are in equilibrium with halothane molecules located in the phospholipid methylene chain near the polar head groups (B). Following membrane-disruptive events, a further environment (C) is created for halothane and may be located toward the terminal methyl end of the hydrocarbon chains of the phospholipid near the center of the bilayer.

In related ^{19}F NMR studies of halothane and sonicated egg lecithin vesicles (Trudell and Hubbell, 1976), it was found that on the NMR scale, there is rapid exchange between the interior of the bilayer and the aqueous medium. By way of contrast, the halothane ^{19}F NMR signal in multilayer vesicles constituted from bovine heart or egg lecithin reveals significant broadening and the absence of a doublet resonance, which suggests that halothane molecules more readily partition into the interior of the bilayer in these lipids (Koehler *et al.*, 1977b).

5.3.2. Tetracaine and Butacaine.   Tetracaine is one member of a class of local anesthetics which block nerve conduction, and its influence upon

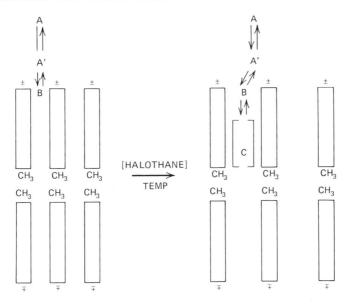

Figure 14. Schematic illustration of the hypothesis relating the effects of temperature and halothane concentration to the distribution of halothane in the system. A = halothane in the bulk aqueous phase; A' = halothane associated with the aqueous-phase–phospholipid interface; B = halothane near the polar head groups of the phospholipid, but in the hydrocarbon chain; C = halothane in the hydrocarbon chain closer to the terminal methyl group.

membrane functions has been studied by ^{1}H NMR (Hauser *et al.*, 1969; Cerbon, 1972; Fernandez and Cerbon, 1973). More detailed physicochemical information was subsequently obtained by combined ^{13}C and ^{31}P NMR analyses of the drug and phospholipid vesicles (Yeagle *et al.*, 1977). When tetracaine was added to PC vesicles without and with cholesterol (30%), the ^{13}C absorptions of the aromatic carbons are broadened while the T_1 values show little change. With regard to the vesicle, addition of equimolar tetracaine had no measurable effect upon ^{13}C T_1 relaxation time and ^{31}P linewidth, T_1 relaxation time, and NOE enhancement. These results suggested a hydrophobic interaction between hydrocarbon portions of the drug and phospholipid bilayer. Analysis of the T_1 values for the proton-bearing aromatic carbons using an isotropic vesicle rotational correlation time of about 10^{-6} s and a much faster anisotropic axial rotation yielded a rotational correlation time of $10^{-10.3}$ s about the long molecular axis of tetracaine, which showed that this parameter is little effected by the vesicle.

$$CH_3(CH_2)_3\overset{H}{N} - \langle \bigcirc \rangle - \overset{O}{\overset{\|}{C}} - OCH_2CH_2N\overset{CH_3}{\underset{CH_3}{<}}$$

tetracaine

In order to assess the rate at which tetracaine might undergo "flip flop" across the two surfaces of a bilayer, vesicles were prepared with Pr^{3+} shift reagent trapped in the interior (Yeagle et al., 1977). In the ^{31}P NMR spectra, the interior lipids give rise to a downfield-shifted signal, relative to the exterior lipids, which therefore allows for the measurement of an outside/inside ratio of lipids. Addition of tetracaine had no influence upon this ratio over a 4 h period, which indicates that the drug was *not* transferred to the inside of the vesicle bilayer and that "flip flop" of the anesthetic molecule must be slow. If such migration had taken place, Pr^{3+} would presumably have been displaced from phosphates of the interior lipids and would have thus increased the outside/inside ratio of ^{31}P NMR resonances.[9]

$$H_2N\!-\!\!\bigcirc\!\!-\overset{\overset{\displaystyle O}{\|}}{C}\!-O(CH_2)_3N\!\!\underset{(CH_2)_3CH_3}{\overset{(CH_2)_3CH_3}{<}}$$

butacaine

Butacaine, like tetracaine, is a local anesthetic of the *p*-aminobenzoic acid family and has been shown to cause marked changes in the rate of adenine nucleotide translocation across the inner membrane of rat liver mitochondria (Spencer and Bygrave, 1974; Fayle et al., 1975). Since no effect of butacaine on the binding of adenine nucleotides to the atractyloside-sensitive sites on the adenine nucleotide translocase was detected, the effect of the local anesthetic is most likely mediated through its interaction with specific phospholipids in the environment of the adenine nucleotide translocase, rather than through a direct interaction with the translocase protein (Fayle et al., 1975). In order to gain further information about the nature of these phenomena, the interaction of butacaine with isolated rat liver mitochondria has been investigated by ^1H NMR (Crompton et al., 1976). At 10° in 0.12 M KCl/D_2O, the addition of mitochondria (1 mg protein/0.3 μmol butacaine) reduced the intensity of *each* proton resonance by ~50%, in a reversible fashion, and caused no chemical shift changes. Under conditions similar to these NMR experiments, ADP translocation was inhibited by ~40%. It was concluded that the *entire* butacaine molecule, rather than specific portions of it, binds to mitochondria, and that this strong immobilization leads to inhibition of the translocation. However, from these

---

[9] ^{31}P NMR spectra of benzene solutions of hydrated dipalmitoyl-lecithin-inverted micelles with and without the incorporation of paramagnetic lanthanide ions have indicated that individual resonances for micelles containing none, one, and two ions can be resolved and simultaneously observed (Chen and Springer, 1979). Moreover, these NMR studies of mixtures of preformed micelles containing different numbers of ions per micelle have indicated equilibration via selective fusion of multi-ion with ion-free micelles, and have also placed constraints on the lifetimes of metal ions, lipid, and water molecules within a micelle before transfer to another.

data alone it was not possible to make a distinction between association of the anesthetic molecule with inner mitochondrial membrane, where the translocase is located, as opposed to the outer membrane region.

5.4. Drugs and Enzymes

5.4.1. Acetylcholinesterase Inhibitors.

As a model for studying drug–receptor interactions by NMR, the nature of atropine and physostigmine (eserine) inhibition of acetylcholinesterase has been studied by ^{1}H NMR (Kato, 1975) using the following generally applicable analysis.

For the equilibrium E + D $\leftrightarrow$ ED describing enzyme (E)–drug (D) association, $K_{dis} = [E][D]/[ED]$, where $[E] = E_0 - [ED]$, $[D] = D_0 - [ED]$, and the terms [ED], E_0 and D_0 refer to the concentration of complex, total enzyme concentration, and total drug concentration, respectively. For reversible binding of D to E, the observed NMR linewidth (ω_{obs}) of a nucleus within D is given by

$$\omega_{obs} = N_{ED}\omega_{ED} + N_D\omega_D + (1/\pi T_2)_{ex},$$

wherein N indicates mole fraction and $(1/\pi T_2)_{ex}$ is the contribution to ω_{obs} due to exchange. In those cases where the exchange lifetime is much less than $1/\omega_{ED}$, and $D_0 \gg E_0$, it follows that $\omega_{obs} = ([ED]/D_0)\omega_{ED} + \omega_D$. Subtraction of ω_D and rearrangement of terms leads to $[ED] = (\omega_{obs}/\omega_{ED})D_0$, which upon substitution in the K_{dis} expression gives, for $\omega_{obs} \ll \omega_{ED}$, $D_0 = (E_0\omega_{ED})/\omega_{obs} - K_{dis}$. Hence, a plot of D_0 versus $1/\omega_{obs}$ gives a line with an intercept equal to $-K_{dis}$ and a slope that can be used to calculate ω_{ED}.

The ^{1}H NMR spectra of atropine and physostigmine feature, for each

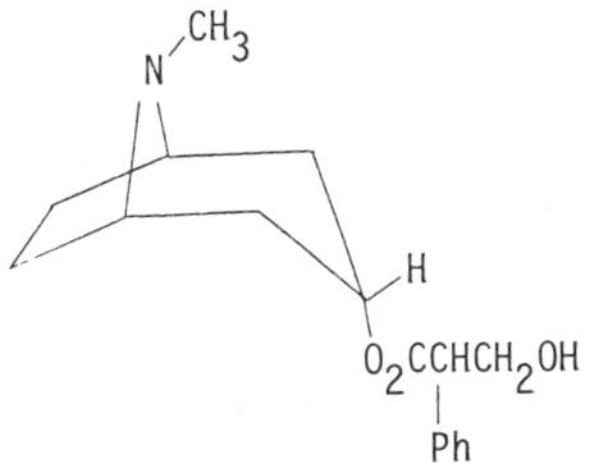

atropine

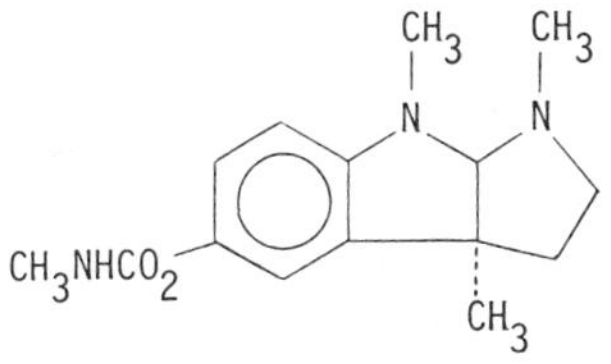

physostigmine

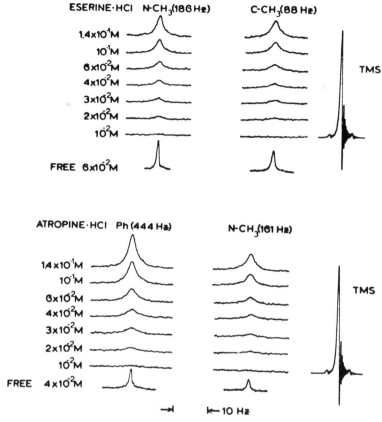

Figure 15. Top half: ^{1}H NMR (60 MHz) spectra of the N-CH$_3$ (left) and C-CH$_3$ (right) resonances of physostigmine hydrochloride, free (lower trace) and in the presence of acetylcholinesterase (1.7×10^{-5} M) (upper traces). Bottom half: ^{1}H NMR (60 MHz) spectra of the C$_6$H$_5$ (left) and N-CH$_3$ (right) resonances of atropine hydrochloride, free (lower trace) and in the presence of acetylcholinesterase (1.7×10^{-5} M) (upper traces). Samples were prepared in 0.1 M phosphate/D$_2$O buffer at pH 7.4, and spectra were recorded at 39°C.

drug, two "sensor" groups (C$_6$H$_5$/N—CH$_3$ and N—CH$_3$/C—CH$_3$) that can be independently used to measure ω_{obs}. Partial 60 MHz spectra obtained for various concentrations of drug (0.14–0.01 M) in the presence of acetylcholinesterase (1.7×10^{-5} M) are presented in Figure 15, and the resultant plots of D$_0$ versus ω_{obs}^{-1} showed good linearity (Kato, 1975). For physostigmine, the average value of $K_{\text{dis}} = 5.1 \pm 3.2 \times 10^{-4}$ M while atropine gave $K_{\text{dis}} = 2.1 \pm 4.9 \times 10^{-4}$ M. The lack of precision was attributed to the fact that the value of K_{dis} was much smaller than D$_0$.

From these data it was suggested that both drugs bind to the enzyme with high affinity ($K_{\text{dis}} < 1$ mM); however, kinetically derived K_i values for the drugs show that physostigmine is an effective inhibitor ($K_i = 1.2$ μM) and that atropine inhibited poorly ($K_i = 6.0$ mM). This set of circumstances was accommodated by proposing that the former compound may bind to one of the two active subsites in acetylcholinesterase, while atropine may be bound at a noncatalytic allosteric site. That the drugs do not bind to the same site on the enzyme was confirmed by adding gallamine triethiodide to enzyme–drug solutions which showed (via ω_{obs}) that gallamine reduced the amount of bound atropine but not physostigmine. It was also possible to monitor, by NMR, the rate of acetylcholine turnover in the presence of either drug and thus determine that atropine does not interfere with the binding of substrate but instead accelerates its hydrolysis, possibly by an allosteric effect (Kato, 1975).

5.4.2. Penicillin Complexation with Carboxypeptidase. *Streptomyces* strain R61 excretes a DD-carboxypeptidase–transpeptidase which appears to be closely related to the corresponding membrane-attached enzyme and serves as a good model for the study of the interaction between the enzyme, its substrates, and penicillin. The soluble DD-carboxypeptidase forms an enzymatically inactive complex with penicillin which slowly degrades to regenerate active enzyme and two penicillin-derived products, N-acyl-glycine and N-formyl-D-penicillamine. Neither the mechanism for cleavage of the C-5—C-6 bond or of the β-lactam amide bond are known yet and, as a starting point, ^{1}H NMR has been used in an attempt to obtain the first direct evidence for identification of the penicillin part of the penicillin–enzyme complex (Degelaen *et al.*, 1979).

penicillin G

High-resolution (270 MHz) ^{1}H NMR comparison of the native R61 enzyme with the benzylpenicillin (penicillin G) complex of interest revealed a downfield shift (0.05 ppm) of one of the eight His C-2 resonances; however, separate pH titration gave no evidence of a pK change for this residue, and the complexity of the overall spectrum was such that no resonances from the penicillin moiety could be clearly identified. To simplify the main region of interest around 1.5 ppm that would be indicative of the expected serine-linked penicilloyl-enzyme, the unfolded penicillin–enzyme complex was examined by use of 6 M guanidinium chloride. Two relatively small resonances at 1.47 and 1.36 ppm were observed and were assigned to penicillin rather than random-coil protein resonances. Supportive evidence for this identification came from the gradual decrease in intensity of these signals which correlated with the decreased proportion of intact complex measured electrophoretically. Signal integrations as a function of time further revealed that the two resonances, which corresponded to a total of only one proton, change from a relative ratio of 2:1 to 1:1 and thus appear to monitor two equilibrating forms for the complex. A chemical shift model for the β-CH$_2$ serine residue of the proposed penicilloyl-enzyme, namely [5R, 6R]-α-methyl-D-benzyl penicilloate, gave a value of 1.42 ppm for the H-5 resonance. Epimerization at the H-5 center to give the [5S, 6R] diastereomer led to the formation of a new signal at 1.32 ppm and thus suggests that a similar epimerization process gives rise to the two forms of the complex: [5R, 6R]- and [5S, 6R]penicilloyl-enzyme diastereomers. While these data and additional double resonance experiments are all consistent with the serine attachment proposal, it was rightly cautioned that covalent bond changes (rearrangement) during denaturation cannot be ruled out.

5.5. Drug Binding to Plasma Proteins

5.5.1. Sulfonamides.   Knowledge of the interactions between a drug and plasma proteins is important from both theoretical considerations and more practical clinical aspects of pharmacology. Details of drug–receptor complexes may be probed by a variety of NMR methods, and differential perturbation of relaxation rates in ^{1}H spectra of sulfonamides complexed with bovine serum albumin (BSA) represents a pioneering example of such methodology (Jardetzky and Jardetzky, 1965).

PABS :   R = H

sulfacetamide :   R = $-\overset{\displaystyle O}{\overset{\displaystyle \|}{C}}-CH_3$

5-methyl-3-sulfanil-
amidoisoxazole :   R =

sulfaphenazole :   R =

Upon addition of BSA to solutions of p-aminobenzenesulfonamide (PABS), sulfacetamide, 5-methyl-3-sulfanilamidoisoxazole, and sulfaphenazole, a larger increment in the relaxation rate was generally found for the p-aminobenzenesulfonamide moiety, relative to the variable N-1 substituents, which indicated that the arylsulfonamide group was the primary binding site. The tricyclic derivative, sulfaphenazole, was found to be an exception in that the arylsulfonamide group *and* the N-phenyl portion appeared to bind independently. The two binding sites were distinguished by addition of phenylpropanol which interferes with binding of the arylsulfonamide and apparently enhances, somewhat, binding of the phenyl substituent.[10]

5.5.2. Penicillin. The differential perturbation of relaxation times has also been used to determine which portion of the penicillin molecule is involved in binding to serum albumin (Fischer and Jardetzky, 1965). A quantitative relationship between the observed broadening and variable penicillin–albumin concentrations was regarded as being incompatible with nonspecific relaxation mechanisms, and the markedly selective effect of binding to the phenyl group indicated its dominant role in the drug–protein complexation process, as opposed to other portions of the penicillin molecule which exhibit greater freedom of motion. Temperature, pH, ionic strength, and inhibition effects were also probed by this NMR technique and provided further details of the binding mechanism.

methotrexate

5.5.3. Methotrexate. Binding between human serum albumin (HSA) and the anticancer drug methotrexate has been recently studied by NMR line broadening and it was reported (Sarrazin *et al.*, 1978) that site exchange is

---

[10]The interaction of variously substituted p-aminosulfonamides with the cyclic polyether "18-crown-6" (1,4,7,10,13,16-hexaoxacyclooctadecane) has been studied in chloroform solution by both UV measurements and measurement of induced ^{1}H chemical shift changes (Takayama *et al.*, 1979). The stability constants determined by NMR compared very well with those determined from UV data. It was concluded that the binding mechanism involved the inclusion of the p-amino group of the sulfonamides in the "hole" of 18-crown-6, and that hydrogen bonding was the primary attractive interaction.

relatively rapid at physiological temperatures since cooling was needed to influence the observed relaxation rate. It was also concluded that a specific interaction is involved, as $1/T_2^{obs} \propto$ [methotrexate]. The value for the association constant obtained from various "sensor" nuclei agreed reasonably well with results obtained by microcalorimetric methods. The H-9 signal was most perturbed while H-3' and H-2' were less influenced, which led to the suggestion that methotrexate is bound to HSA primarily through the methylated nitrogen position and, to a lesser degree, by the aromatic and heterocyclic rings.

5.5.4. Metronidazole. The broad-spectrum antiprotozoal agent metronidazole and its derivatives A–D shown in Table I are known to bind to human plasma protein in a manner that is dependent upon the nature of the alkyl side chain (Sanvordeker *et al.*, 1975). The absence of a linear correlation between the extent of plasma protein binding and lipophilic character (partition coefficient) prompted these investigators to perform SCF–MO calculations regarding the side-chain conformation and frontier electron density (qr) via MINDO/2 and CNDO methods, respectively. It was found that the 3' position of the side chain exhibited the highest value of qr for all of

Table I. Binding and solubility data for metronidazole and metronidazole derivatives A–D

Compound	R	R'	Fraction of Bound Drug (%)[a]	Partition Coefficient[b]
Metronidazole	CH_2OH	CH_3	4.22	0.78
A	$CH_2S_2O_3Na$	CH_3	67.12	0.03
B	CH_2NHAc	CH_3	11.36	0.34
C	CO_2H	H	4.45	0.03
D	CH_2OAc	CH_3	2.36	2.00

[a]Total drug concentration = 4.8×10^{-5} M, as selected on the basis of minimum inhibitory concentration for *Trichomonas vaginalis*.
[b]*n*-Octanol/pH 7.4 phosphate buffer. Increasing values of the partition coefficient correspond to increasing lipophilicity.

the compounds and, moreover, that there was a linear correlation between these values and $\ln P_{mean}$, where P_{mean} is the mean percent fraction of the total drug in the bound state: $\ln P_{mean} = 0.848 + 2.668\ qr$ (correlation coefficient = 0.994). These findings suggested that the plasma protein binding for this series of compounds was nonspecific, that is, drug–protein (mostly serum albumin) interaction occurs through a combination of van der Waals and electrostatic forces, and supportive evidence for participation of the side chain in drug–protein binding was provided by NMR. The proton spectra for metronidazole and its derivatives in the presence of added HSA (1.15 and 2.30%) showed only slight broadening for the CH_2 and CH_3 resonances of metronidazole and derivative D, while the proton absorptions in derivative A were extensively broadened. It was reasoned that the relaxation effects observed for derivative A compared with metronidazole and derivative D were consistent with the relative protein binding data (Table I), and that the specific relaxation of the CH_2 and CH_3 groups was due to their participation in the binding process.

5.6. Drug Interaction with Metal Ions

5.6.1. Metronidazole. In addition to the aforementioned plasma protein binding studies with metronidazole, NMR has been used to assess the feasibility of this drug acting as an "electron sink" and thereby inhibiting the metabolic activity of ferredoxin and hydrogenase systems involved in the phosphoroclastic reaction in anaerobes which succumb to this antimicrobial drug (Chien *et al.*, 1975). As a test for the ability of metronidazole, like other imidazole compounds, to undergo complexation with biological metal ions, the effect of transition metal ions on the drug's proton spectrum was probed by examination of observed linewidths. While Sn^{2+}, Fe^{2+}, Mg^{2+}, Co^{2+}, Ni^{2+}, and Zn^{2+} over the range of 1–$3 \times 10^{-4}\ M$ had no measurable influence on ω values for metronidazole at $6 \times 10^{-2}\ M$. the linewidths were found to be proportional to the concentration of added Cu^{2+}:$\log(1/T_2^{obs}) = \beta \log[Cu^{2+}] +$ const. The effect of cupric ion on the C-4 olefinic proton resonance was greater than that of the C-2 methyl protons ($\beta = 1.30$ compared with 0.26), which was regarded as consistent with metal ion complexation via the sterically less hindered N-3 position. Hückel calculations further indicated that the N-3 position bears a greater negative charge (-1.0012) than the alternative N-1 binding site ($+0.0374$).

The stoichiometry of the presumed cupric ion complex with metronidazole was investigated by ac polarography and the estimated coordination number of 3.63 suggested that four molecules of drug are required for the square-planar tetracoordinate structure pictured in Figure 16 (Chien *et al.*, 1975). In conclusion, it is interesting to note that $10^{-3}\ M$ metronidazole is reported to completely inhibit the activity of uricase, which is a copper-containing enzyme.

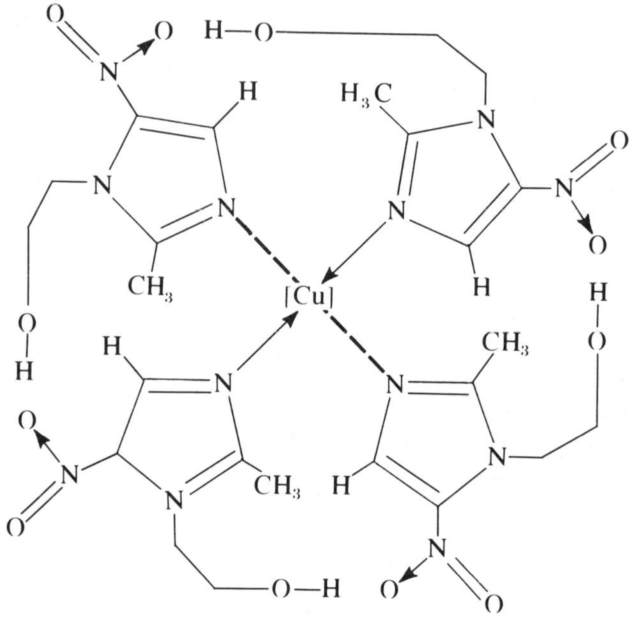

Figure 16. Possible square-planar tetracoordinate complex between Cu^{2+} and four metronidazole ligands.

5.6.2. Bleomycin. The antitumor antibiotic bleomycin A2 has one iron binding site, and a bleomycin–iron complex is believed to cause single-strand breaks in DNA, while other transition metals such as Cu^{2+} and Zn^{2+} inhibit this reaction (Gupta *et al.*, 1979). The bleomycin–Fe^{2+} complex reportedly produces superoxide and hydroxyl radicals which may initiate strand breaks. To further probe the structure of the iron–drug complex, the paramagnetic effects of Fe^{3+} on T_1 of the aliphatic carbon atoms have been

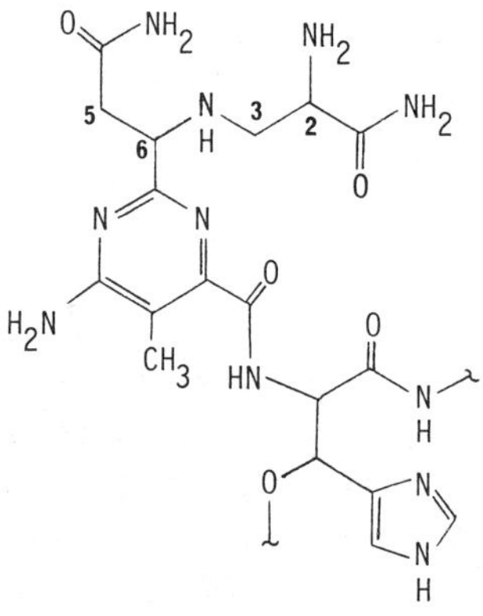

bleomycin A2

studied. The addition of Fe^{3+} (0.7 mM) to a solution of bleomycin (51 mM) enhances T_1 for only four of the aliphatics, C-2, C-3, C-5, and C-6. From the magnitude of these Fe^{3+}-induced effects on ^{13}C relaxation rates, distances of 4.6, 5.3, 5.3, and 4.6 Å from the metal to C-2, C-3, C-5, and C-6, respectively, were calculated. The results were reported to be consistent with the coordination of the α-NH$_2$ group of the terminal diaminopropionic acid residue and the pyridine ring, and apparently exclude additional binding sites.

In contrast to these conclusions, more recent studies (Dabrowiak *et al.*, 1979) of bleomycin by ^{13}C NMR, ESR, titrations, and electrochemistry have led to the suggestion that Fe^{2+}/Fe^{3+} "binds to the amine–pyrimidine–imidazole region" of the antibiotic.

5.6.3. Antibiotics. Specific complexation of the macrocyclic antibiotic rifamycin S with calcium and sodium ions has been investigated by ^{13}C NMR in CDCl$_3$/CH$_3$OH (1:1) solution and, on the basis of chemical shift arguments, the structure shown in Figure 17 was proposed (Leibfritz, 1974). The application of ^{23}Na and ^{43}Ca NMR can provide further details regarding such complexes. Relaxation measurements of ions containing quadrupolar moments have long been recognized (Jardetzky and Jardetzky, 1971) as sensitive indicators of weak ionic interactions. Relatively small distortions of the electronic shell of the ion from spherical symmetry, as a result of binding, introduces an additional relaxation mechanism through coupling of the nuclear quadrupole moment in the electric field gradients in the distorted shell.

^{1}H and ^{13}C NMR studies (Toeplitz *et al.*, 1979) of the polyether antibiotic ionomycin have been used in conjunction with X-ray analysis to establish topological details regarding the binding of this lipid-soluble Ca^{2+} and Cd^{2+} ion transporter. Dynamical aspects of such ionophores have been studied by

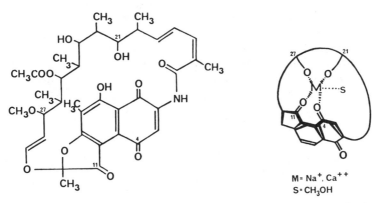

Figure 17. Structure of refamycin S and a schematic representation of a generalized complex between refamycin S and either Na$^+$ or Ca^{2+}.

the "controllable hyperfine induced time-scale (CHITS)" method, which employs the isotropic hyperfine shifts caused by paramagnetic metal ions and is used to vary the NMR time scale of exchange phenomena in a controlled manner. The lifetimes of the ionophoric antibiotic lasalocid A (X-537A, HX) complexes with various metal ions have thus been determined (Krishnam *et al.*, 1978) in methanol solution: BaX^+, 132 μs; SrX^+, 34 μs; CaX^+, 24 μs; and KX, 15 μs. The method depends upon the broadness of the H-29, H-5, and H-4 resonance lines of X^- due to reactions such as $X^- + Ba^{2+} \leftrightarrow BaX^+$. when the resonance of the proton in the unbound X^- has been shifted from its normal frequency by Pr^{3+} which acts via the exchange $PrX_3 \leftrightarrow PrX_2^+ + X^-$. To determine the aforementioned lifetimes, the equations for a three-state spin system give rise, under relevant conditions, to Swift–Connick equations for a two-state system, with the characteristics of one of the states dependent upon the PrX_3 concentration which allows for extraction of the lifetimes in a relatively simple way.

X-537A (lasalocid)

In order to gain insight into mechanisms by which metal cations are transported across biological membranes, the di- and trivalent-specific ionophore A23187 has been widely used to study Ca^{2+} transport and function in membranes. Further mechanistic understanding has been sought by ^{1}H, ^{31}P, and ^{13}C NMR approaches which utilize lanthanide ions such as Pr^{3+} and Eu^{3+} to distinguish inner and outer surfaces of phosphatidylcholine vesicles. Initial studies (Hunt, 1975) demonstrated that the rate of A23187 (or X-537A)-assisted Pr^{3+} transport across dipalmitoylphosphatidylcholine was measureable by time-dependent ^{1}H NMR line broadening; however, more recent proton measurements using chemical shift data have demonstrated a considerable improvement in the method and have afforded both rate constants and activation parameters for Pr^{3+} transport by A23187 (Hunt *et al.*, 1978). The rates were found to be proportional to A23187 concentration and suggested that the rate-limiting step involves $[Pr(A23187)]^{2+}$. A detailed theoretical analysis was made of various possible mechanisms for the initial stages of transport in view of the experimental NMR data, and good agreement obtains for diffusion of $[Pr(A23187)]^{2+}$ over the "image potential barrier" (Hunt *et al.*, 1978).[11]

A physicochemical technique which probes the functional features of amino glycoside antibiotics such as kanamycin A and ribostamycin has been described (Hanessian and Patil, 1978). The technique, which involves the effect of increasing Cu^{2+} concentrations on line broadenings, provides for the simplification of ^{13}C NMR spectra and complements the study of shifts induced by acidification.

5.7. Drug Interaction with Other Small Molecules

The association or complexation of drugs with other relatively small molecules by means of electrostatic forces, hydrogen bonding, and/or π-orbital interactions can be considered as models for extension to drug transport phenomena and drug–receptor specificity. As an example of these kinds of studies, consider complexes of theophylline, caffeine, and other xanthines

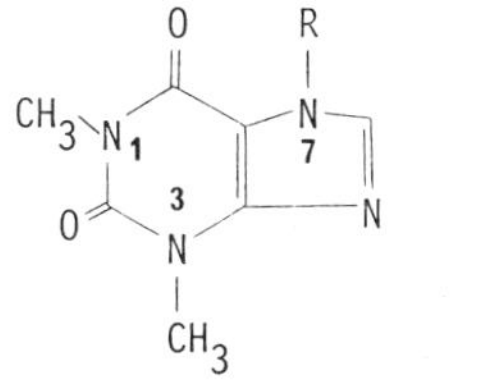

caffeine : R = CH_3

theophylline : R = H

with various drug molecules, wherein the mode of association has remained a matter of speculation, and concern exists for the relevance of *solid-state* crystallographic structural data to the disposition of interacting partners in *solution* (Thakkar and Tensmeyer, 1974). Theophylline–sodium benzoate complexation has been the subject of some debate and has been more recently investigated by ^1H NMR. Theophylline, like caffeine, undergoes self-association in solution, which is reflected by the upfield shift of all three types of proton resonances with increasing concentration; hence, relatively dilute samples (4–20 mM) have been studied (Thakker and Tensmeyer, 1974) to minimize this potentially complicating feature. Plots of the induced

---

[11]As a somewhat related example of the utility of ^{23}Na NMR, a saline emulsion of nonionic surfactant and higher alcohol was found to exhibit single T_1 and T_2 values for sodium nuclei, while in the presence of gramicidin D the ^{23}Na relaxations were greatly accelerated and T_2 was the sum of two decaying exponentials (Monoi and Uedaira, 1979). It was concluded that ^{23}Na NMR may be useful for studying the nature of ion-permeable channels in biological membranes even when the channel has no ionizable group.

chemical shift change ($\Delta\delta$) for both methyl groups versus the concentration of sodium benzoate (0.1–1.0 M) revealed substantial *upfield* shifts, which suggested that these substituents come under the shielding region of the anisotropic benzoate ring. Since the conical shielding regions of the aromatic ring are perpendicular to the plane of the benzene ring, it was suggested that the complex is formed by "vertical stacking." If side-by-side or "horizontal stacking" were the case, it was reasoned that downfield changes in chemical shift would have been observed. The somewhat larger value of the upfield shift for 3-CH$_3$ (0.71 ppm) compared with 1-CH$_3$ (0.60 ppm) was further taken to indicate that, averaged over time, 3-CH$_3$ resides nearer the axis of the benzene ring; however, such conclusions must be regarded as highly speculative. In any event, the observed $\Delta\delta$ values were treated by three different literature methods for calculating the apparent formation constant, K, assuming 1:1 stoichiometry, and an average value of $K = 1.20 \pm 0.13\ M^{-1}$ was obtained.

A combination of IR and ^{1}H NMR has been applied to structural elucidation of the 1:1 complex obtained from ephedrine and phenobarbital and, in summary, chemical-shift arguments were used to propose a structure involving a combination of salt formation together with hydrogen bonding and/or "horizontal stacking" of the aromatic rings (Mrtek *et al.*, 1977).

Additional studies of drug–small-molecule interactions include ^{1}H NMR analysis of the hydrogen bonding between minor tranquilizers diazepam and nitrazepam and a number of nucleoside bases in CDCl$_3$ solution (Paul *et al.*, 1978). Interactions of diazepam ($\Delta G_0^{25} = -0.2$ to -7.4 kJ/mol) were stronger than those of nitrazepam ($\Delta G_0^{25} = 6.0$ to 6.4 kJ/mol) and were rationalized according to various hydrogen bonding models.

```
diazapam  :  X = Cl,  R = CH₃

nitrazapam :  X = NO₂,  R = H
```

6. CONFORMATIONAL ASPECTS OF DRUG ACTIVITY

6.1. Anticancer Agents

6.1.1. Comformational Stereochemistry of Cyclophosphamide (CP). Variable temperature ^{1}H, ^{13}C, and ^{31}P NMR investigations of CP (Egan and Zon,

1976) and a number of closely related heterocycles, as well as force-field calculations with CP and ultrasonic relaxation measurements, are all consistent with a relatively low energy barrier for ring inversion which interconverts the two diastereomeric chair conformations of CP, that is, **1** and **2**

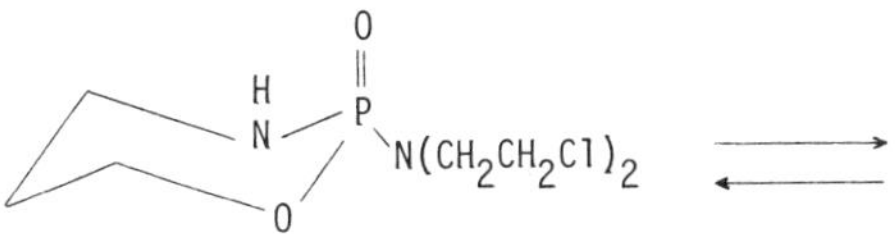

(Egan, Govil, Sharpless, Hemmes and Zon, unpublished work). (In the diagrams, the boldfaced numerals are represented as lightfaced numbers with a tilde underneath.) The conformational stereochemistry of CP in solution has been more recently elaborated by NMR comparisons with "conformationally biased" model compounds **3** and **4**, which for steric reasons exist predominantly in the conformations shown (White *et al.*, 1979). Lanthanide-induced shift (LIS) analysis of proton spectra of **3** and **4** with Eu(dpm)$_3$ involving computerized comparison of observed LIS values with calculated values for specified geometries of the shift-complex was used to confirm the relative stereochemical identity of these conformational model compounds. When the C-5 protons of CP were decoupled from the adjacent POCH$_2$ ring protons and the resulting ABX spectrum (X = ^{31}P) was analyzed, it was found that $^3J_{POCH_{A,B}}$ = 17.7 and 4.7 Hz. From the dependence of such coupling on the POCH dihedral angle and the significantly smaller values of $^3J_{POCH}$ for **3** (1.8 Hz) and **4** (2.5 Hz), it was concluded that the values of $^3J_{POCH_A}$ and $^3J_{POCH_B}$ for CP stem from weighted averages of conformers **1** and **2**. If the assumptions are made that $^3J_{POCH_{ax}}$ = 2.2 Hz (average from **3** and **4**) and that

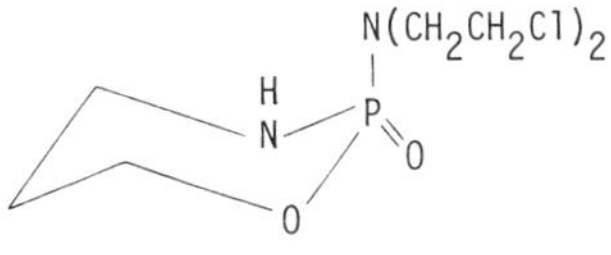

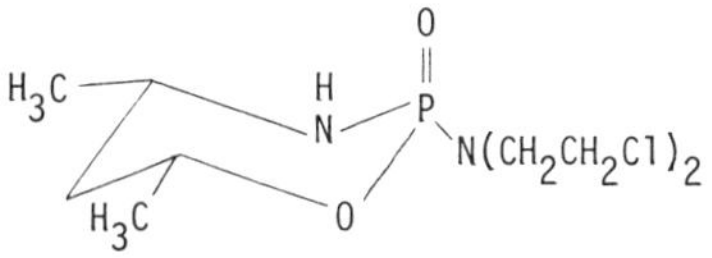

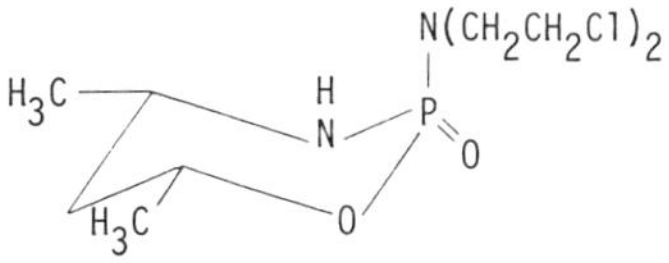

$$^3J_{POCH_{eq}} = {}^3J_{POCH_A} + {}^3J_{POCH_B} - {}^3J_{POCH_{ax}} = 20.2 \text{ Hz},$$

then the observed values of $^3J_{POCH_{A,B}}$ for CP in water indicate that **1** : **2** = 6 : 1. This estimate was confirmed by LIS measurements in CCl$_4$, which gave a 99% confidence limit for conformer **1** and suggested that this diastereomer was somewhat destabilized in aqueous media. It was noted that **3** (which possesses the configuration matching the dominant form of CP, namely, **1**) shows activity against KB tumor cell cultures whereas **4** does not, and that this could be taken as tentative evidence for **1** being a more active conformer than **2**.

6.1.2. Hydroxyureas. Possible relationships between molecular conformation and mechanism of action have been studied in antileukemic hydroxyureas (**5–8**), which may inhibit growth by ligand formation with enzymatic octahedral iron (Parker *et al.*, 1976). Internal hydrogen bonding and hindered rotation in hydroxyureas are related to their complexing abilities, and an attempt was made to characterize the intramolecular dynamics/bonding by ^{1}H NMR analyses of a large series of compounds related to **5–8**. However, for each of the hydroxyurea derivatives, there are a deceivingly large number of diastereomers to consider, and it is a rather formidable (if not impossible) task to unambiguously assign structures from the sets of absorption signals which may be observed for a given compound. From the spectral changes induced by solvent variations it was nevertheless suggested that conformational isomerism in hydroxyurea systems may be due to hydrogen bonding between the carbonyl oxygen and the hydroxyl proton, which appeared to be stabilized in nonhydrogen-bonding media. The significance of these findings to the mechanism of action and/or biological activity of these hydroxyureas remains to be established.

$$
\begin{array}{c}
O \\
\parallel \\
R \diagdown \ C \diagup OH \\
N \qquad N \\
H \qquad R'
\end{array}
$$

5 : R = Et, R' = H

6 : R = n-Pr, R' = H

7 : R = H, R' = Me

8 : R = H, R' = Et

6.1.3. Tenuazonic Acids. Hydrogen bonding has also been suggested as a biologically significant factor for tenuazonic acid analogs such as 3-acetyltetramic acids (**9**), which are of interest as potential anticancer agents (Yamaguchi *et al.*, 1976). The existence of different "external tautomers" (**9a**, **9b** ↔ **9c**, **9d**) and "internal tautomers" (**9a** ↔ **9b**, **9c** ↔ **9d**) had been previously discussed (Forsén and Nilsson, 1970) in connection with ^{1}H NMR data which suggested that the "external tautomers" are interconverted at a relatively slower rate. A combination of ^{13}C and ^{1}H NMR was subsequently used (Yamaguchi *et al.*, 1976) for structural assignments within these tautomer distributions; however, the relationship of these equilibria to biological activity has not been determined as yet.

6.2. Antibiotics

Puromycin, a nucleoside antibiotic bearing structural resemblance to the 3'-O-aminoacyl-adenylyl end of charged tRNA, inhibits the elongation cycle in protein biosynthesis by accepting the carboxyl function of a growing peptidyl chain to form a peptidylpuromycin (Nathans, 1964; Traut and Monro, 1964). The conformation of puromycin has been studied by ^{1}H NMR and it was concluded that this antibiotic prefers the Ng^+ combination of ribose and the exocyclic CH_2OH group, together with g^+ orientation of the C_α—C_β bond (De Leeuw et al., 1977). These conformational findings suggest homology with the structural situation in a charged tRNA and may therefore be related to the efficacy of the drug's inhibitory action.

Virginiamycins S and S_4, as well as veramycin B_α, are closely related cyclic hexadepsipeptide antibiotics and have been studied by combined ^{1}H and ^{13}C NMR techniques to characterize their solution conformations (Anteunis et al., 1975). It was found that the depside bond can rotate and thus provides the backbone with some conformational mobility; moreover, the orientation of the depsicarbonyl was found to be dependent upon solvent effects, and apparent discrepancies in the nature of hydrogen bonding were rationalized.

The relatively large cyclic polypeptide bacitracin (F.W. = 1420) has been studied by a pulsed triple resonance method to avoid excitation of the water resonance, which thereby allowed for the observation and assignment of exchangeable hydrogens (Campbell et al., 1974).

More recently, the conformation of the antibiotic iturin A has been elaborated by measurement of the temperature dependence of peptide NH chemical shifts and their exchange rates (Garbay-Jaureguiberry et al., 1978). ^{13}C T_1 values were determined and the correlation time for tumbling was $\sim 8 \times$

10^{-10} s. All of the data supported a relatively rigid backbone structure having a small cavity into which the backbone carbonyls are directed. On the other hand, a 270 MHz ^{1}H NMR investigation of an ion-binding cyclic peptide analog of valinomycin, *cyclo*-(L—Val—Gly—Gly—L—Pro)$_3$ and its cation complexes has been carried out in CD$_2$Cl$_2$ and CDCl$_3$ (Easwaran *et al.*, 1979) and the peptide is proposed to exist in a C_3-symmetry conformer which lacks a cavity. Cations did not bind, or bound only weakly, to the peptide in these solvents. The uncomplexed trimer in acetonitrile appeared to be averaging among several conformations with no evidence for preferred intramolecular hydrogen bonds.

6.3. Enkephalins and Analgesics

The enkephalins are an intriguing class of small endogenous peptides that exhibit morphine-like action (Hughes *et al.*, 1975). In contradistinction to morphine and other opiates having relatively inflexible structures, the pentapeptide Met-enkephalin (H—Tyr—Gly—Gly—Phe—Met—OH) may adopt one of many possible conformations as the preferred (active) solution structure; consequently, it is of importance to determine what if any structural analogy exists between the two classes of compounds. ^{1}H NMR analyses of a DMSO solution of Met-enkephalin at 300 MHz have utilized temperature variation, coupling constant correlations, and conformational energy steric maps to deduce a preferential conformation characterized by a highly folded secondary structure (Garbay-Jaureguiberry *et al.*, 1978). More importantly, the relatively free N-terminal Tyr—Gly segment exhibited steric requirements found in stereospecific opiate–receptor interaction, which suggested that it functions as a primary attachment point. The existence of one predominant conformation of Leu-enkephalin in DMSO has also been deduced from ^{13}C and ^{1}H NMR data obtained with ^{13}C-enriched carbonyl carbons (Stimson *et al.*, 1979).

Somewhat different conclusions have come from recent studies of a ^{2}H-labeled enkephalin (Cowburn, 1978). The chemical shifts and coupling constants of the amide and α protons of the labeled glycine residues in H—Tyr—X—X′—Phe—Leu—OH (X—X′ = HNCD$_2$CONHCD$_2$CO, HNCD$_2$CONHCHDCO) were determined in DMSO, H$_2$O, and D$_2$O. The observed coupling constants for the glycine residues were very close to the calculated averaged coupling constants and, apparently, a favored conformation of the backbone region for residues 2 and 3 of enkephalin does not exist in aqueous solution. Hence, it was suggested that a possible conformational resemblance of enkephalin to opiates might arise *after* binding of the effector to its receptor.

The structure–activity considerations for these materials is further complicated by NMR studies (Gacel *et al.*, 1979) of the conformation and biological activities of hexapeptide analogs of enkephalins which led to the suggestion that the characteristic conformation of the enkephalin backbone was retained in these analogs and thus ruled out the possibility that the observed biological activity changes were related to a modified conformation of this portion of the molecule. An interaction mode for the enkephalins with their receptors was proposed and involved a combination of electrostatic and van der Waals forces.

Circumstances *formally* analogous to the aforementioned Met-enkephalin analysis may be found in NMR conformational studies of *threo-* and *erythro*-5-methyl-methadone (Henkel *et al.*, 1976). Based on the magnitude of the vicinal $^1H-^1H$ coupling constant, it was suggested that the inactive threo diasteromer, as the hydrochloride salt, exists predominantly in an intramolecular hydrogen bonded conformation while the erythrohydrochloride, which exhibits five times the analgesic activity of racemic methadone, exists as a conformational mixture. Since the inactive threo isomer has the same configurations as that found in the active enantiomers of methadone and isomethadone, it was suggested that (1) the chiral centers do not behave independently, (2) conformational factors play an important role, and (3) the "pharmacophoric" conformations of the diphenylpropylamine analgesics involve antiperiplanar-like disposition of the Ph_2CCOEt and $N\overset{+}{H}(CH_3)_2$ groups.[12]

threo-5-methyl-methadone

erythro-5-methyl-methadone

In the search for clinically useful narcotic agonists and antagonists, a large number of oripavine and thebaine derivatives have been synthesized, and C-19 diastereomers were found to exhibit significantly different potencies (Lewis *et al.*, 1971). Recent conformational studies of a series of oripavine compounds by semiempirical quantum mechanical calculation have addressed this structure–activity problem, and low-energy conformers of carbinol substituents at the C-19 position have been found with and without intramolecular hydrogen bonding to the C-6 OCH_3 group (Loew and Berkowitz, 1979). The relative energies of these conformers depends upon the nature of the R_1 and R_2 groups and the alcohol configuration, which is also in

[12]Correlation of chemical shift data for somatostatin analogs with biological activity has led to the conclusion that the biologically active conformation of somatostatin at the receptor controlling insulin release is *not* the major conformation of this hormone in solution (Arison *et al.*, 1978).

accord with earlier variable temperature ^{1}H NMR data in the thebain series that indicated the presence of more than one conformation in solution (Fulmor *et al.*, 1967).

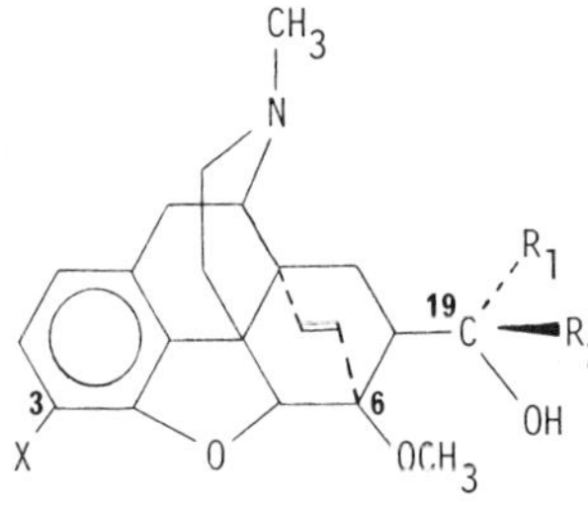

oripavine : X = OH

thebaine : X = OCH$_3$

6.4. Miscellaneous

Conformational studies of the antiradiation agents cysteamine (HSCH$_2$CH$_2$ NH$_2$), 2-aminoethanethiosulfuric acid (HO$_3$SSCH$_2$CH$_2$NH$_2$), and 2-substituted thiazolidines have been carried out (Kulkarni and Govil, 1977) for comparison with theoretical calculations and solid-state data in an attempt to further define their mechanism of action, which had been previously ascribed to either mixed disulfide formation with proteins, chelation of vital metal ions, and/or binding to DNA.

    The room-temperature proton spectrum of cysteamine is an AA′BB′ type, which is the weighted average resulting from relatively rapid interconversion of staggered *trans* and *gauche* rotamers, T, G, and $\bar{\text{G}}$, where G and $\bar{\text{G}}$ are

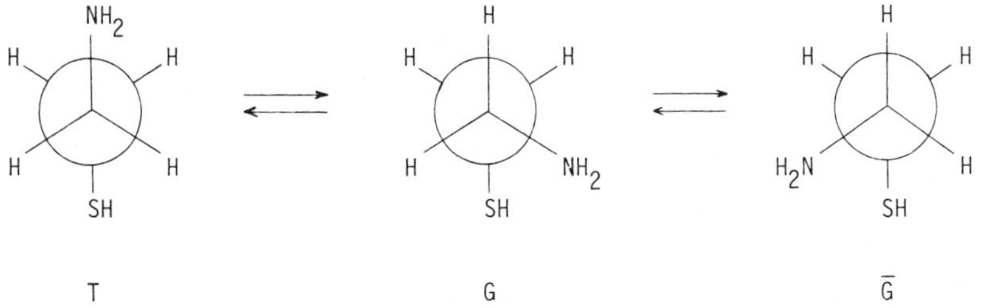

enantiomers. The observed coupling constants (J_{AB} and $J_{AB'}$) were used to extract quantitative information about relative conformer populations, and it was found that the *gauche:trans* ratio was rather insensitive to pH over the range studied: pH 3, 56:44; pH 7, 75:25; and pH 11, 66:34. At pH 7, the slightly higher population of the *gauche* conformation (which is found in the solid state) was rationalized in terms of electrostatic attractive forces and

N$\cdots$H$\cdots$S hydrogen bonding. It was further suggested that the biologically active conformation may be the *gauche* form, as this arrangement facilitates metal ion chelation and prevents *in vivo* oxidation of sulfur and nitrogen by water radiolysis.

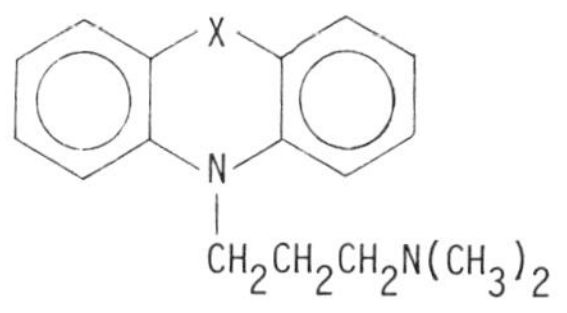

imipramine : X = CH_2CH_2

promazine : X = S

Clinically active tricyclic compounds such as the antidepressant imipramine and its analogs. and the antipsychotics or neuroleptics akin to promazine have been studied by ^1H NMR in order to assess the importance of conformational factors (Abraham *et al.*, 1974). A relatively large effect of benzyl alcohol and other aromatic solutes on the induced chemical shifts ($\Delta\delta$) of these drug molecules has also been found in ^1H spectra and has been interpreted in terms of ion pairs (Abraham *et al.*, 1977): a specific complex is formed in which the aromatic ring of the benzyl alcohol interacts with the positively charged ammonium ion of the side chain and, in addition, the OH group of benzyl alcohol forms a hydrogen bond with the chloride counterion of the hydrochloride salt. Phenol and aniline, but not 2-phenylethanol, may substitute for benzyl alcohol, while tetraphenylborate substitution for chloride in this general type of complex is not possible. Increasing the dielectric constant (ϵ) of the solvent from 4.6 with chloroform to 46.6 with DMSO results in a gradual decrease in $\Delta\delta$, which was rationalized on the basis of increasing dissociation of the ion pair and competitive hydrogen bonding with solvent. Aside from providing details on ion pairing of psychotropic drugs, extension of these studies to cognate systems can be used to check for similar associative phenomena related to, for example, membrane passage or transportation of drugs or drug metabolites.

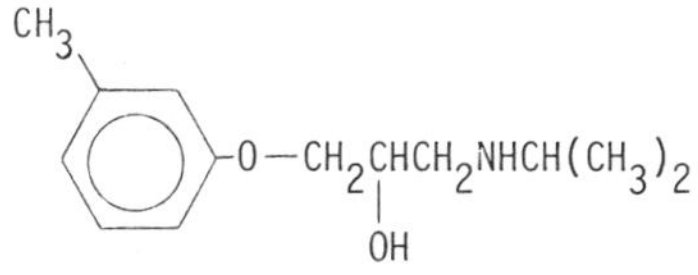

toliprolol

Another example of ion pairing has been provided by conformational analysis of hydrohalide salts of the β-adrenoceptor antagonist toliprolol (Zaagsma, 1979). In chloroform solution, ^1H NMR evidence was consistent with a halide ion being intramolecularly bonded to both the hydroxyl and ammonium ions via a rather unusual 7-membered ring, which was *not* formed in an aqueous medium. It was noted, however, that such an interaction is conceivable for anionic sites in the hydrophobic microenvironment of the plasma membrane carrying β-adrenoceptors.

acridans

The acridan derivatives represent a related class of CNS-active psychotropic compounds that have also been the subject of structure–activity studies (Shambhu *et al.*, 1974). The carboxamide system pictured features restricted rotation about the CO—N amide bond and diastereoisomerism associated with the central "puckered boat" ring. Conclusions drawn from ^{1}H NMR analysis regarding the degree of puckering resulting from the combined steric bulk at C-9 and C-10 were extrapolated to pharmacologically active analogs bearing a $(CH_2)_3N(CH_3)_2$ side chain as opposed to the carboxamido moiety; however, the validity of this extension and the use of $CHCl_3$ as the most appropriate NMR solvent for this conformational analysis are reasonable points for debate.

ACKNOWLEDGMENTS

The author gratefully acknowledges financial support from the National Institutes of Health through research grants CA-21345 and CA-18366, which provided partial funding for some of the unpublished work that has been referenced herein. The assistance of Mrs. Elizabeth Pohlhaus and Dr. Susan M. Ludeman during manuscript preparation is also sincerely appreciated.

REFERENCES

Abraham, R. J., L. J. Kricka, and A. Ledwith (1974), *J. Chem. Soc. Perk. II*, 1648.

Abraham, R. J., K. Lewtas, and W. A. Thomas (1977), *J. Chem. Soc. Perk II*, 1964.

Abraham, R. J., and P. Loftus (1978), *Proton and Carbon-13 NMR Spectroscopy, An Integrated Approach*, Heyden, London.

Anhoury, M. L., P. Crooy, R. DeNeys, and A. Laridant (1976), *J. Pharm. Sci.* **65**, 590.

Anteunis, M. J. O., R. E. A. Callens, and D. K. Tavernier (1975), *Eur. J. Biochem.* **58**, 259.

Arison, B. H., R. Hirschmann, and D. F. Veber (1978), *Bioorg. Chem.* **7**, 447.

Arnold, H., F. Bourseaux, and N. Brock (1968), *Naturwiss.* **45**, 64.

Bartlett, P. D., S. D. Ross, and C. G. Swain (1949), *J. Am. Chem. Soc.* **71**, 1415.

Benckhuijsen, C. (1968), *Biochem. Pharmacol.* **17**, 55.

Berner, D., and J. F. McGarrity (1979), *J. Am. Chem. Soc.* **101**, 3135.

Beroza, M., and A. B. Borkovec (1964), *J. Med. Chem.* **7**, 44.

Browner, S. M., A. F. Cockerill, R. J. Maidment, D. M. Rackham, and G. F. Snook (1976), *J. Pharm. Sci.* **65**, 1305.

Brundrett, R. B., J. W. Cowens, M. Colvin, and I. Jardine (1976), *J. Med. Chem.*, **19**, 958.

Campbell, I. D., C. M. Dobson, G. Jeminet, and R. J. P. Williams (1974), *FEBS Letters* **49**, 115.

Campbell, I. D., and D. M. Dobson (1979), *Meth. Biochem. Anal.* **25**, 1.

Case, D. E. (1973), *Xenobiotica* **3**, 451.

Case, D. E. (1976), *Chem. Ind.*, 391.

Cavalieri, P., P. De Santis, S. Morosetti, and M. Savino (1978), *Gazz. Chim. Ital.* **108**, 509.

Cerbon, J. (1972), *Biochim. Biophys. Acta* **290**, 51.

Chang, C. J., and C. G. Lee (1976), *Arch. Biochem. Biophys.* **176**, 801.

Chang, C. J., and C. G. Lee (1978), *Cancer Res.* **38**, 3734 (1978).

Chen, S. T., and C. S. Springer, Jr. (1979), *Chem. Phys. Lipids* **23**, 23.

Chien, Y. W., H. J. Lambert, and D. R. Sandvordeker (1975), *J. Pharm. Sci.* **64**, 957.

Chiu, F. T., F. P. Tsui, and G. Zon (1979), *J. Med. Chem.* **22**, 802.

Chow, S. T., and R. B. Martin (1974), *Inorg. Nucl. Chem. Letters* **10**, 1131.

Chu, G. Y. M., S. Mansy, R. E. Duncan, and R. S. Tobias (1978), *J. Am. Chem. Soc.* **100**, 593.

Cowburn, D. (1978), *FEBS Letters* **94**, 236.

Cox, P. J., P. B. Farmer, M. Jarman, M. Jones, W. J. Stec, and R. Kinas (1976), *Biochem. Pharmacol.* **25**, 993.

Crompton, M., G. J. Barritt, J. H. Bradbury, and F. L. Bygrave (1976), *Biochem. Pharmacol.* **25**, 2461.

Dabrowiak, J. C., F. T. Greenaway, and F. S. Santillo (1979), 178th Natl. ACS Meeting Abstracts, ORGN, No. 12.

Davidson, J. P., P. J. Faber, R. G. Fisher, S. Mansy, H. J. Peresie, B. Rosenberg, and L. VanCamp (1975), *Cancer Chemoth. Rep.* **59**, 287.

Degelaen, J., J. Feeney, G. C. K. Roberts, A. S. V. Burgen, J. M. Freve, and J. M. Ghuysen (1979), *FEBS Letters* **98**, 53.

Dehand, J., and J. Jordanov (1977), *J. Chem. Soc., Dalton*, 1588.

De Leeuw, H. P. M., J. R. De Jager, H. J. Koeners, J. H. Van Boom, and C. Altona (1977), *Eur. J. Biochem.* **76**, 209.

Dornberger, K., H. Thrum, and G. Engelhardt (1976), *Tetrahedron Letters*, 4469.

Easwaran, K. R. K., L. G. Pease, and E. R. Blout (1979), *Biochemistry* **18**, 61.

Egan, W., and G. Zon (1976), *Tetrahedron Letters*, 813.

El-Fatatry, H. M., and H. Y. Aboul-Enein (1978), *Spectr. Letters* **11**, 921.

Engle, T. W., G. Zon, and W. Egan (1979), *J. Med. Chem.* **22**, 897.

Fayle, D. R. H., G. J. Barritt, and F. L. Bygrave (1975), *Biochem. J.* **148**, 527.

Fernandez, M., and J. Cerbon (1973), *Biochim. Biophys. Acta* **298**, 8.

Fischer, J. J., and O. Jardetzky (1965), *J. Am. Chem. Soc.* **87**, 3237.

Forlani, L., L. Lunazzi, D. Macciantelli, and B. Minguzzi (1979), *Tetrahedron Letters*, 1451.

Forsén, S., and M. Nilsson (1970), in *The Chemistry of the Carbonyl Group*, Vol. 2, Wiley, New York, NY, p. 157.

Friedman, O. M., and A. M. Seligman (1954), *J. Am. Chem. Soc.* **76**, 655.

Friedman, O. M., and E. Boger (1961), *Anal. Chem.* **33**, 906.

Friedman, O. M., E. Boger, V. Grubliauskas, and H. Sommer (1963), *J. Med. Chem.* **6**, 50.

Friedman, O. M. (1967), *Cancer Chemoth. Rep.* **51**, 347.

Friedman, O. M., S. Bien and J. K. Chakrabati (1965), *J. Am. Chem. Soc.* **87**, 4978.

Friedman, O. M., A. Myles, and M. Colvin (1979), in "Advances in Cancer Chemotherapy" (Marcel Dekker, New York), 143–204.

Fulmor, W., J. E. Lancaster, G. D. Morton, J. J. Brown, C. F. Howell, C. T. Nora, and R. A. Hardy, Jr. (1967), *J. Am. Chem. Soc.* **89**, 3322.

Gacel, G., M. C. Fournie-Zaluski, E. Fellion, B. P. Roques, B. Senault, J. M. Lecomte, B. Malfroy, J. P. Swertz, and J. C. Schwartz (1979), *Life Sci.* **24**, 725.

Garbay-Jaureguiberry, C., B. P. Roques, L. Delcambe, F. Peypoux, and G. Michel (1978), *FEBS Letters* **93**, 151.

Gent, M. P. N., and J. H. Prestegard (1976), *Biochim. Biophys. Acta* **426**, 17.

Golumbic, C., J. S. Fruton, and M. Bergmann (1946), *J. Org. Chem.* **11**, 518.

Granot, J., and J. Reuben (1978), *J. Am. Chem. Soc.* **100**, 5209.

Gupta, R. K., J. A. Ferretti, and W. J. Caspary (1979), *Biophys. J.* **25**, 236a.

Hanessian, S., and G. Patil (1978), *Tetrahedron Letters*, 1031.

Hauser, H., S. A. Penkett, and D. Chapman (1969), *Biochim. Biophys. Acta* **183**, 466.

Henkel, J. G., E. P. Berg, and P. S. Portoghese (1976), *J. Med. Chem.* **19**, 1308.

Hill, D. L. (1975), *A Review of Cyclophosphamide*, Charles C. Thomas Publishing Company, Springfield, Ill.

Hohorst, H. J., U. Draeger, G. Peter, and G. Voelcker (1976), *Cancer Treat. Rep.* **60**, 309.

Hughes, J., T. W. Smith, H. W. Kosterlitz, L. A. Fothergill, B. A. Morgan, and H. R. Morris (1975), *Nature* **258**, 577.

Hunt, G. R. A. (1975), *FEBS Letters* **58**, 194.

Hunt, G. R. A., L. R. H. Tipping, and M. R. Belmont (1978), *Biophys. Chem.* **8**, 341.

Inagaki, K., and Y. Kidani (1978), *Bioinorg. Chem.* **9**, 157.

Jardetzky, O., and N. G. Jardetzky (1965), *Mol. Pharmacol.* **1**, 214.

Jardetzky, O., and N. G. Jardetzky (1971), *Ann. Rev. Biochem.* **40**, 605.

Jarman, M., P. B. Farmer, A. B. Foster, and P. J. Cox (1978), in *Stable Isotopes, Applications in Pharmacology, Toxicology, and Clinical Research*. T. A. Baillie, Ed., University Park Press, Baltimore, MD, pp. 85–95.

Jarman, M., R. A. V. Milsted, J. F. Smyth, R. W. Kinas, K. Pankiewicz, and W. Stec (1979), *Cancer Res.* **39**, 2762.

Jeffrey, A. M., J. S. H. Blobstein, I. B. Weinstein, F. A. Beland, R. G. Harvey, H. Kasai, I. Miwra, and K. Nakanishi (1976), *J. Am. Chem. Soc.* **98**, 5714.

Jordanov, J., and R. J. P. Williams (1978), *Bioinorg. Chem.* **8**, 77.

Kato, G. (1975), *J. Pharm. Sci.* **64**, 488.

Kaufman, R. D. (1977), *Anesthesiology* **46**, 49.

Kawashima, T., R. D. Kroshefsky, R. A. Kok, and J. G. Verkade (1978), *J. Org. Chem.* **43**, 1111.

Kearns, D. R., D. J. Patel, and R. G. Shulman (1971), *Nature* **229**, 338.

Koehler, L. S., E. T. Fossel, and K. A. Koehler (1977a), *Biochemistry* **16**, 3700.

Koehler, L. S., W. Curley, and K. A. Koehler (1977b), *Mol. Pharmacol.* **13**, 113.

Kong, P. C., and T. Theophanides (1974a), *Inorg. Chem.* **13**, 1167.

Kong, P. C., and T. Theophanides (1974b), *Inorg. Chem.* **13**, 1981.

Kong, P. C., and T. Theophanides (1975), *Bioinorg. Chem.* **5**, 51.

Kram, T. C. (1977), *J. Pharm. Sci.* **66**, 443.

Krishman, C. V., H. L. Friedman, and C. S. Springer, Jr. (1978), *Biophys. Chem.* **9**, 23.

Krugh, T. R. (1978), in *Nuclear Magnetic Resonance Spectroscopy in Molecular Biology* B. Pullman, Ed., D. Reidel Publishing Company, Dordrecht, Holland, pp. 137–146.

Kulkarni, V. M., and G. Govil (1977), *J. Pharm. Sci.* **66**, 483.

Lanouette, M., D. Legault, and B. A. Lodge (1976), *J. Pharm. Sci.* **65**, 1214.

Leibfritz, D. (1974), *Tetrahedron Letters*, 4125.

Levins, P. L., and Z. B. Papanastassiou (1965), *J. Am. Chem. Soc.* **87**, 826.

Levins, P. L., and W. I. Rogers (1965), *Cancer Chemoth. Rep.*, No. 44, 15.

Lewis, J. W., K. W. Bentley, and A. Cowan (1971), *Ann. Rev. Pharmacol.* **11**, 241.

Loew, G., and D. S. Berkowitz (1979), *J. Med. Chem.* **22**, 603.

Lown, J. W., L. W. McLaughlin, and Y. M. Chang (1978), *Bioorg. Chem.* **7**, 97.

Ludeman, S. M., D. L. Bartlett, and G. Zon (1979), *J. Org. Chem.* **44**, 1163.

Mansy, S., G. Y. H. Chu, R. E. Duncan, and R. S. Tobias (1978), *J. Am. Chem. Soc.* **100**, 607.

Marker, A., L. G. Paley, and T. M. Spotswood (1978), *Chem. Phys. Lipids* **22**, 39.

Marzilli, L. G. (1978), *Biochem. Biophys. Res. Commun.* **84**, 70.

Maxwell, J., D. S. Kaushik, and C. G. Butler (1974), *Biochem. Pharmacol.* **23**, 168.

Monoi, H., and H. Uedaira (1979), *Biophys. J.* **25**, 535.

Mrtek, M. B., J. E. Gearien, and M. I. Blake (1977), *J. Pharm. Sci.* **66**, 1019.

Muller, N., P. C. Lauterbur, and G. F. Svatos (1957), *J. Am. Chem. Soc.* **79**, 1807.

Munchausen, L. L., and R. O. Rahn (1975), *Biochim. Biophys. Acta* **414**, 242.

Nathans, D. (1964), *Proc. Natl. Acad. Sci. U.S.A.* **51**, 585.

Nelson, D. J., P. L. Yeagle, T. L. Miller, and R. B. Martin (1976), *Bioinorg. Chem.* **5**, 353.

Neville, G. A., F. B. Hasan, and I. P. Smith (1977), *J. Pharm. Sci.* **66**, 638.

Olah, G. A., D. J. Donovan, and L. K. Keefer (1975), *J. Natl. Cancer Inst.* **54**, 465.

Omura, S., A. Neszmelyi, M. Sangare, and G. Lukacs (1975), *Tetrahedron Letters*, 2939.

O'Neill, I. K., and M. A. Pringuer (1975), *Chem. Ind.*, 494.

Parfitt, R. T., G. H. Dewar, and J. K. Kwakye (1978), *J. Pharm. Pharmacol.* **30**, 62P.

Parker, G. R., N. K. Hilgendorf, and J. G. Lindberg (1976), *J. Pharm. Sci.* **65**, 585.

Patel, D. J., and A. E. Tonelli (1974), *Biopolymers* **13**, 1943.

Patel, D. J. (1979a), *Accounts Chem. Res.* **12**, 118.

Patel, D. J. (1979b), *Biopolymers* **18**, 553.

Paul, H. H., H. Sapper, and W. Lohmann (1978), *Z. Naturforsch., C: Biosci.* **33c**, 870.

Pettit, G. R., and R. L. Smith (1964), *Can. J. Chem.* **42**, 572.

Pettit, G. R., J. A. Settapani, and R. A. Hill (1965), *Can. J. Chem.* **43**, 1792.

Pirkle, W. H., and M. S. Hoekstra (1976), *J. Am. Chem. Soc.* **98**, 1832.

Rackham, D. M. (1976), *Talanta* **23**, 269.

Roberts, J. J., and A. J. Thomson (1979), *Prog. Nucl. Acid Res. Mol. Biol.* **22**, 71.

Rodriguez, M. A., M. T. Pizzorno, and S. M. Albonico (1977), *J. Pharm. Sci.* **66**, 121.

Rosenberg, B. (1978), *Adv. Exp. Med. Biol.* **91**, 129.

Sandvordeker, D. R., Y. W. Chien, T. K. Lin, and H. J. Lambert (1975), *J. Pharm. Sci.* **64**, 1797.

Sarrazin, M., M. Chauvet-Deroudhile, M. Bourdeaux-Pontier, and C. Briand (1978), in *Proc. European Conf. on NMR of Macromolecules*, Lerici, Italy, p. 503.

Seydel, J. K., and O. Wassermann (1973), *Naunyn-Schmiedeberg's Arch. Pharmacol.* **279**, 207.

Seydel, J. K., and O. Wassermann (1976), *Biochem. Pharmacol.* **25**, 2357.

Shambhu, M. B., R. R. Koganty, and G. A. Digenis (1974), *J. Med. Chem.* **17**, 805.

Singer, B. (1975), *Prog. Nucl. Acid Res. Mol. Biol.* **15**, 219.

Spencer, T. L., and F. L. Bygrave (1974), *Biochem. J.* **140**, 413.

Stimson, E. R., Y. C. Meinwald, and H. A. Scheraga (1979), *Biochemistry* **18**, 1661.

Stone, P. J., A. D. Kelman, F. M. Sinex, M. M. Bhargava, and H. O. Halvorson (1976), *J. Mol. Biol.* **104**, 793.

Takayama, K., N. Nambu, and T. Nagai (1979), *Chem. Pharm. Bull.* **27**, 715.

Thakkar, A. L., and L. G. Tensmeyer (1974), *J. Pharm. Sci.* **63**, 1319.

Toeplitz, B. K., A. I. Cohen, P. T. Funke, W. L. Parker, and J. Z. Gongoutas (1979), *J. Am. Chem. Soc.* **101**, 3344.

Traut, R. R., and R. E. Monro (1964), *J. Mol. Biol.* **10**, 63.

Trudell, J. R., and W. L. Hubbell (1976), *Anesthesiology* **44**, 202.

Tsui, F. P., J. A. Brandt, and G. Zon (1979), *Biochem. Pharmacol.* **28**, 367.

Turczan, J. W., and T. Medwick (1976), *J. Pharm. Sci.* **65**, 235.

Vanderkooi, J. M., R. Landesberg, H. Selick, and G. G. McDonald (1977), *Biochim. Biophys. Acta* **464**, 1.

Vinson, J. A., and D. M. Kozak (1978), *Am. J. Pharm. Ed.* **42**, 290.

Warren, R. J., J. E. Zarembo, D. B. Staiger, and A. Post (1976), *J. Pharm. Sci.* **65**, 738.

Wehrli, F. W., and T. Wirthlin (1978), *Interpretation of Carbon-13 NMR Spectra*, Heyden, London.

Weinberg, E. D. (1974), in *Principles of Medicinal Chemistry*, W. O. Foye, Ed., Lea and Febiger, Philadelphia, PA, pp. 761–765.

White, D. W., D. E. Gibbs, and J. G. Verkade (1979), *J. Am. Chem. Soc.* **101**, 1937.

Wiemer, D. F., D. I. C. Scopes, and N. J. Leonard (1976), *J. Org. Chem.* **41**, 3051.

Wilson, W. L., H. W. Avdovich, D. W. Hughes, and G. W. Buchanan (1977), *J. Pharm. Sci.* **66**, 1079.

Wood, A., L. G. Paleg, and T. M. Spotswood (1974), *Austral. J. Plant Physiol.* **1**, 167.

Workman, P., and J. A. Double (1978), *Biomedicine* **28**, 255.

Yamaguchi, T., K. Saito, T. Tsujimoto, and H. Yuki (1976), *J. Heterocyclic Chem.* **13**, 533.

Yeagle, P. L., W. C. Hutton, and R. B. Martin (1977), *Biochim. Biophys. Acta* **465**, 173.

Yeh, H. J. C., R. S. Wilson, W. A. Klee, and A. E. Jacobson (1976), *J. Pharm. Sci.* **65**, 902.

Zaagsma, J. (1979), *J. Med. Chem.* **22**, 441.

Zon, G., W. Egan, and J. B. Stokes (1976), *Biochem. Pharmacol.* **25**, 989.

Zon, G., S. M. Ludeman, and W. Egan (1977), *J. Am. Chem. Soc.* **99**, 5785.

Zwelling, L. A., T. Anderson, and K. W. Kohn (1979), *Cancer Res.* **39**, 365.

Four

Biosynthesis and ^{19}F NMR Characterization of Fluoroamino Acid Containing Proteins

Brian D. Sykes and Joel H. Weiner

Department of Biochemistry and
MRC Group on Protein Structure and Function
University of Alberta
Edmonton, Alberta T6G 2H7, Canada

1. INTRODUCTION

High-resolution nuclear magnetic resonance (NMR) methods offer a powerful tool for the elucidation of the structure and function of proteins. When

171

resonances in the NMR spectrum of a protein can be resolved and assigned to individual nuclei in the protein, these resonances can then serve as probes of the structure, chemical state, and dynamic properties of the protein in solution. To date, most applications of NMR in the study of proteins have used ^{1}H NMR since the proton is the most sensitive nucleus with respect to signal intensity, ^{1}H NMR instrumentation is most commonly available, and ^{1}H chemical shifts, coupling constants, and relaxation times have been widely studied. However, ^{1}H resonance frequencies are confined to a narrow 10 parts per million range, exclusive of protons in unique environments (such as near a paramagnetic center). This means that even using the highest magnetic fields presently attainable with superconducting magnets (approximately 94 kG), which result in ^{1}H NMR frequencies near 400 MHz, the resonances from the several thousand protons in a protein fall within a spectral width of 4000 Hz. Since the widths of individual resonances increase with increasing protein molecular weight ($\Delta\nu$ ~10 Hz for MW = 10,000), the ^{1}H NMR spectra of proteins larger than MW ~20,000 are generally unresolved with few resolved components. Several approaches have been attempted to simplify ^{1}H NMR spectra of proteins: deuteration of the protein in all but a selected few proton positions (Putter et al., 1969), various resolution-enhancement techniques including difference NMR (Campell et al., 1973a), and the use of NMR shift reagents (Campell et al., 1973b). The usefulness of these methods, however, is still limited by the intrinsic narrow chemical shift range of ^{1}H NMR, especially for larger proteins where the NMR linewidths are correspondingly larger.

Each of the other spin $\frac{1}{2}$ nuclei commonly found in proteins (e.g., ^{13}C, ^{15}N, ^{31}P) have specific advantages making them attractive for NMR studies. For example, ^{13}C has a wide chemical shift range of ~250 ppm, easily interpretable linewidths and relaxation times, and relatively narrow linewidths for unprotonated carbons. However, the overriding consideration for biochemical studies is that these other nuclei, even if the protein is specifically enriched to 100% in ^{13}C or ^{15}N, *are considerably less sensitive in terms of signal intensity than ^{1}H*. A practical lower limit in concentration for NMR observation of individual resonances in a protein, with time averaging limited to approximately 5 h, ranges from $10^{-4} M$ for ^{1}H to about $5 \times 10^{-3} M$ for natural abundance ^{13}C with even higher concentrations required for ^{15}N. These requirements are even more striking in terms of amounts of sample. For ^{1}H NMR typically 350 μl of protein solution is required in a 5 mm NMR tube. At a concentration of 0.5 mM, and for a protein of molecular weight 10,000 daltons, this means that 1.7 mg of protein is required for a ^{1}H NMR spectrum. On the other hand, for natural abundance ^{13}C NMR, 1.0 ml of protein solution is required to fill a 10 mm NMR tube. For a 10 mM solution of a protein of molecular weight 10,000 daltons, this means that 100 mg of protein is required for a single NMR spectrum.

An alternative approach is to prepare a protein specifically labeled with another nucleus, such as ^{19}F. The advantages of ^{19}F as a label are severalfold. First, the sensitivity of ^{19}F NMR is only slightly lower than that of ^{1}H NMR. This sensitivity loss (approximately 17%) is offset in part by the elimination of the dynamic range problems associated with ^{1}H Fourier Transform NMR that are caused by the signal from solvent H_2O or residual HDO in D_2O solutions (Patt and Sykes, 1972). Second, ^{19}F NMR chemical shifts are spread over a much larger range than ^{1}H chemical shifts and are much more sensitive to the environment of the nucleus. Third, ^{19}F is 100% naturally abundant, so that specific reagents for fluorine labeling are not as expensive as those for specific enrichment in ^{13}C or ^{15}N. The ^{19}F NMR spectrum of a specifically labeled protein will therefore be simpler in terms of the number of resonances present, more spread out in chemical shift, more sensitive to conformational changes in the protein, and free from background signals from other components of the system (such as lipids) if they are present. The major disadvantage and *caution* is that one must *demonstrate that the labeled protein is identical in conformation and activity to the native protein.* The van der Waals radius of fluorine is 1.35 Å compared with 1.2 Å for hydrogen, so that size should not be a major factor in the incorporation of fluorine into a protein. However, fluorine is considerably more electronegative than hydrogen so that a substituted fluorine could participate in abnormal hydrogen bonds or influence the pK values of neighboring acid–base groups.

The first applications of ^{19}F NMR to biochemical problems generally examined the interaction of ^{19}F-labeled substrates or inhibitors with enzymes (Atler and Magnuson, 1974; Dwek, 1971; Zeffren, 1968, 1970; Ashton and Capon, 1971; Tayler *et al.*, 1971; Hunkapiller and Richards, 1972; Sykes, 1969; Robillard and Wishnia, 1972; Spotswood *et al.*, 1967; Gammon *et al.*, 1972; Raftery *et al.*, 1971), or studied proteins labeled by reaction with organic reagents containing fluorine (Bittner and Gerig, 1970; Paselk and Levy, 1974; Huestis and Raftery, 1971, 1973; Bode *et al.*, 1975). Recently, however, considerable success has been obtained incorporating fluorinated amino acid analogs into bacterial proteins (Sykes *et al.*, 1974; Hull and Sykes, 1974, 1975a, b, 1976; Browne and Otvos, 1976; Robertson *et al.*, 1977; Anderson *et al.*, 1975; Coleman *et al.*, 1976; Coleman and Armitage, 1977; Lu *et al.*, 1976; Lu *et al.*, 1980; Opella *et al.*, 1979; Kimber *et al.*, 1977, 1978; Hagen *et al.*, 1978, 1979a, 1980; Gally *et al.*, 1978; Lee *et al.*, 1977; Richmond, 1963; Pratt and Ho, 1975; Browne *et al.*, 1970). This clearly is the best choice for monitoring the properties of the protein since the amino acid residues to be labeled can be chosen and studied at will. Although fluorine analogs are generally toxic to the cell, conditions can often be found where limited growth and biosynthetic incorporation of analogs into proteins can take place. It is then possible to determine the effects of such incorporation on individual proteins and to isolate them for ^{19}F NMR studies. The proteins which have been prepared containing ^{19}F-labeled amino acids and for which ^{19}F NMR spectra have been recorded are listed in Table I. In this chapter we will focus our discussion on these proteins. We will discuss their biosyn-

Table I. Partial list of proteins prepared containing ¹⁹F-labeled amino acid analogs where ¹⁹F NMR spectrum is known

Protein	Analog[a]	Source	Incorporation of Analog (%)	¹⁹F NMR Frequency, ν_0 (MHz)	Reference
Alkaline phosphatase	3-F-Tyr	*E.coli* W3747	73	94.1 235.2	Sykes *et al.* (1974); Hull and Sykes (1974; 1975a,b; 1976)
	(3-F-Tyr) 4-F-Trp	*E.coli* H677	75	94.1	Browne and Otvos (1976)
Histidine BP J	5-F-Trp	*Salmonella typhimurium* TA1010	80	235.2	Robertson *et al.* (1977)
fd Gene 5	3-F-Tyr	*E.coli* AT2741		84.67 94.1	Anderson *et al.* (1975); Coleman *et al.* (1976); Coleman and Armitage (1977)
Lac repressor	3-F-Tyr	*E.coli* CSH46	90	94.1	Lu *et al.* (1976); Lu *et al.* (1980); Opella *et al.* (1979)
Dihydrofolate reductase	3-F-Tyr 6-F-Trp	*Lactobacillus casei* MTX/R		94.1 254	Kimber *et al.* (1977, 1978)
M13 Gene 8	3-F-Tyr	*E.coli* AT2471	~50	84.67 254	Hagen *et al.* (1978, 1979, 1980)
Acyl carrier protein	3-F-Tyr	*E.coli* K12		84.67	Gally *et al.* (1978)
Lipoprotein	3-F-Tyr	*E. coli* X'121		94.1	Lee *et al.* (1977)

[a]Tyr:

Trp:

Phe:

thesis and the various details about their structure that have been elucidated from the ^{19}F NMR spectra. Our main purpose is to highlight the general conclusions that can be made from analysis of ^{19}F NMR spectra of proteins. The principles and practical aspects of the techniques for acquiring the spectrum of a fluorine-labeled protein will not be emphasized here since they have been discussed in detail elsewhere (Sykes and Hull, 1978).

2. BIOSYNTHESIS OF FLUORINE-LABELED PROTEINS

In this section we will describe the biochemical basis of the procedures which have been developed to prepare fluorine-labeled proteins. The discussion is almost exclusively limted to the incorporation of fluorine-labeled aromatic amino acids into the bacterium *Escherichia coli* because this system has been the most utilized and is the most amenable to manipulation. For a more detailed discussion of fluoroamino acids, readers should consult the review by Marquis (1970).

2.1. The Use of Auxotrophs to Prepare Fluorine-Labeled Proteins

Wild type *E. coli* strains are termed prototrophic and are capable of endogenously producing all the amino acids necessary to sustain growth. These growing prototrophic bacterial cultures have a profound ability to discriminate against fluorinated amino acid analogs, utilizing instead endogenously produced amino acids or amino acids available from the medium. The factors primarily responsible for discrimination are the high affinity of the specific amino acid transport systems and amino acyl tRNA synthetases for normal amino acids compared with various fluoroamino acid analogs (see Sections 2.3 and 2.4). Thus, fluorinated analogs are generally utilized only after endogenous supplies of the natural amino acids have been severely depleted. In fact, this forms the basis of one replacement technique (Lu *et al.*, 1976). However, with auxotrophic mutants unable to synthesize specific amino acids, the amino acid supply can be exogenously controlled, and fluoroamino acids supplied in large excess. Addition of a large excess of a fluorinated amino acid analog to an auxotrophic bacterial culture grown on a limiting concentration of the required amino acid results in an immediate slowing of growth. As measured by increases in cell mass, viable counts, and protein biosynthesis, growth continues at this slow rate for one to several generations (depending on the analog employed) before the culture becomes nonviable.

Fluoroamino acid analogs are toxic for two reasons. First, they are incorporated into all protein, thereby altering enzyme activity and upsetting metabolism (Cohen and Munier, 1959). Secondly, they act as structural analog inhibitors (corepressors and allosteric inhibitors) (Richmond, 1962). Of course the former reason, although harmful to the bacteria, forms the basis of the *in vivo* replacement technique for incorporating fluorine probes at *defined positions* in the protein sequence. It is a simple matter to partially starve auxotrophic cells for a particular amino acid and replace it with the fluorinated analog. The most efficient way to deplete an auxotrophic culture of a natural amino acid is to harvest the cells, wash them free of amino acid and resuspend them in fresh medium. Unfortunately, this is not feasible with the large cultures required for preparative biochemistry (often 10–300 l). Instead, it is necessary to adjust the levels of normal amino acid and fluorinated amino acid, such that a reasonable compromise between maximum incorporation and cell viability is attained. This is particularly relevant for "nonregulated" or "constitutive" proteins. In certain cases, if a protein can also be induced when the fluorinated analog is added it may be possible to simultaneously achieve high levels of replacement with analog and reasonable yield of protein (Pratt and Ho, 1975; Lu *et al.*, 1976; Sykes *et al.*, 1974).

Addition of the most commonly used analog, 3-fluorotyrosine (3-F-Tyr), to a tyrosine auxotroph growing on a nearly depleted supply of tyrosine results in a marked slowing of growth which none the less continues for two to five generations (Browne and Otvos, 1976; Cohen and Munier, 1959). This period of continued growth is sufficient to permit replacement of tyrosine by 3-F-Tyr in a variety of proteins.

Similarly addition of the fluorotryptophan analogs (10^{-4} M) to a tryptophan auxotroph growing on limiting tryptophan results in a change from exponential to linear growth. Growth continues for 2 generations on 4-F-Trp, 1.3 generations on 5-F-Trp and 1.6 generations on 6-F-Trp (Browne *et al.*, 1970; Pratt and Ho, 1975).

The fluorophenylalanine analogs present a slightly different picture. Addition of 2-F-Phe or 4-F-Phe to a phenylalanine auxotroph growing on limiting phenylalanine results in linear growth (Cohen and Munier, 1959) which continues for about two generations. 3-F-Phe, however, slows growth only slightly and the cells continue to grow and divide at nearly normal rates for at least 18 h (Dettman, Weiner, and Sykes, unpublished observations).

The marked slowing of growth observed in the presence of most fluoroamino acids is reflected in a difficulty in obtaining reasonable yields of protein. This has led to the development of procedures to differentially label proteins of interest for NMR studies. The most fruitful procedure has been to induce the synthesis of a specific protein at the same time as the fluoroamino acid is added to the culture. For example, Sykes *et al.* (1974) and Browne and Otvos (1976) induced bacterial alkaline phosphatase by devising conditions such that a culture of a tyrosine auxotroph ran out of both phosphate and tyrosine simultaneously. 3-F-Tyr was then added and the low phosphate concentration resulted in enhanced synthesis of 3-F-Tyr labeled alkaline phosphatase. An identical approach was employed to prepare 4-F-Trp labeled alkaline phosphatase (Browne and Otvos, 1976). Induction has also proved useful to differentially label lac-operon proteins

(Pratt and Ho, 1975). In an analogous manner, bacteriophage infection has been initiated at the same time as analog addition in order to prepare fluorolabeled bacteriophage-coded proteins (Coleman *et al.*, 1976; Hagen *et al.*, 1978).

A novel approach to prepare a fluorolabeled protein for NMR studies was devised by Lee *et al.* (1977). The major lipoprotein of *E. coli* lacks histidine in its primary sequence; thus, they employed a histidine plus tyrosine double auxotroph and depleted the culture of both amino acids. Addition of only 3-F-Tyr resulted in substantial synthesis of the labeled lipoprotein, while other protein synthesis was impaired for lack of histidine. Although most of the fluorolabeled proteins prepared to date are inducible or bacteriophage coded, the technique is not limited to these proteins. For example, Robertson *et al.* (1977) produced 5-F-Trp labeled histidine binding protein J (a constitutive protein) by balancing the relative concentrations of tryptophan and 5-F-Trp (10^{-5} M and 5×10^{-4} M, respectively) such that growth could proceed at a diminished rate in the presence of excess 5-F-Trp. Sufficient 5-F-Trp was incorporated into the protein to provide a useful probe for NMR studies.

2.2. The Use of Repression and Feedback Inhibition to Produce Fluorine-Labeled Proteins

Addition of an amino acid to a growing prototrophic *E. coli* culture results in repression of synthesis of the amino acid biosynthetic enzymes, coupled with immediate allosteric product inhibition of the biosynthetic pathway. These effects are reflected in a severe depression of endogenous amino acid supplies in favor of the exogenously added amino acid. In general, amino acid analogs which are incorporated into protein act as repressors of biosynthetic enzyme synthesis (Maas, 1961; Freundlich, 1967). Thus Pauley *et al.* (1978) found that 0.25 mM 5-F-Trp caused an 85–90% repression of biosynthesis of the tryptophan operon enzymes and 6-F-Trp caused a 99% repression. Similarly, 4-F-Phe functioned as a repressor of synthesis of the aromatic pathway enzymes (Smith *et al.*, 1964).

One exception to this rule is the fluorine analog 5′,5′,5′-trifluoroleucine. Formation of the leucine biosynthetic enzymes is repressed by leucine, which must be activated to the leucyl tRNA to act as a corepressor. 5′,5′,5′-trifluoroleucine could not act as a corepressor of the leucine enzymes even though it is charged to leucyl tRNA by the leucyl-tRNA synthetase (Freundlich and Trela, 1969). This is one of the few examples of total discrimination against a fluorinated analog. It is interesting however that 5′,5′,5′-trifluoroleucine could replace leucine for repression of the isoleucine and valine biosynthetic enzymes (Freundlich and Trela, 1969).

Inhibition of endogenous amino acid utilization by repression is a relatively slow process. Of more immediate and probably exclusive importance is direct product or feedback inhibition of the amino acid biosynthetic

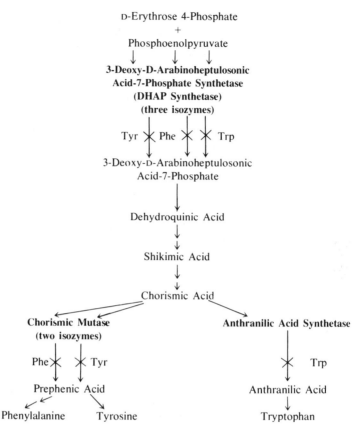

Figure 1. Schematic representation of the aromatic biosynthetic pathway of *Escherichia coli* showing the enzymatic reactions sensitive to feedback inhibition.

pathways by the fluorinated analogs. In general it has been found that the fluorinated amino acids are effective feedback inhibitors.

The aromatic amino acid biosynthetic pathway is diagrammed schematically in Figure 1. The first enzyme, called 3-deoxy-D-arabinoheptulosonic acid-7-phosphate synthetase (DHAP synthetase), combines phosphoenolpyruvate with erythrose-4-phosphate to form 3-deoxy-D-arabinoheptulosonic acid-7-phosphate, the common precursor of all three aromatic amino acids. This key enzymatic reaction is inhibited by tyrosine, phenylalanine, and tryptophan. DHAP synthetase exists as three isozymic forms, each sensitive to a different aromatic amino acid. Smith *et al.* (1964) and Previc and Binkley (1964) showed that the fluorophenylalanine analogs acted as feedback inhibitors of the crude phenylalanine-sensitive DHAP synthetase in the order 2-F-Phe > Phe > 3-F-Phe > 4-F-Phe. Simpson and Davidson (1976) arrived at essentially the same result with highly purified enzyme. Similarly 3-F-Tyr inhibits the tyrosine-sensitive DHAP synthetase (Marquis, 1970).

The biosynthesis of the aromatic amino acids subsequently follows a common pathway to the formation of chorismic acid, at which point the pathways branch. Phenylalanine and tyrosine arise from the chorismate mutase branch, while tryptophan arises from the anthranilate synthetase branch. The branch-point enzyme chorismate mutase exists as two isozymic forms; one sensitive to allosteric inhibition by tyrosine, the other by phenylalanine and presumably to the fluoroamino acids as well. At least in the case of anthranilate synthetase from *Brevibacterium flavum,* inhibition by 5-F-Trp has been documented (Shio and Sugimoto, 1978).

Detailed studies have been carried out with the leucine biosynthetic pathway. The enzyme susceptible to feedback inhibition is α-isopropylmalate synthetase; $5',5',5'$-trifluoroleucine has been shown to be an effective inhibitor (Calvo *et al.,* 1969).

Lu *et al.* (1976) have recently developed a procedure for incorporating certain fluorinated amino acids into proteins, which is based on the results presented above and eliminates the need for auxotrophs. The addition of 10 mM 3-F-Tyr, 1 mM phenylalanine and 1 mM tryptophan to the medium of a growing prototrophic culture resulted in a 96% drop in endogenous tyrosine utilization. A fluorolabeled protein was thus produced as one of the aromatic amino acids in the growth medium (tyrosine) was replaced by its fluorinated analog 3-F-Tyr. In this case, 90% replacement of the tyrosine residues of *E. coli* lac repressor protein with 3-F-Tyr was achieved, with a 50% yield of protein. In a similar manner, it should be possible to incorporate the other aromatic fluoroamino acids into proteins. In fact, encouraging results have already been obtained with the fluorinated trytophan analogs and it appears that this technique will have general applicability.

2.3. Transport of Fluoroamino Acids

Active transport is the initial point at which the cell discriminates against the fluoroamino acid analogs, and it has been found that although these analogs compete with the natural amino acids, the carrier proteins have rather poor affinities for them.

In *E. coli,* 80–90% of aromatic amino acid accumulation occurs via the general aromatic transport system, which is coded for by a gene termed *aro* P (Ames, 1964). This system exhibits a K_m for the natural aromatic amino acids of about $5 \times 10^{-7} M$. 4-F-Phe is a competitive inhibitor of the uptake of all three aromatic amino acids although it requires $2 \times 10^{-4} M$ 4-F-Phe to obtain 90% inhibition (Brown, 1970, 1971). The remaining 10–20% of aromatic amino acid uptake is mediated by specific aromatic transport systems, which have K_m's for the natural amino acids in the range of $2 \times 10^{-6} M$. As might be expected, 4-F-Phe is a competitive inhibitor of the phenylalanine-specific transport system. It is also a weak inhibitor of the tyrosine-specific system but not of the trytophan-specific system.

2.4. Aminoacyl tRNA Charging of Fluoroamino Acids

Before it can be incorporated into protein, the fluorinated amino acid must be activated to an enzyme-bound aminoacyl adenylate by the appropriate aminoacyl tRNA synthetase and then transferred to the correct tRNA. It is at this stage that the cell exerts its major discrimination against fluorinated analogs.

Santi and Denenberg (1971) carried out a study of the kinetics of purified phenylalanine tRNA synthetase with the fluorophenylalanine analogs. They found that while the V_{max} for phenylalanine and the three fluorine analogs was the same, the K_m's varied greatly. Phenylalanine had a K_m of 0.05 mM; 4-F-Phe, 0.56 mM; 3-F-Phe, 1.4 mM; and 2-F-Phe, 5 mM. These results are of critical importance and indicate that the cell must be very rigorously depleted of the natural amino acid if high levels of replacement are to be achieved. For example, a 3% contamination of 3-F-Phe by the natural analog would reduce incorporation of the fluorine amino acid by 50%. Hennecke and Böch (1975), who carried out essentially similar studies, also isolated a mutant strain resistant to 4-F-Phe, which had an altered phenylalanine tRNA synthetase. The mutated proteins had dramatically reduced turnover with 4-F-Phe as well as reduced affinity. It is interesting that the sensitivity to 3-F-Phe and 2-F-Phe was not affected by the mutation.

Dunn and Leach (1967) carried these studies further by using the polynucleotides poly (UUU) and poly (UUC) (the codons for phenylalanine) as message in an *in vitro* protein-synthesizing system. They found that UUU allowed for the incorporation of both phenylalanine and 4-F-Phe whereas UUC permitted only polyphenylalanine synthesis. They believed this was because the AAA codon species of phenylalanine tRNA could be charged by both amino acids while the GAA species could be charged only with phenylalanine. This result is hard to rationalize in view of the finding by Richmond (1963) that all 16 phenylalanines in alkaline phosphatase could be labeled with 4-F-Phe, because it is unlikely that all 16 phenylalanines are coded by UUC. In fact, the results obtained by Sanger *et al.* (1977) from the total sequence of ϕX174 indicate that both codons are used to specify phenylalanine. Although the observations of Dunn and Leach may be an artifact of the *in vitro* system or the ionic composition of their experimental system, it may indicate that the replacement of phenylalanine residues by 4-F-Phe is not equivalent at all positions in a protein sequence.

Calender and Berg (1966) carried out studies with the purified tyrosine tRNA synthetase. They found that the K_m for L-tyrosine was 0.006 mM; for L-3-F-Tyr 0.039 mM and for DL-3-F-Tyr 0.13 mM. As the D-analogs are poorly accumulated by cells (Brown, 1970) it would appear that a 6- to 10-fold excess of the fluorine analog is needed inside the cell to get about

50% replacement. Calender and Berg also showed that 4-F-Phe was not a substrate for the tyrosine tRNA synthetase, thus eliminating the possibility that this analog could replace tyrosine residues in a protein.

Fenster and Anker (1969) measured the ability of 5′,5′,5′-trifluoroleucine to be charged to tRNA in crude extracts of *E. coli.* They observed that it was charged more efficiently than leucine indicating that it was charged to more tRNAs than just that for leucine. From competition studies they found that it was also charged to isoleucine and phenylalanine tRNAs. This would result in the incorporation of this analog in place of isoleucine, phenylalanine, and leucine residues in a protein sequence and would therefore severely restrict its utility as an NMR probe.

2.5. Extent of Labeling and Activity of Labeled Proteins

An examination of the proteins which have been labeled with fluoroamino acids indicates that the extent of replacement of the natural amino acid by the fluorinated analog varies from less than 10% to 100% but usually is in the range of 50–75% (Table II). As discussed above, the extent of labeling will depend on adequate depletion of the natural amino acid, on the relative affinities of the tRNA synthetase for the fluorinated amino acid compared with the natural amino acid, and on the concentration of fluoroamino acid supplied to the culture.

As stated at the outset, the conclusions drawn from fluorine NMR are contingent on the protein having an activity identical to that of the unlabeled protein. Unfortunately this cannot always be achieved and will depend on the relative importance of the particular amino acid to the conformation and catalytic activity of the protein. The results reported to date are tabulated in Table II and are as varied as the proteins studied. Replacement of tyrosine by 3-F-Tyr often results in retention of activity, indicating that at least for this amino acid, the change of a hydrogen to a fluorine is usually tolerated. It would also appear from the small sample of proteins which have been examined that incorporation of fluorotryptophan usually leads to some loss of activity. This may be a reflection of the observation that tryptophan residues are located at many substrate binding sites. In at least one case, the D-lactate dehydrogenase (Pratt and Ho, 1975), replacement of tryptophan by 4-F-Trp results in a marked enhancement of substrate turnover.

The yield of enzyme also varies widely, from a few percent to more than double the normal amount. Obviously the aim is to obtain as high a yield of active protein as possible. The reduced yields observed in many cases may result from a combination of several factors. It is now becoming apparent that proteins labeled with fluorinated analogs may be thermodynamically slightly less stable than the natural proteins (Goldberg *et al.*, 1975). In this respect Munier and Sarazin (1964) observed that β-galactosidase labeled with 3-F-Tyr was more susceptible to denaturation by urea or heating than normal enzyme. *E. coli* cells have also evolved proteolytic enzymes which

Table II. Proteins labeled with fluoroamino acids

Protein	Analog	Replacement (%)	Amount of Protein (%)	Activity (%)	Reference
Acyl carrier protein	3-F-Tyr	56	—	100	Gally et al. (1978)
Alkaline phosphatase	4-F-Phe	73	10	100	Richmond (1963)
	3-F-Tyr	75	100	100	Sykes et al. (1974)
	3-F-Tyr	75	100	100	Browne and Otvos (1976)
	4-F-Trp	75	100	115	Browne and Otvos (1976)
	5-F-Trp	—	Low	—	Browne and Otvos (1976)
	6-F-Trp	—	Low	—	Browne and Otvos (1976)
α-amylase	4-F-Phe	15	—	70	Yoshida (1960)
Dihydrofolate reductase	3-F-Tyr	—	—	—	Kimber et al. (1977, 1978)
	6-F-Trp	—	—	—	Kimber et al. (1977, 1978)
Exopenicillinase	4-F-Phe	75	—	50	Richmond (1960)
	4-F-Phe	20	—	70	Yoshida (1960)
β-galactosidase	4-F-Trp	75	—	28	Browne et al. (1970)
	5-F-Trp	—	—	5.6	Browne et al. (1970)
	6-F-Trp	—	—	0.7	Browne et al. (1970)
	4-F-Trp	73	60	60	Pratt and Ho (1975)
	5-F-Trp	50–60	20	10	Pratt and Ho (1975)
	6-F-Trp	50–60	50	10	Pratt and Ho (1975)

Protein	Analog				Reference
Gene 5 protein	3-F-Tyr	—	—	normal DNA binding	Coleman et al. (1976)
Gene 8 protein	3-F-Tyr	~50	100	normal infectivity	Hagen et al. (1978)
	4-F-Phe	~10	100	infectivity	Dettman, Weiner, and Sykes (unpublished)
	3-F-Phe	—	10	—	Dettman, Weiner, and Sykes (unpublished)
Histidine binding protein J	5-F-Trp	80	62	normal binding	Robertson et al. (1977)
D(−)lactate dehydrogenase	4-F-Trp	~70	—	7200	Pratt and Ho (1975)
	5-F-Trp	~50	—	50	Pratt and Ho (1975)
	6-F-Trp	~50	—	50	Pratt and Ho (1975)
Lac permease	4-F-Trp	60	—	35	Pratt and Ho (1975)
	5-F-Trp	~50	—	10	Pratt and Ho (1975)
	6-F-Trp	~50	—	100% then falls to 35%	Pratt and Ho (1975)
Lac repressor	3-F-Tyr	90	50	30	Lu et al. (1976)
Major lipoprotein	3-F-Tyr	100	240	—	Lee et al. (1977)

detect and destroy aberrant proteins and Goldberg *et al.* (1975) has observed a 50% increase in the rate of protein degradation in cells exposed to 4-F-Trp. Mutants which lack at least one of these proteases (the deg protein) are available (Goldberg *et al.*, 1975) and it may prove fruitful to utilize these strains when attempting to incorporate fluoroamino acids. Decreased protein yields may also reflect the fact that the analogs are being incorporated into the enzymes responsible for protein biosynthesis and such modified enzymes may have reduced catalytic ability.

3. ¹⁹F NMR SPECTRA OF FLUORINE-LABELED PROTEINS

3.1. ¹⁹F Chemical Shifts

The large range of ^{19}F chemical shifts and in particular the sensitivity of ^{19}F chemical shifts to environment has led in the proteins studied to well-resolved spectra for the individual amino acid residues of a given type (i.e., tyrosine) which have been substituted by their fluorinated analog (i.e., 3-F-Tyr). For example, the 94.1 MHz ^{19}F NMR spectrum of 3-F-Tyr alkaline phosphatase, which contains 11 tyrosines per monomer and has a total MW ~86,000, has resolved resonances for all 11 fluorotyrosyl residues (see Figure 2). The corresponding ^1H NMR spectrum shows a featureless envelope for the aromatic region. All of the other proteins from Table I show similarly well-resolved spectra. Possibly the best resolved is the ^{19}F NMR spectrum of 6-F-Trp dihydrofolate reductase, recorded either at 94.1 or 254 MHz (Figure 3).

From a biochemical standpoint, the major problem is the interpretation of the range of chemical shifts occurring for a given type of substituted amino acid in a protein. No detailed theory is available which allows us to understand the chemical shifts in terms of the microenvironments experienced by individual fluorines at particular sites in the protein (see Kimber *et al.*, 1977, for discussion of theory of ^{19}F chemical shifts). From studies of fluorinated compounds in solvents of varying polarity (Robertson *et al*, 1977), downfield shifts have generally been taken as indicative of the amino acid residue being situated in a hydrophobic (buried) environment. This empirical conclusion has been supported by the studies of fluorinated proteins. Detailed analysis of the relaxation of the fluorines of fluorotyrosine alkaline phosphatase (Hull and Sykes, 1976) yielded a parameter indicated as Σr^{-6} which was found to be correlated with the chemical shifts observed in the spectrum. This parameter expresses the weighted average distance to neighboring protons; that is, a larger number indicates that more and/or nearer protons surround the fluorine in the protein. The F-Tyr resonances which had chemical shifts near that of the free amino acid were found to make little contact with neighboring protons appropriate for a residue near the surface of the protein. Resonances which were shifted well downfield of the free amino acid, on the other hand,

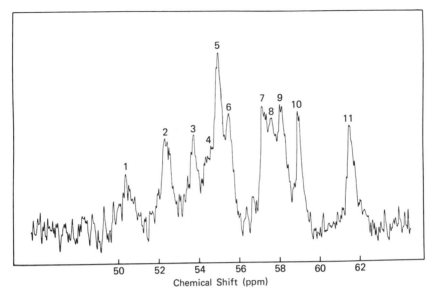

Figure 2. ¹⁹F NMR spectrum of native fluorotyrosine alkaline phosphatase at 94.16 MHz. The enzyme concentration was approximately 25 mg/ml in 0.5 *M* Tris-D₂O buffer (pH 7.85), 25–30°C. This represents a concentration of about 0.6 mM for each fluorotyrosine. The chemical shift scale is in parts per million upfield of a capillary of trifluoroacetic acid. This Fourier transform spectrum was obtained with a spectral width of 2500 Hz, acquisition time of 0.2 s, and a pulse delay of 0.2 s. The number of transients per block was 128, and 520 blocks were averaged with a sensitivity enhancement time constant of 0.1 s. Total time was about 7.5 h. From Hull and Sykes (1974).

were near more protons on the average. A second approach involved recording the ¹⁹F NMR spectrum in H₂O versus D₂O as a solvent. Then the solvent induced isotope shift (SIS) of resonance position (Lauterbur *et al.,* 1978) was determined for each resonance and compared with the chemical shift. It was found that the resonances with chemical shift near that of the free amino acid experienced the full SIS appropriate for a residue exposed on the surface of the protein. Correspondingly, the downfield-shifted resonances experienced little or no SIS, as is appropriate for residues buried within the protein and not interacting with the solvent.

Similar conclusions about the influence of environment on the ¹⁹F chemical shifts of individual resonances have also been drawn for the 3-F-Tyr *fd* gene 5 protein (Anderson *et al.,* 1975). For this protein, various chemical and biochemical evidence such as nitration of tyrosines indicate two buried and three exposed tyrosines. The ¹⁹F NMR spectrum is grouped with two downfield shifted resonances and three resonances near the position of the free amino acid (Anderson *et al.,* 1975).

All of the above notwithstanding, caution is advised in the interpretation of ¹⁹F chemical shifts. Indeed, the F-Tyr alkaline phosphatase data upon

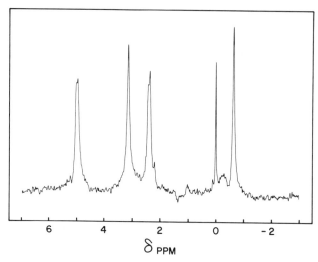

Figure 3. The proton noise-decoupled ¹⁹F NMR spectrum at 254 MHz of a 1 m*M* solution of 6-F-Tyr-labeled dihydrofolate reductase in the presence of 1 equivalent of methotrexate and 1 equivalent of NADPH. The sample was examined as a D_2O solution containing 50 m*M* potassium phosphate, 500 m*M* potassium chloride, pH 6.5 (meter reading); sample temperature was 28°C. The sharp signal at zero frequency arises from a small amount of denatured enzyme which appears on standing. The spectrum was obtained with a 90° pulse angle (10 μs), 8000 points, sweep width = ±4000Hz, 0.5 s acquisition time, repetition time = 2 s, number of scans = 21,400, and a 2.5 Hz line broadening. From Kimber *et al.* (1978).

which much of the above is based is not perfect, with significant scatter in several individual points, and exceptions or conflicts with the general picture of downfield shifts indicating buriedness have been reported in the literature. For example, the single tyrosine of the *E. coli* acyl carrier protein is not susceptible to nitration, implying buriedness, but the ¹⁹F NMR chemical shift of the F-Tyr acyl carrier protein is similar to that of the free amino acid (Gally *et al.*, 1978). Also the two fluorotyrosines of the labeled M13 gene 8 protein have been shown to be buried from solvent when the protein is incorporated into synthetic phospholipid vesicles (Hagen *et al.*, 1979a) and yet neither F-Tyr has a chemical shift significantly different from the free amino acid (Hagen *et al.*, 1979b).

A final point relative to the interpretation of ¹⁹F chemical shifts for 3-F-Tyr is that the chemical shift change upon titrating the free amino acid or the denatured protein is small (~0.5 ppm) while the chemical shift range observed for fluorotyrosine alkaline phosphate is ~11 ppm. Therefore, different pK_a's for individual fluorotyrosines, and correspondingly different contributions of the protonated and unprotonated forms at a given pH, cannot explain the chemical shift differences observed.

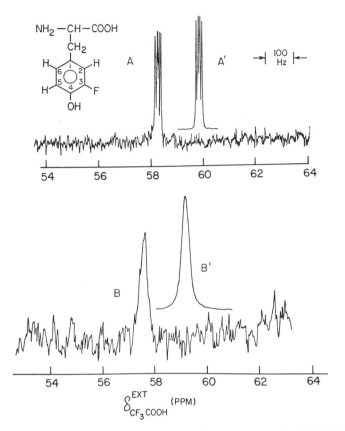

Figure 4. (A) 94.1 MHz ¹⁹F NMR spectrum of 1.7×10^{-2}m m-fluoro-DL-Tyr in 0.3 M Tris·HC1 buffer (in D_2O), pD 9.9; 256 transients, acquisition time = 0.8 s. (A') Simulated ¹⁹F NMR spectrum of m-fluoro-DL-Tyr, $^3J(F_3,H_2)$ = 13 Hz, $^1J(F_3,H_5)$ = 8 Hz, $\Delta\nu$ =3 Hz. (B) ¹⁹F NMR spectrum of 6 mg/ml of fluorotyrosine alkaline phosphatase in 6 M guanidine·HC1, pD = 7.3; 8633 transients, acquisition time = 0.3 s, sensitivity enhancement filter function = 0.1 s. (B') Simulated spectrum of denatured fluorotyrosine alkaline phosphatase, parameters are the same as (A') except that $\Delta\nu$ = 20 Hz. From Sykes et al. (1974). Chemical shift standard same as Figure 2.

3.2. ¹⁹F–¹H Spin–Spin Coupling

The spin–spin coupling to protons seen in the spectrum of the individual fluorinated amino acid (Figure 4, spectrum A) is not observed in the spectrum of the protein (Figure 4, spectrum B). This is because the widths of the individual lines are larger in the protein and the coupling is obscured. The observed linewidth will have some contribution, however, from the spread in line positions due to ¹⁹F–¹H spin–spin coupling. Experience with small-molecule NMR would lead one in this situation to attempt proton decoupling to narrow the observed lineshape. In all cases observed to date, this has led to a large decrease in the intensity of the observed resonances due to a negative nuclear Overhauser enhancement (NOE) which is discussed below. The alternative is to use gated decoupling with the proton decoupler gated on during the acquisition of the free induction decay [equivalent to acquisition of spectrum, see Sykes and Hull (1978) for details of Fourier transform NMR], to collapse the proton–fluorine spin–spin splitting, and gated off during a time period inserted between successive free induction decays to allow the fluorine spin system to return to its equilibrium intensity before each pulse. This technique was used for the spectrum presented in Figure 3 and allowed the observation of an unexpected fluorine–fluorine homonuclear spin–spin coupling which has been taken as indicative of two of the fluorotryptophans being stacked in dihydrofolate reductase (Kimber *et al.*, 1977, 1978). However, the technique is time consuming since the extra time period must be of a length equal to several times the spin-lattice relaxation time T_1 of the resonance. If the T_1 and the acquisition time are of the same order ($\sim$0.3 s, Sykes and Hull, 1978), then the time required to acquire the spectrum is considerably lengthened. Also the decrease in observed linewidth may not be great enough to warrant the increased time, since the natural linewidth could easily be as large as 50–100 Hz, which is much greater than the observed coupling constants ($J \sim$ 8–10 Hz).

3.3. $^{19}F-\{^1H\}$ Nuclear Overhauser Enhancement

The nuclear Overhauser enhancement (NOE) is the change in the intensity of the fluorine resonance when the resonances of nearby protons are saturated (as in a decoupling experiment). The NOE is defined as NOE = $(A_{obs} - A_0)/A_0$ where A_{obs} is the area of the ^{19}F resonance in the presence of irradiation at the ^1H NMR frequency and A_0 the area in the absence of this irradiation. There are two major differences between small molecules and proteins with respect to this experiment. The first is that for small molecules the $^{19}F-\{^1H\}$ NOE is positive, and can be as large as +0.5, meaning that upon saturation of the nearby protons the fluorine signal intensity can be increased by as much as 50%. When this occurs concomitant with the collapse of the spin–spin multiplet, the height of the resonance can be very much increased. However, for slowly tumbling proteins with rigidly affixed amino acids, the NOE approaches −1.0; that is, complete *disappearance* of the observed signal in the presence of proton irradiation (see Figure 5) (Sykes *et al.*, 1974; Hull and Sykes, 1975a, b). The criterion of slowly tumbling is defined relative to the frequency of the ^{19}F NMR spectrometer: $(\tau_{rot}\omega_0)^2 > 1$, where τ_{rot} is the rotational tumbling time of the protein and ω_0 the spectrometer frequency. Even for a protein as small as MW $\sim$6000, τ_{rot}

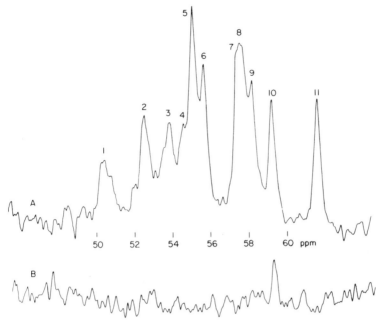

Figure 5. Experimental observation of the ¹⁹F NOE at 94 MHz in fluorotyrosine alkaline phosphatase. The sample contained 0.28 mM protein in 0.01 M Tris [pH 7.8, 0.1 M NaCl (D$_2$O)]. The ¹⁹F NMR spectrum (A) is the control; (B) is in the presence of continuous broad-band decoupling of ¹H. From Hull and Sykes (1976). Chemical shift standard same as Figure 2.

$\sim 2 \times 10^{-9}$ s and $(\tau_{\text{rot}}\omega_0)^2 \sim 1.4$ for $\omega_0 = (2\pi)(94.1 \times 10^6)$ rad s^{-1}. However, the individual amino acid can rotate faster than the protein as a whole by internal motions within the protein. If the internal motions are over a significant range and fast enough, the amino acid will appear as a small molecule and the NOE will be less negative or even positive. The NOE measurement can therefore serve as a very sensitive indicator of the rate of internal motions of amino acid residues in proteins (Hull and Sykes, 1975a) and has been used for example to observe the decreased mobility of surface fluorotyrosines of the *fd* gene 5 protein upon interaction with oligonucleotides (Coleman and Armitage, 1977) and to measure the mobility of fluorotyrosines in the membrane-bound M13 gene 8 protein (Hagen *et al.*, 1978). The exact theoretical treatment (Hull and Sykes, 1975a, b) involves consideration of the rate of internal motion and the angle between the motion axis and the vector connecting the interacting proton and fluorine nuclei, as well as other relaxation mechanisms such as chemical shift anisotropy (see below).

3.4. Interpretation of ^{19}F Linewidths and Spin-Lattice Relaxation Times

The relaxation times for fluorinated amino acids at lower fields (^{19}F NMR frequency $\lesssim 94.1$ MHz) can be adequately understood in terms of the dipolar relaxation of the fluorines by the neighboring protons (Hull and Sykes, 1974, 1975a, b). A difficulty of importance here is that relaxation may occur from protons other than those on the amino acid in question. In this case the geometry relative to the fluorine nucleus is not known. Therefore the dipolar relaxation can really only be used to set limits on the internal motion of the substituted amino acid and the degree of interaction with its neighbors. (These considerations also apply, but less critically, to the NOE measurements.) It can be safely assumed that interaction between fluorines on different tyrosine residues is not important, that relaxation due to dissolved oxygen is not significant (this is generally valid for $T_1 < 1$ s), and that the effect of solvent need not be considered if the experiments are performed in D_2O.

The analysis of the linewidth ($\Delta \nu = 1/\pi T_2$), spin-lattice relaxation time T_1, and NOE in terms of standard formulas has been treated in detail elsewhere (Hull and Sykes, 1974, 1975a, b; Doddrell $et\ al.$, 1972; Woessner, 1962) and those formulas will not be elaborated here. The parameters involved in the first-order relaxation calculation, including only the dominant nearby proton(s) on the amino acid residue, are: τ_{rot}, the overall rotational correlation time of the protein; τ_{int}, the correlation time(s) for internal motion(s); r, the distance to neighboring proton(s); and θ, the angle between the internuclear dipole–dipole vector(s) and the axis of internal rotation. The parameters for the $\alpha\beta$ internal motion are shown below for fluorotyrosine. For this residue,

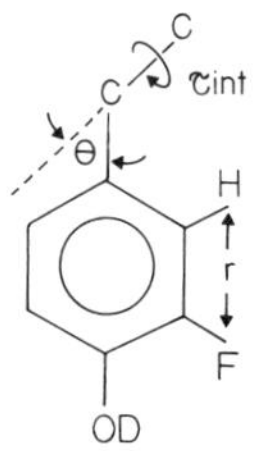

the β–γ rotation does not reorient the H–F internuclear vector and is therefore invisible in the measurements, which therefore become most sensitive to the $\alpha\beta$ internal motion. Coupled with the sensitivity of the chemical shift anisotropy linebroadening (present at higher fields, see below) to the $\beta\gamma$ motion, a complete analysis of the internal mobility of a given amino acid can be achieved. The analysis requires knowledge of τ_{rot} (which can generally be estimated from the Stokes–Einstein equation or obtained from other measurements such as fluorescence depolarization), T_1, $\Delta\nu$ as a function of ω_0, and the NOE. This has been accomplished for F-Tyr alkaline phosphatase (Hull and Sykes, 1974, 1975a) and for the F-Tyr gene 8 protein from the coliphage M13 when incorporated into synthetic phospholipid vesicles of known size (Hagen *et al.*, 1978).

The spin-lattice relaxation time can be measured using either the inversion recovery or progressive saturation methods (Hull and Sykes, 1975b; Sykes and Hull, 1978). The result is formally dependent upon whether or not the measurement is made in the presence of irradiation of the proton spin system. In general, the result is simple only in the presence of proton irradiation. However, the NOE as noted is negative and the measurement of the ¹⁹F T_1 in the presence of ¹H irradiation becomes impossible. For proteins, Hull and Sykes (1975b) have shown that the efficient cross relaxation within the proton spin system simplifies the interpretation of the measurement in the absence of proton decoupling. Opella *et al.* (1979) have verified this conclusion in practice and have shown that the value of T_1 determined by these two methods in the absence of ¹H irradiation is equivalent to the value of T_1 derived from the transient development of the NOE, a method which takes advantage of the negative NOE (Kuhlman and Grant, 1971; Opella *et al.*, 1976).

3.5. Higher Fields and Chemical Shift Anistropy

With the development of spectrometers operating at higher and higher magnetic field strengths, with corresponding greater sensitivity, it would seem obvious that studies of fluorinated proteins should be pushed to higher fields. However, a different linebroadening mechanism (chemical shift anisotropy, CSA) comes into play because of its dependence upon ω_0^2, and can actually lead to a *decrease* in resolution between individual resonances (whose separation is increasing proportional only to ω_0) at higher fields. This has been observed for F-Tyr alkaline phosphatase (Hull and Sykes, 1975a), F-Trp dihydrofolate reductase (Kimber *et al.*, 1978), and F-Tyr M13 gene 8 protein in phospholipid vesicles (Hagen *et al.*, 1978). Consequently some compromise between sensitivity and resolution must be reached. At the present moment, for proteins of moderate size (MW ~20,000–100,000), this compromise would seem to be in the range of 100–200 MHz. At higher fields the CSA linebroadening becomes too severe (Hull and Sykes, 1975a; Sykes and Hull, 1978).

The CSA relaxation mechanism involves the averaging of the orientation-dependent chemical shift of the ¹⁹F nucleus by the overall and internal motions of the labeled amino acid and can be used to provide additional information on the internal motions of the amino acid. For F-Tyr, the CSA contribution to the linewidth is very sensitive to the β–γ motion of the F-Tyr ring and has been used to quantitate this motion for the 11 tyrosines of alkaline phosphatase (Hull and Sykes, 1975a) and for the tyrosines of the M13 gene 8 protein in phospholipid vesicles (Hagen *et al.*, 1978). These calculations have an advantage over the corresponding interpretation of the dipole–dipole contribution to the relaxation times in that the geometry of the chemical shift anisotropy is fixed within the amino acid residue, and the sensitivity of the linewidth to the various internal motions can be calculated directly without worry about interresidue effects such as the influence of nearby protons from other amino acid residues.

3.6. Spectral Assignments

A great deal more specific information on the structure of a protein can be obtained when the ^{19}F NMR resonances can be *assigned* to individual amino acids within the protein. Apart from the trivial cases where only one fluorotyrosine is present, as in the acyl carrier protein (Gally *et al.*, 1978) or two are present but overlapping as in the M13 gene 8 protein (Hagen *et al.*, 1978, 1979, 1980), this problem has been tackled in only two proteins using two different approaches. Lu *et al.* (1976, 1979) have used elegant genetic techniques to assign the fluorotyrosine resonances of the fluorotyrosine lac repressor protein. In their experiments, mutants are used to replace one fluorotyrosine at a time at a known position in the amino acid sequence with another amino acid. The peak missing in the resultant spectrum is thereby assigned to the "deleted" fluorotyrosine. Matthews (1979), on the other hand, has attempted to assign the fluorotyrosine and fluorotyrptophan residues of the correspondingly labeled dihydrofolate reductases by rationalizing the chemical shifts of the individual peaks and their changes upon adding substrates and inhibitors in terms of the known X-ray structure of the protein and the expected correlation (discussed above) of the fluorine chemical shift with buriedness within the protein.

Neither of these approaches is perfect, however. The first approach suffers if fluorotyrosines other than the one "deleted" are perturbed by the substitution of a different amino acid. The second approach depends too heavily on the chemical shift criterion when the complete theoretical basis for the chemical shifts is not understood. In general, peak assignment remains the biggest problem to solve if one is to extract the maximum amount of information from nuclear magnetic resonance studies of fluorine-labeled proteins.

4. FUTURE TRENDS

The future applications of the approach discussed in this chapter depend upon the success of the biosynthesis of other fluoroamino acid containing proteins and upon advances in ^{19}F NMR technology which will permit the observation of the spectra of proteins available only in limited quantities. The rapidly evolving field of recombinant DNA technology is making it possible to construct *E. coli* strains harboring plasmids which greatly overproduce the protein(s) specified by the recombinant DNA insert (Clarke and

Carbon, 1976; Weiner *et al.*, 1978). The use of such strains will result in greatly improved protein yields and permit extension of the fluorine labeling approach to proteins normally available in only limited quantities. Further, one is not limited to bacterial proteins since recent success has been obtained in the isolation of proteins such as procollagen containing 4-fluoroproline from nonbacterial sources (Uitto and Prockop, 1977).

Potential sources of improved sensitivity for ^{19}F NMR are also at hand. Raising the spectrometer frequency should increase sensitivity but also results in a loss of resolution as discussed in Section 3.5. Increases in sample

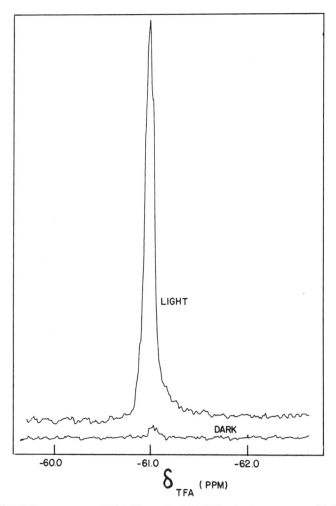

Figure 6. ^{19}F NMR spectrum at 254 MHz at 4 mM 3-F-Tyr in the presence of 0.1 mM flavin mononucleotide taken with and without irradiation of the sample with a 4 W argon ion laser. Chemical shifts from 10 mM TFA in D_2O, pH 7 (J.H. Baldo, D.S. Hagen, and B.D. Sykes, unpublished results).

size are also possible if larger amounts of the fluorine labeled protein are also available. However, Kaptein and coworkers (Kaptein *et al.*, 1978a, b) and Muszkat and coworkers (Muszkat and Gilon, 1978) have recently demonstrated a new laser CIDNP (chemically induced dynamic nuclear polarization) technique which greatly enhances the signal intensity of the ^{1}H NMR resonances of tyrosine, tryptophan, and histidine if these residues are exposed on the surface of the protein. We have extended this technique, which involves the addition of a flavin dye to the solution and irradiation of the solution with an argon ion laser, to fluorotyrosine and fluorotryptophan. The result of the laser irradiation for fluorotyrosine is shown in Figure 6; the irradiation produces an approximately 40-fold increase in intensity of the ^{19}F NMR resonance (J. H. Baldo, D. S. Hagen, and B. D. Sykes, unpublished results). This means that the same spectrum could be obtained in constant time for 40× less protein, or 1600× faster for the same concentration of protein (Sykes and Hull, 1978). This technique can thus be used in two ways; to increase sensitivity and to discriminate between buried and exposed residues.

ACKNOWLEDGMENTS

The authors would like to thank Dr. John H. Baldo and D. Scott Hagen for permission to quote the experimental results shown in Figure 6; and D. Scott Hagen for a very helpful reading and constructive criticisms of the manuscript. This work was supported by the MRC Group on Protein Structure and Function and MRC Grant MA-5838. JHW is a scholar of the MRC.

REFERENCES

Ames, G. F. (1964), *Arch. Biochem. Biophys.* **104**, 1.

Anderson, R. A., Y. Nakashima, and J. C. Coleman (1975), *Biochemistry* **14**, 907.

Ashton, H., and B. Capon (1971), *J. Chem. Soc., Chem. Commun.*, 513.

Atler, G. M., and J. A. Magnuson (1974), *Biochemistry* **13**, 4038.

Bittner, E. W., and J. T. Gerig (1970), *J. Amer. Chem. Soc.* **92**, 5001.

Bode, J., M. Blumenstein, and M. A. Raftery (1975), *Biochemistry* **14**, 1153.

Brown, K. D. (1970), *J. Bacteriol.* **104**, 177.

Brown, K. D. (1971), *J. Bacteriol.* **106**, 70.

Browne, D. T., G. L. Kenyon, and G. D. Hegeman (1970), *Biochem. Biophys. Res. Commun.* **39**, 13.

Browne, D. T., and J. D. Otvos (1976), *Biochem. Biophys. Res. Commun.* **68**, 907.

Calender, R., and P. Berg (1966), *Biochemistry* **5**, 1690.

Calvo, J. M., M. Freundlich, and H. E. Umbarger (1969), *J. Bacteriol.* **97**, 1272.

Campell, I. D., C. M. Dobson, R. J. P. Williams, and A. V. Xavier (1973a), *J. Magn. Reson.* **11**, 172.

Campell, I. D., C. M. Dobson, R. J. P. Williams, and A. V. Xavier (1973b), *Ann. N. Y. Acad. Sci.* **222**, 163.

Clarke, L., and J. Carbon (1976), *Cell* **9**, 91.

Cohen, G. N., and R. Munier (1959), *Biochem. Biophys. Acta* **31**, 347.

Coleman, J. E., R. A. Anderson, R. G. Ratcliffe, and I. M. Armitage (1976), *Biochemistry* **15**, 5419.

Coleman, J. F., and I. M. Armitage (1977), *NMR in Biology*, R. A. Dwek, I. D. Campbell, R. E. Richards, and R. J. P. Williams, eds., Academic Press, London, p. 171.

Doddrell, D., V. Glushko, and A. Allerhand (1972), *J. Chem. Phys.* **56**, 3683.

Dunn, T. F., and F. R. Leach (1967), *J. Biol. Chem.* **242**, 2693.

Dwek, R. A. (1971), *Carbon–Fluorine Compounds: Chemistry, Biochemistry and Biological Activity*, Ciba Foundation Symp., 13–15 September 1971, Associated Scientific Publishers, Amsterdam, p. 239.

Fenster, E. D., and H. S. Anker (1969), *Biochemistry* **8**, 269.

Freundlich, M. (1967), *Science* **157**, 823.

Freundlich, M., and J. M. Trela (1969), *J. Bacteriol.* **99**, 101.

Gally, H. V., A. K. Spencer, I. M. Armitage, J. H. Prestegard, J. E. Cronan, Jr. (1978), *Biochemistry* **17**, 5377.

Gammon, K. L., S. H. Smallcombe, and J. H. Richards (1972), *J. Am. Chem. Soc.* **94**, 4573.

Goldberg, A. L., K. Olden, and W. F. Prouty (1975), in *Intracellular Protein Turnover*, R. T. Schimke, and N. Katunuma, Eds., Academic Press, New York, NY.

Hagen, D. S., J. S. Weiner, and B. D. Sykes (1978), *Biochemistry* **17**, 3860.

Hagen, D. S., J. S. Weiner, and B. D. Sykes (1979), *Biochemistry* **18**, 2007.

Hagen, D. S., J. S. Weiner, and B. D. Sykes (1980), in *NMR and Biochemistry*, S. J. Opella, and P. Lu, Eds., Marcel Dekker, New York, NY, p. 51.

Hennecke, H., and A. Böch (1975), *Eur. J. Biochem.* **55**, 431.

Huestis, W. H., and M. A. Raftery (1971), *Biochemistry* **10**, 1181.

Huestis, W. H., and M. A. Raftery (1973), *Biochemistry* **12**, 2531.

Hull, W. E., and B. D. Sykes (1974), *Biochemistry* **13**, 3431.

Hull, W. E., and B. D. Sykes (1975a), *J. Mol. Biol.* **98**, 121.

Hull, W. E., and B. D. Sykes (1975b), *J. Chem. Phys.* **63**, 867.

Hull, W. E., and B. D. Sykes (1976), *Biochemistry* **15**, 1535.

Hunkapiller, M. W., and J. H. Richards (1972), *Biochemistry* **11**, 2829.

Kaptein, R., K. Dijkstra, F. Müller, C. G. Van Schagen, and A. J. W. G. Visser (1978a), *J. Magn. Reson.* **31**, 171.

Kaptein, R., K. Dijkstra, and K. Nicolay (1978b), *Nature* **274**, 293.

Kimber, B. J., I. Feeney, G. C. K. Roberts, B. Birdsall, D. V. Griffiths, A. S. V. Burgen, and B. D. Sykes (1978), *Nature* **271**, 184.

Kimber, B., D. V. Griffiths, B. Birdsall, R. W. King, P. Scudder, J. Feeney, G. C. K. Roberts, and A. S. V. Burgen (1977), *Biochemistry* **16**, 3492.

Kuhlman, K., and D. M. Grant (1971), *J. Chem. Phys.* **55**, 2998.

Lauterbur, P. C., B. V. Kaufman, and M. K. Crawford (1978), in *Biomolecular Structure and Function*, P. F. Agris, R. N. Loeppky, and B. D. Sykes, Eds., Academic Press, New York, NY, p. 329.

Lee, N., M. Inouye, and P. C. Lauterbur (1977), *Biochem. Biophys. Res. Commun.* **78**, 1211.

Lu, P., M. C. Jarema, H. R. Rackwitz, R. L. Friedman, and J. H. Miller (1980), in *NMR and Biochemistry*, S. J. Opella, and P. Lu, Eds., Marcel Dekker, New York, NY, p. 59.

Lu, P., M. Jarema, K. Mosser, and W. E. Daniel, Jr. (1976), *Proc. Natl. Acad. Sci. U.S.A.* **73**, 3471.

Maas, W. K. (1961), *Cold Spring Harbor Symp. Quant. Biol.* **26**, 183.

Marquis, R. E. (1970), in *Handbook of Experimental Pharmacology*, Vol. XX/2, Springer-Verlag, Berlin, Germany.

Matthews, D. A. (1979), *Biochemistry* **18**, 1602.

Muszkat, K. A., and C. Gilon (1978), *Nature* **271**, 685.

Munier, R., and G. Sarrazin (1964), *Compt. Rend.* **259**, 677.

Opella, S. J., R. A. Friedman, M. C. Jarema, and P. Lu (1979), *J. Mag. Res.* **36**, 81.

Opella, S. J., D. J. Nelson, and O. Jardetsky (1976), *J. Chem. Phys.* **64**, 2533.

Paselk, R. A., and D. C. Levy (1974), *Biochemistry* **13**, 3340.

Patt, S. L., and B. D. Sykes (1972), *J. Chem. Phys.* **56**, 3182.

Pauley, J., W. W. Fredricks, and O. H. Smith (1978), *J. Bacteriol.* **136**, 219.

Pratt, E. A., and C. Ho (1975), *Biochemistry* **14**, 3035.

Previc, E., and S. Binkley (1964), *Biochem. Biophys. Res. Commun.* **16**, 162.

Putter, I., A. Baretto, J. L. Markley, and O. Jardetzky (1969), *Proc. Natl. Acad. Sci. U.S.A.* **64**, 1396.

Raftery, M. A., W. H. Huestis, and F. Millet (1971), *Cold Spring Harbor Symp. Quant. Biol.* **36**, 541.

Robertson, D. E., P. A. Kroon, and C. Ho (1977), *Biochemistry* **16**, 1443.

Robillard, K. A., and A. Wishnia (1972), *Biochemistry* **11**, 3841.

Richmond, M. H. (1960), *Biochem. J.* **77**, 121.

Richmond, M. H. (1962), *Bacteriol. Rev.* **26**, 398.

Richmond, M. H. (1963), *J. Mol. Biol.* **6**, 284.

Sanger, F., F. M. Air, B. G. Barrell, N. L. Brown, A. R. Coulson, J. C. Fiddes, C. A. Hutchison, P. M. Slocombe, and M. Smith (1977), *Nature* **265**, 687.

Santi, D. V., and P. V. Danenberg (1971), *Biochemistry* **10**, 4812.

Shio, I., and S. Sugimoto (1978), *J. Biochem.* **83**, 879.

Simpson, R. J., and B. E. Davidson (1976), *Eur. J. Biochem.* **70**, 509.

Smith, L. C., J. M. Ravel, S. R. Lax, and W. Shive (1964), *Arch. Biochem. Biophys.* **105**, 424.

Spotswood, T. M., J. M. Evans, and J. H. Richards (1967), *J. Am. Chem. Soc.* **89**, 5052.

Sykes, B. D. (1969), *J. Am. Chem. Soc.* **91**, 949.

Sykes, B. D., and W. E. Hull (1978), *Meth. Enzymol.* **49**, Part G, 270.

Sykes, B. D., H. I. Weingarten, and M. J. Schlesinger (1974), *Proc. Natl. Acad. Sci. U.S.A.* **71**, 469.

Tayler, P. W., J. Feeney, and A. S. V. Burgen (1971), *Biochemistry* **10**, 3866.

Uitto, J., and D. J. Prockop (1977), *Arch. Biochem. Biophys.* **181**, 293.

Weiner, J. H., E. Lohmeier, and A. Schryvers (1978), *Can. J. Biochem.* **56**, 611.

Woessner, D. E. (1962), *J. Chem. Phys.* **36**, 1.

Yoshida, A. (1960), *Biochem. Biophys. Acta* **41**, 98.

Zeffren, E. (1968), *Biochem. Biophys. Res. Commun.* **33**, 73.

Zeffren, E. (1970), *Arch. Biochem. Biophys.* **137**, 291.

Five

Structure of the Capsular Polysaccharide Antigens from *Haemophilus influenzae* and *Neisseria meningitidis* by ^{13}C NMR Spectroscopy

William Egan

Division of Bacterial Products
Bureau of Biologics
Food and Drug Administration
8800 Rockville Pike
Bethesda, MD 20205

1. INTRODUCTION

The presence of extracellular capsular polysaccharides on both gram-negative and gram-positive bacteria is related to their ability to cause invasive disease in humans, with septicemia generally noted as a common factor (Davis *et al.*, 1973; Mims, 1977); noncapsulated bacterial strains are usually nonvirulent.[1] The best explanation at present for the association of virulence with encapsulation is that capsules protect the bacterial organism against ingestion and digestion by host phagocytic cells (Davis *et al.*, 1973; Mims, 1977). The cell-wall components of gram-positive and gram-negative noncapsulated bacteria can initiate the complement[2] sequence leading to the elimination of these organisms (Schreiber *et al.*, 1978; Cooper and Morrison; 1978; Winkelstein and Tomasz, 1977, 1978). The capsular polysaccharide associated with pathogenic strains permits survival in the presence of complement; the mechanism whereby the capsule functions in this manner is not understood. In the presence of antibodies specific for these capsules, complement is activated and bacterial clearance readily takes place.

Polysaccharide capsules are not necessarily associated with virulence; of the hundreds of known encapsulated bacteria, only a few dozen are commonly associated with invasive disease in humans (Robbins, 1978; Mims, 1977; Davis *et al.*, 1973). The majority of encapsulated bacteria are either noninvasive by nature, cleared via the alternate complement pathway, or cleared without the aid of complement. The structural features and physical properties associated with, or common to, the few virulent organisms have yet to be elucidated.

Resistance and immunity to encapsulated pathogens is associated with the presence of serum antibodies specific for the polysaccharide capsule [for examples and references, see Robbins *et al.*, 1976]. From the time of the initial finding that capsule-derived polysaccharides could be immunogenic (Francis and Tillet, 1930), numerous attempts were made to actively immunize with purified polysaccharide preparations. The first unequivocal demonstration of the efficacy of a polysaccharide vaccine was provided by MacCleod and associates (1945).

Success with polysaccharide vaccines has, in many senses, been extraor-

---

[1]An introductory discussion of bacterial architecture may be found in, for example, Zinsser (1964).

[2]Complement is a major component of the body's immunological defense system. It consists of a set of proteins which, when activated, leads to the destruction and elimination of foreign entities [for reviews, see Fearon *et al.*, (1976) and Müller-Eberhardt (1975)].

dinary. Aside from being virtually nontoxic (millions of doses of polysaccharide vaccines have been administered worldwide without serious incident due directly to the polysaccharide) they are highly efficacious; single 25 to 50 μg doses per individual of the polysaccharide often furnish lifelong protection against the homologous organism. This success, however, has not been total. As examples, polysaccharide vaccines are generally not effective in children under 3 years of age, the group for whom they are often most needed, and it has not yet been possible to produce an efficacious vaccine against *Neisseria meningitidis* Group B organisms (Robbins, 1978).

The interactions of encapsulated organisms with the body's defense system (both in the presence and absence of antibodies directed against the capsule) and the interactions of isolated polysaccharide antigens with antibody-forming lymphocytes are only recently beginning to be understood [see, for example, the book edited by Bellanti and Dayton (1975)]. It is hoped that studies of capsule structure will furnish further insight into the fascinating area of immunological chemistry.

The development of nuclear magnetic resonance (NMR) spectroscopy, and in particular ^{13}C NMR spectroscopy, has greatly facilitated and stimulated structural studies of bacterial polysaccharides. This review concerns such studies. In particular, it deals with (1) general characteristics of the ^{13}C NMR spectra of sugars and polysaccharides, (2) structural studies of *Haemophilus influenzae* and *N. meningitidis* type-specific polysaccharides, (3) structural studies of *Escherichia coli* capsular polysaccharides that cross react with *N. meningitidis* and *H. influenzae* pathogen-derived capsules, and (5) studies of the microdynamic behavior of meningococcal Groups B and C capsular polysaccharides.

2. GENERAL FEATURES OF THE ^{13}C NMR SPECTRA OF SUGARS AND POLYSACCHARIDES; A FEW USEFUL CHEMICAL REACTIONS

The capsules from *N. meningitidis, H. influenzae,* and *E. coli* are polymers based on simple di- or trisaccharide repeating units. In a number of instances, the repeating units are linked by phosphoric diester groups; additionally, O-acetyl groups are often found. Structural analysis of a capsular polysaccharide involves determination of (1) the overall composition and stoichiometry, (2) the linear sequence and linkage sites, (3) the chirality of the anomeric carbon (if a reducing sugar is present), and (4) the nature and placement of appended groups if they are present.

The sugar components of a polysaccharide are generally determined by hydrolyzing the polymer and then analyzing for the released monosaccharides employing either an automated sugar analyzer (for neutral reducing sugars), an amino acid analyzer (for amino sugars), or a gas chromatograph (for all sugars); see Boykins and Liu (1980). Since hydrolysis of the polysaccharide simultaneously removes O- and N-acetyl groups, these must be quantitated separately, either by a chemical or spectral method. Phosphorus, when present, is readily determined either by chemical or spectroscopic methods.

Several review articles have dealt in detail with the ^{13}C NMR spectra of sugars and polysaccharides (Jennings and Smith, 1978; Nunez *et al.*, 1977; Perlin, 1976; Friebolin *et al.*, 1976). In this Section, therefore, we will only mention some of the more salient features of the spectra of these compounds.

^{13}C NMR spectra are generally recorded with broad-band ^{1}H decoupling, resulting in single resonances for distinct carbon atoms (coupling to ^{31}P, if present, can lead to splitting of the ^{13}C resonances; see later). The position of the ^{13}C NMR signals are relatively characteristic of carbon type: carboxyl, ester, and amide carbonyl carbon resonances are generally found between 170 and 180 parts per million (ppm)[3]; anomeric carbon resonances are generally found between 90 and 110 ppm; secondary hydroxyl group carbons (including those that are part of a ketal or acetal linkage) between 70 and 90 ppm; hydroxymethyl group carbons between 60 and 65 ppm, acetamido (instead of hydroxyl) substituted carbons between 50 and 60 ppm; deoxy (i.e., —CH$_2$—) carbons between 40 and 45 ppm; methyl group carbons from N-acetyl and O-acetyl groups between 24 and 25, and 22 and 23 ppm, respectively. In general, corresponding resonances from furanose sugars are to lower field of the pyranose sugars.

Substitution of the hydroxyl hydrogen atom with a carbon or phosphorus atom causes shifts of the contiguous and adjacent carbon atoms. In general, O-alkylation causes an approximately 5 to 10 ppm downfield shift of the contiguous carbon atom and a small (~2 ppm) upfield shift of the neighboring carbon atoms. O-acetylation or O-phosphorylation generally causes a ~3 ppm downfield shift of the contiguous carbon and a 1 to 3 ppm upfield shift of the neighboring carbon atoms.

Since many of the polysaccharides that are to be discussed in this Chapter liberate inorganic phosphate on hydrolysis, a short digression regarding the nature and characterization of phosphate esters is in order. (It can be noted at the outset that, to date, only phosphoric diesters have been observed in capsular polysaccharides.) Phsophoric triesters are not acidic and thus are easily characterized. Phosphoric mono- and diesters are distinguishable by titration, possessing, respectively, two and one ionizable protons. Phosphoric monoesters generally exhibit pK_a's of ~1 and ~6 and diesters ~1. The pK_a of a phosphate ester can be conveniently determined by ^{31}P NMR from a plot of chemical shift versus pH [see, for example, Cozzone and Jardetzky, (1976)]. A typical plot for a sugar monophosphate and sugar

---

[3]In this Review, chemical shifts are referenced to internal sodium 3-trimethylsilylpropionate-2,2,3,3-d_4 (TSP). In a mixture of water, acetone, TSP and tetramethylsilane (TMS), the TSP resonance appears 1.6 ppm to higher field of TMS; we have employed this conversion factor in transforming literature data referenced to external TMS.

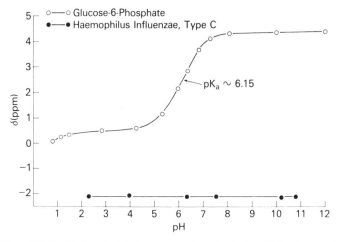

Figure 1. Plot of ^{31}P NMR chemical shifts as a function of solution pH for glucose 6-phosphate (o—o—o) and the *H. influenzae*, type c capsular polysaccharide (•—•–•); the chemical shifts are referenced to an internal capillary containing 25% H_2PO_4. The titration curve was carried out in a H_2O/D_2O mixture (95:5) at a temperature of ~25°C. The pK_a for glucose 6-phosphate is indicated. The ^{31}P NMP spectra were determined at 40.25 MHz.

phosphodiester is shown in Figure 1. Large changes in ^{31}P chemical shift are observed for the phosphate monoester in the pH 4 to 7 region; in contrast, no changes are observed for the diester in this region (Figure 1).

Phosphate mono- and diesters may be distinguished enzymatically through their susceptibility to alkaline phosphatase catalyzed hydrolysis; this enzyme is specific for monoesters. The liberation of inorganic phosphate following the addition of alkaline phosphatase to a solution of the polysaccharide may be taken as evidence for the presence of phosphate monoesters (providing the polysaccharide is stable to the reaction conditions in the absence of the active enzyme). Chemically, mono- and diesters generally differ dramatically in their susceptibility to acid- or base-catalyzed hydrolysis; diesters hydrolyze rapidly whereas monoesters hydrolyze slowly. However, the interpretation of hydrolysis rate constants is very difficult and hydrolytic stability should not be advanced as the sole argument for the degree of phosphate esterification (see, for example, Hudson, 1965). ^{31}P NMR, because it distinguishes chemically distinct species, can be used to demonstrate the stepwise nature (or nonstepwise nature) of phosphate hydrolyses and thereby furnish evidence for the degree of phosphate esterification in the intact polysaccharide. This is illustrated in Figure 2.

One final note with regard to ^{31}P concerns the coupling of phosphorus to carbon. In phosphodiesters, ^{31}P coupling is usually observed at the contiguous carbon atom; this is a coupling through two bonds and its magnitude is generally between 5 and 10 Hz. Three-bond P—O—C—C couplings are also commonly observed. The range of values for this coupling is approximately

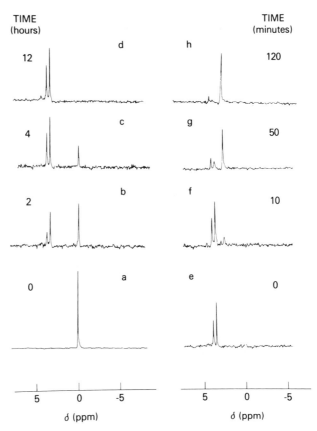

Figure 2. ^{31}P NMR spectra (40.25 MHz) of the base-catalyzed hydrolysis of the *E. coli* K100 capsular polysaccharide as a function of time (*a–d*) and subsequent alkaline phosphatase catalyzed hydrolysis (*e–h*). A 0.05 *M* solution of NaOH was utilized in *a–d*; the alkaline phosphatase reactions was carried out at pH 10.0. Both reactions were carried out at 25°C. The spectra in *b–d* indicate that a migration of phosphate is accompanying the initial hydrolysis.

2–10 Hz, the exact value depending, *inter alia*, on the dihedral angle (Bundle *et al.*, 1974); for a dihedral angle of ~180°, a coupling of approximately 8–10 Hz is expected while, for a *gauche* disposition (dihedral angle of approximately 60°), a coupling of approximately 2 Hz is expected.

^{13}C NMR spectra of polysaccharides are usually not recorded under conditions which lead to quantitative results as one is generally more interested in obtaining resonance positions than in integrated absorption intensities. It is generally differences in spin-lattice relaxation times and nuclear Overhauser effect (NOE) enhancements between the various kinds of carbons in the polysaccharide repeating unit that lead to nonquantitative effects. (The effect of being off-resonance is generally insignificant, at least for a $\frac{1}{2}\pi$ pulse.) When quantitative results are needed, the Overhauser enhancement is generally suppressed by a ^{1}H gating technique, and a long delay period between pulses is used. Quantitative aspects of ^{13}C NMR are discussed in Shaw (1976) and Wehrli and Wirthlin (1978).

Assignments of sugar resonances in monomers are generally made on the basis of (1) empirical correlations; (2) nondecoupled or off-resonance decoupled spectra; (3) specific isotopic substitution; and (4) specific ^{1}H proton decoupling (Jennings and Smith, 1978; Wehrli and Wirthlin, 1978; Perlin, 1976). Many other techniques are applicable to ^{13}C assignment; the above-mentioned four are among the most common. Other valuable techniques include (1) perturbations in chemical shifts brought about by changes in ionization state, (2) perturbations in shift due to isotopic substitution at a neighboring atom, usually replacing O—H with O—D, (3) use of lanthanide ion induced shifts and broadenings, (4) specific heteronuclear (not ^{1}H) scalar couplings. Among the "newer" techniques, we might single out (1) use of two-dimensional Fourier transformation [reviewed recently by Freeman and Morris (1979)], (2) use of $T_{1\rho}$ relaxation methods to locate carbon atoms bearing quadrupolar nuclei, and (3) selective excitation (Bodenhausen *et al.*, 1976). Assignments in polysaccharides are generally made from comparison with monomers and related sugars as well as a knowledge of the effects of substitution on chemical shift. It should be emphasized at the outset that structural determinations of capsular polysaccharides do not require the assignment of all resonances; in most instances, only a few resonances need be assigned.

In addition to magnetic resonance spectroscopy, several chemical reactions have proved extremely useful in structural elucidations. The reaction of most concern in this Review is periodate oxidation. Periodate is specific for vicinal hydroxyl groups and the number of equivalents of periodate consumed is equal to the number of C(OH)—C(OH) groupings present. NH_2 groupings may take the place of hydroxyl groups; in these cases, however, the reaction is generally slower. A C(OH)—CH_2OH grouping will liberate formaldehyde on periodate oxidation and a C(OH)—CH(OH)—C(OH) grouping will liberate formic acid on periodate oxidation. The periodate method is discussed in Sharon (1975).

In addition to periodate oxidations, many investigators have relied extensively on permethylation analyses. The analysis involves methylation of all hydroxyl groups in the polymer, hydrolysis of the polymer, and analysis of the resulting methylated saccharides, usually by gas chromatography and mass spectroscopy (Lindberg, 1972). We will not have occasion to use the permethylation analysis in this Review, except incidentally. Methylation analysis is usually indispensable in the analysis of branched polysaccharides. Permethylation methods applicable to phosphorylated polysaccharides have not yet been devised.

3. *HAEMOPHILUS INFLUENZAE*

3.1. Introduction

Bacteria of the genus *Haemophilus* (blood-loving), species *influenzae* (of the influenza), are indigenous to the upper respiratory tract of man, the only known natural host for this organism. It is a small, gram-negative bacillus, often difficult to stain. Its colonial morphology is variable, tending to be cocco-bacillary under optimal culture conditions, otherwise markedly pleomorphic. Many *H. influenzae* strains display a characteristic irridescence when obliquely lighted and this is due to capsular polysaccharide. It was originally noted by Pittman (1931) that invasive *H. influenzae* disease in humans was caused by encapsulated organisms, that there were six encapsulated serotypes (designated by the letters "a" through "f"), and that virtually all serious invasive disease was due to type b.

   H. influenzae type b is the etiologic agent of a variety of invasive diseases in children, most commonly meningitis (Smith and Robbins, 1974). In the United States, for example, the annual incidence of type b caused meningitis is estimated to be approximately 15,000 cases (Parke *et al.*, 1972; Fraser *et al.*, 1974). Although antimicrobial treatment has reduced the mortality from *H. influenzae* type b meningitis from nearly 100% to less than 10%, the morbidity in "cured" patients remains high (Sell *et al.*, 1972). This, coupled with the emergence of antibiotic-resistant strains (Elwell *et al.*, 1975; Ward *et al.*, 1978), stress the need for disease prevention.

   Human immunity to *H. influenzae* type b was first elucidated by Fothergill and Wright (1933) who showed that bactericidal antibodies ("bactericidal power of blood") present at birth, declined to nondetectable levels by age 3 months, reappeared by age 3 years, and rose to adult levels by age 8 years. The high incidence of type b meningitis in children between the age of 3 months and 3 years was associated with the lack of bactericidal antibodies; conversely, the presence of these antibodies implied protection. The age dependence of type b antibodies and the inverse relationship between antibody level and incidence of disease has been confirmed (Norden, 1974; Anderson *et al.*, 1972; Schneerson *et al.*, 1971; Alexander, 1943). These latter studies showed that most of the bactericidal antibodies were directed against the capsule.

   The relationship between type b capsule structure and type b virulence is, as yet, unknown. To increase understanding of the pathogenic role of the type b capsule, comparative studies of the other, nonvirulent encapsulated serotypes have been undertaken. Structural studies of the six serotypes constitute the subject matter of the remainder of this Section.

3.2. *H. influenzae*, type b

The capsular structure from the pathogenic type b organism was the first to be studied by NMR methods. The structure was elucidated by a group of

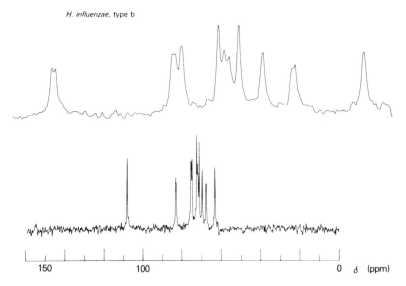

H. influenzae, type b

150   100   0   δ (ppm)

Figure 3. ^{13}C NMR spectrum (25.05 MHz) of the *H. influenzae* type b capsular polysaccharide in H_2O/D_2O (90:10) at pH 7.0 and 25°C. Shifts are in ppm relative to internal TSP. A solution of ∼50 mg of polysaccharide dissolved in 0.5 ml of water was placed in a 5 mm NMR tube and the spectrum recorded under continuous broad-band decoupling conditions. Approximately 25,000 transients were collected using a 90° pulse angle and a 2.5 s pulse repetition time.

workers from Eli Lilly and Company (Crisel *et al.*, 1975). Compositional analysis showed that the polymer was composed of equimolar amounts of sodium, phosphate, ribose, and ribitol; other sugar constituents were not detected.

The ^{13}C NMR spectrum of the polymer (Figure 3) revealed the presence of ten different carbon atoms; of these, five displayed observable scalar couplings to phosphorus. The NMR spectrum was thus in accord with the compositional analysis. Several of the resonances were readily assigned by comparison to model compounds and by use of gated decoupling techniques. The resonance at 109.4 ppm is in the anomeric carbon region and can be assigned to C-1 of ribose; by comparison of the resonance positions of the C-1 carbons in α and β O-methyl ribofuranosides, the chirality at C-1 in the type b polysaccharide is determined to be β. The resonance at 84.7 ppm is assignable to the ribose C-4 by comparison to model compounds. The 65.07 ppm resonance is a primary hydroxyl group carbon resonance and that at 69.5 ppm is a —CH_2O—P(O)(O^-)O— carbon (the resonances at 65.07 and 69.5 ppm were triplets under coupled conditions).

Mild hydrolysis with 0.1 *M* NaOH followed by alkaline phosphatase catalyzed hydrolysis resulted in a neutral ten-carbon disaccharide; the anomeric resonance in this disaccharide was unaltered in position and a new —CH_2OH resonance was observed in the ^{13}C NMR spectrum (Egan and

Tsui, unpublished findings). The emergence of a new hydroxymethyl group resonance and the observation of an unaltered anomeric resonance establishes the ribosyl—ribitol (1 → 1) linkage.

Crisel *et al.* (1975) showed that the polysaccharide consumed two equivalents of periodate and liberated formic acid, but not formaldehyde; this evidences a —CH(OH)—CH(OH)—CH(OH)— fragment. It was also shown that ribose was not degraded by periodate. The above fragment must therefore correspond to C-2, C-3, and C-4 of ribitol. The C-1 carbon atom of ribitol is linked to the C-1 carbon atom of ribose, and by elimination, the site of phosphate attachment to ribitol must be C-5. This is consistent with the above chemical and spectroscopic findings.

The C-1 carbon atom of ribose is linked to ribitol and the C-5 hydroxyl is unsubstituted. Therefore, phosphate must be linked to either the C-2 or C-3 of ribose (this is also in accord with the periodate oxidation studies). The presence of a scalar ^{31}P coupling to C-4 of ribose and the absence of a scalar coupling to C-1 establishes the site of phosphate linkage to ribose as C-3. The structure of the *H. influenzae* type b capsular polysaccharide is shown below. This structure or portions thereof have been confirmed by synthesis (Garegg *et al.*, 1977) and by spectroscopic techniques (Branefors-Helander *et al.*, 1976; Fraser *et al.*, 1979; Egan and Tsui, unpublished findings). The chirality of ribitol in the polymer remains to be unequivocally demonstrated, although the work of Branefors-Helander *et al.* (1976) indicate D-ribitol.

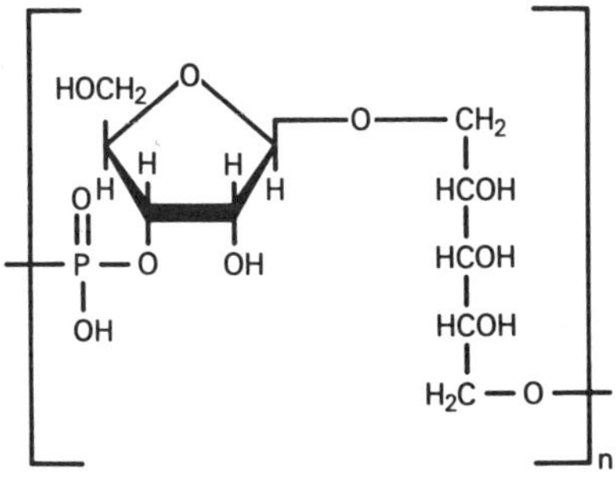

H. influenzae type b

^{13}C NMR spectral data for *H. influenzae* type b polysaccharide and model compounds are collected in Table I. With the exception of resonance assignments given above in the text, other assignments are tentative. Their correctness, or incorrectness, however, does not alter the already presented structural conclusions.

3.3. *H. influenzae*, type a

Shortly following the report of the type b capsular polysaccharide structure, the type a structure was communicated by the Swedish group—Branefors-

Table I. ^{13}C NMR chemical shifts (ppm) and $^{31}P-^{13}C$ coupling constants (Hz) for *Haemophilus influenzae*, type b capsular polysaccharide and model compounds[a]

Compound	C-1	C-2	C-3	C-4	C-5	C-1'	C-2'	C-3'	C-4'	C-5'
Ribitol						65.2	75.95	75.05	74.95	65.2
α-O-methyl-ribose	107.4	73.7	72.4	88.7	64.2					
β-O-methyl-ribose	110.6	76.9	73.5	85.5	65.5					
H. influenzae, type b	109.4	77.0 (4.9)	76.5 (3.1)	84.7 (6.7)	65.1	71.4	74.2	72.8	73.6 (7.9)	69.5 (6.1)

[a]Unprimed carbons refer to the ribose residue in the type b polysaccharide and primed carbons refer to the ribitol residue; phosphorus-carbon scalar coupling constants are given in parentheses.

207

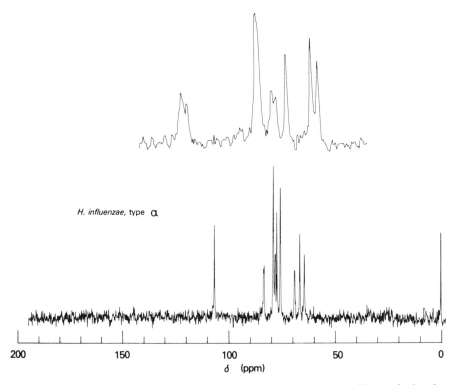

Figure 4. ^{13}C NMR spectrum of *H. influenzae* type a capsular polysaccharide acquired under conditions similar to those described in Figure 3.

Helander, Lindberg, Kenne, and coworkers (Branefors-Helander *et al.*, 1977). These workers showed that the type a polysaccharide was a phosphoric diester linked polymer of D-glucose and ribitol. The ^{13}C NMR spectrum was in accord with this composition; 11 resonances, accounting for 12 carbons, were discernible. Three signals showed measurable scalar couplings to ^{31}P. Spectral data for the type a polysaccharide and model compounds are collected in Table II; the ^{13}C NMR spectrum is shown in Figure 4.

The anomeric carbon resonance of glucose is observed at 104.9 ppm, establishing the chirality at C-1 as being β; see Table II. Two —CH$_2$OH groupings are observed at 63.4 and 65.5 ppm. A signal at 67.7 ppm shows coupling to ^{31}P and most likely derives from a —CH$_2$O—P(O)(O$^-$)O— fragment (see following); gated decoupling shows that this carbon bears two hydrogen atoms.

Hydrolysis of the polysaccharide with base and then alkaline phosphatase yielded a material that was identified as a 2-O-hexopyranosyl-pentitol by mass-spectroscopic analysis of the trimethylsilyl derivative (Branefors-Helander *et al.*, 1977). This disaccharide was then shown to be identical to

Table II. ^{13}C NMR chemical shifts (ppm) and $^{31}P-^{13}C$ coupling constants (Hz) for *Haemophilus influenzae*, type a capsular polysaccharide and model compounds[a]

Compound	Carbon										
	C-1	C-2	C-3	C-4	C-5	C-6	C-1'	C-2'	C-3'	C-4'	C-5'
Me-O-β-Glc	105.5	75.4	78.1	72.0	78.2	63.1	59.9				
4-O-β-Glc-ribitol	104.7	75.6	78.0	72.0	78.2	63.1	65.1	75.3	73.8	83.3	62.7
H. influenzae, type a	104.5	75.7	78.2	76.4 (6.0)	77.4	63.0	65.1	74.2	74.0	81.7 (6.5)	76.3 (5.0)
β-Glc-3-phosphate	98.0	75.7 (3.3)	84.6 (6.5)	71.1 (3.1)	77.8	63.1					

[a]Glc is to be equated with glucose or glucosyl. Scalar phosphorus to carbon coupling constants are given in parentheses.

authentic 4-O-β-D-glucopyranosyl-D-ribitol (comparison was made as the respective octa-acetates).

Periodate oxidation was performed and three equivalents were reported to be consumed; in addition, formaldehyde and formic acid were liberated. This periodate data indicated the presence of a —CH(OH)—CH(OH)—CH$_2$OH grouping, deriving from ribitol.

Following periodate oxidation, the material was reduced with sodium borohydride, hydrolyzed with acid, and reduced with sodium borodeuteride; the reaction products were then per-acetylated and examined by gas–liquid chromatography and mass spectroscopy. Analysis showed equivalent amounts of glycerol triacetate and erythritol tetraacetate, neither of which contained a deuterium label. In order to account for the production of erythritol from the oxidized and reduced material, phosphate must be linked to either O-4 of glucose or O-3 of ribitol in the untreated polymer. This is outlined in Scheme I on page 211. Phosphate attachment to O-3 of ribitol may be excluded by the observation of the formation of formaldehyde as well as by the ^{13}C NMR spectrum of the starting polymer.

It was concluded that the type a polysaccharide was composed of 4-O-β-D-glucopyranosyl-D-ribitol residues joined through phosphoric diester linkages between O-4 of glucose and O-5 of ribitol. The structure proposed by the Swedish group is presented below.

Haemophilus influenzae, type a

Our group at the Bureau of Biologics was also working on the structure of the type a capsule. Periodate oxidation experiments (Egan and Tsui, unpublished findings) revealed that (1) two equivalents of periodate were consumed (not three), (2) formaldehyde and formic acid were liberated, and (3) glucose was not consumed in the reaction. These results with periodate oxidation indicated that the site of phosphate attachment to glucose was C-3 and not the proposed C-4. The ^{13}C NMR spectrum of the starting polysaccharide was, however, strongly indicative of C-4 attachment. Thus, if phosphate were attached at C-3, the glucose C-3 and C-4 resonances would be expected to occur at about 81 and 70 ppm, respectively; resonances from glucose at these positions were not observed. The ^{13}C NMR spectrum was in accord with C-4 attachment; see Table II.

SCHEME I

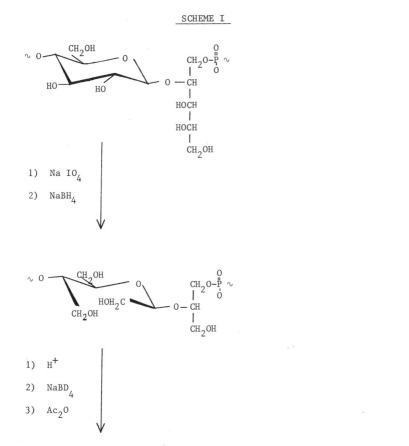

1) Na IO$_4$

2) NaBH$_4$

1) H$^+$

2) NaBD$_4$

3) Ac$_2$O

R = C-CH$_3$

The observation that two equivalents of periodate were consumed in the oxidation reaction was confirmed by Lindberg and Kenne (unpublished findings). However, Lindberg and Kenne went on to note that periodate oxidation, followed by sodium borohydride reduction, and reoxidation by periodate oxidation resulted in the uptake of three equivalents of periodate and destruction of glucose.

This raises the distinct possibility that a stable hemiacetal is formed (a possible structure is given below) following the initial oxidation of the ribitol moiety. The hemiacetal would be susceptible to sodium borohydride reduction and this reaction product would then be capable of consuming the third equivalent of periodate.

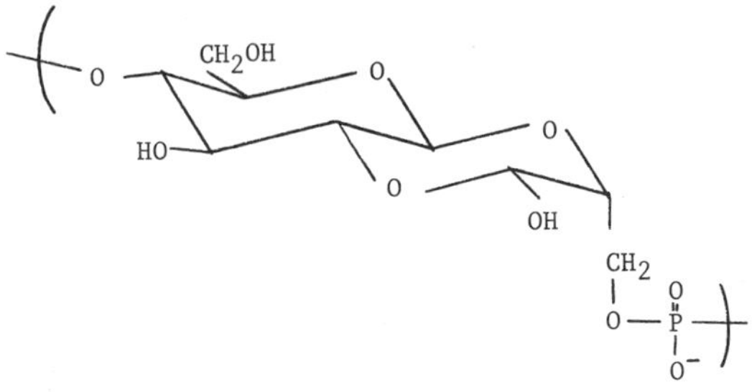

Haemophilus influenzae, type a hemiacetal

The structure of the type a polysaccharide is therefore that shown above. These results indicate that extreme caution must be placed in the over-reliance on periodate oxidation studies. It has not yet been determined whether glucose is linked to the pro-**R** or pro-**S** carbon atom of ribitol; we are currently pursuing this final aspect of the problem.

3.4. *H. influenzae,* type c

The type c structure has been elucidated by our group at the Bureau of Biologics and by the Swedish group in Stockholm; both studies are in agreement as to all structural details (Egan *et al.*, 1980a; Branefors-Helander *et al.*, 1979a).

Following acid hydrolysis, galactose, glucosamine, and phosphate were found to be present in equimolar ratios. The D configuration at C-1 for glucosamine and galactose was established by specific rotation and confirmed enzymatically. Analysis by the method of Hestrin for the presence of O-acetyl groups indicated that they were present, but not in an equimolar ratio.

The ^{13}C NMR spectrum of the type c polysaccharide is presented in Figure 5; spectral data from the type c polysaccharide and from model compounds are collected in Table III. Twelve major resonances, four of which exhibited ^{31}P–^{13}C scalar couplings, were distinguished in the 50–100 ppm region; additionally, the spectrum showed resonances characteristic of N- and O-acetyl groups.

The resonances at 24.8 and 177.1 ppm correspond to the respective C(O)$\underline{C}$H$_3$ and $\underline{C}$(O)CH$_3$ carbon atoms of the N-acetyl group; the resonances at 23.5 and 175.7 ppm correspond to the C(O)$\underline{C}$H$_3$ and $\underline{C}$(O)CH$_3$ carbon atoms of the O-acetyl group. The resonance at 56.83 ppm corresponds to the C-2 of 2-acetamido-2-deoxyglucose (N-acetyl glucosamine). The resonances at 63.56 and 62.91 ppm are characteristic of primary alcohol carbon atoms and correspond to the C-6 carbon atoms of N-acetyl glucosamine and galactose. The resonances at 104.58 and 98.6 ppm correspond to the anomeric carbon atoms of N-acetyl glucosamine and galactose.

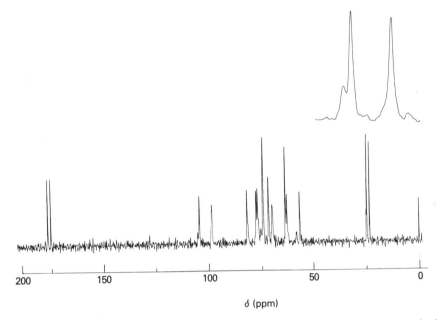

δ (ppm)

Figure 5. ^{13}C NMR spectrum of the *H. influenzae* type c capsular polysaccharide acquired under conditions similar to those described in Figure 3. The acetyl region is shown in expanded form.

A number of minor resonances were also discernible in the spectrum, the most telling of which was the shoulder at 25.1 ppm on the N-acetyl methyl carbon resonance. These minor resonances were shown to derive from repeating units that were not O-acetylated (approximately 80% of the repeating units were O-acetylated). The O-deacetylated polymer spectrum is shown in Figure 6. It can be noted in this spectrum that among the changes effected by complete O-deacetylation, the C-2 resonance of N-acetylglucosamine has shifted upfield by ~1.6 ppm; this establishes the site of O-acetylation as C-3 of this residue.

Dephosphorylation resulted in a disaccharide whose ^{13}C spectrum is shown in Figure 7. The N-acetyl glucosamine C-1 resonance was not altered by dephosphorylation, whereas the galactose anomeric linkage was altered (it was replaced by two resonances at 99.0 and 94.93 ppm, and the ^{31}P scalar coupling to the erstwhile C-1 resonance disappeared). Upon treatment of the disaccharide with sodium borohydride, these two, new anomeric resonances disappeared (see Figure 7). Sugar analysis of the borohydride reduced material revealed glucosamine but not galactose. Phosphate is therefore attached to C-1 of galactose.

The type c polymer did not consume periodate, even after prolonged treatment at room temperature; accordingly it contains no vicinal hydroxyl groups.

Table III. ^{13}C NMR chemical shifts (ppm) and ^{31}P–^{13}C coupling constants (Hz) for *H. influenzae*, type c capsular polysaccharide and related compounds[a]

Carbon	α-GlcNAc-4-P	β-GlcNAc-4-P	*H. influenzae,* type c	α-Gal	β-Gal
C-1	93.2	97.6	104.6		
C-2	56.5	59.1	56.8		
C-3	72.8(1.7)	76.0(1.9)	76.8		
C-4	76.7(5.9)	76.4(6.2)	74.1(7.0)		
C-5	73.7(6.1)	77.8(6.1)	77.4(4.9)		
C-6	63.1	63.3	63.6		
NC(O)CH$_3$	177.3	177.3	177.1		
NC(O)CH$_3$	24.9	25.1	24.8		
C-1′			98.6(6.7)	95.2	99.3
C-2′			69.7(8.5)	71.4	74.9
C-3′			81.8	72.2	75.8
C-4′			71.5	72.2	71.7
C-5′			74.3	73.3	77.9
C-6′			62.9	64.1	63.9
OC(O)CH$_3$			177.7		
OC(O)CH$_3$			23.4		

[a]GlcNAc-4-P is to be equated with 2-acetamido-2-deoxy-glucose 4-phosphate; Gal is to be equated with galactose. The primed carbon atome of the type c polysaccharide refer to the galactose residue; the unprimed carbon atoms refer to the GlcNAc residue. Phosphorus scalar couplings to carbon are given in parentheses.

Since the anomeric carbon of N-acetylglucosamine is linked to galactose, O-3 of N-acetylglucosamine is acetylated, and O-6 is unsubstituted, phosphate must be attached to O-4.

Galactose is linked at C-1 to N-acetylglucosamine via a phosphoric-diester group. Since the galactose C-6 hydroxyl is unsubstituted, and phosphate is attached to C-1, and the polymer does not consume periodate, N-acetyl glucosamine must be linked to the C-3 carbon of galactose.

The chirality at C-1 for both moities was established by observation of the H-1–H-2 scalar couplings, in conjunction with the Karplus relationship [for use of the ^{1}H NMR technique, see, for example, Lemieux and Stevens (1971)]. Glucosamine was shown to possess the β configuration and galactose the α configuration. The resulting structure of the type c *H. influenzae* polysaccharide is shown on page 216.

3.5. *H. influenzae*, type f

The structure of the type f capsule has been established by workers at the Bureau of Biologics (Egan *et al.*, 1980b) and the University of Stockholm

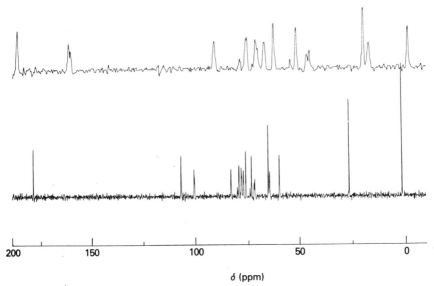

Figure 6. ¹³C NMR spectrum of the O-deacetylated *H. influenzae* type c capsular polysac-charide acquired under conditions similar to those described in Figure 3. The polymer was O-deacetylated with ammonium hydroxide at pH 10.5 at 20°C for 12 h. The region between approximately 50 and 110 ppm is shown on top in expanded form.

(Branefors-Helander *et al.*, 1979b). Compositionally, the polysaccharide consists of 2-acetamido-2-deoxy-galactose and phosphate in a molar ratio of 2:1. In addition, 1 equivalent of O-acetyl is present per 2 equivalents N-acetyl galactosamine. Galactose was present as the D enantiomer as determined by its optical rotation.

The ¹³C NMR spectrum of the type f polysaccharide is shown in Figure 8. Twelve resonances are observed in the region 50–110 ppm; additionally,

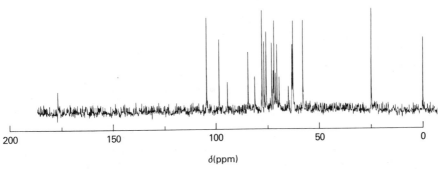

Figure 7. ¹³C NMR spectrum of the *H. influenzae* type c capsular polysaccharide following mild treatment with acid (0.5 M HCl) and alkaline phosphatase acquired under conditions similar to those described in Figure 3.

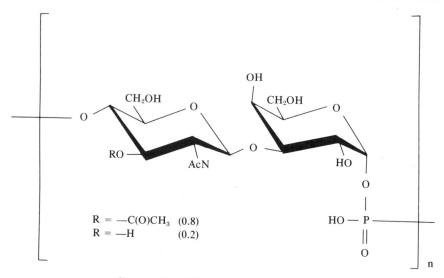

Haemophilus Influenzae type c Polysaccharide

three lie in the region 22–25 ppm and three in the region 175–177 ppm. ^{13}C NMR data for the type f polymer and model compounds are assembled in Table IV. The six resonances in the ranges 22–25 and 175–177 ppm are characteristic of two N-acetyl and one O-acetyl group. In conjunction with compositional data, the ^{13}C NMR spectrum provides evidence for a repeating unit in which two galactosyl groups are linked by a phosphoric diester; one of the galactosyl groups is O-acetylated.

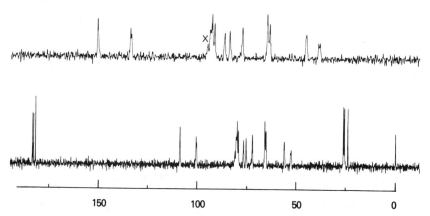

Figure 8. ^{13}C NMR spectrum of the capsular polysaccharide from *H. influenzae* type f acquired under conditions similar to those described in Figure 3. The upper tracing is an expansion of the region between approximately 50 and 85 ppm; an impurity (unidentified) is marked with an "x" in the expansion.

Table IV. ^{13}C NMR chemical shifts (ppm) and ^{31}P–^{13}C coupling constants (Hz) for the *H. influenzae* type f polysaccharide and 2-acetamido-2-deoxy-galactose[a]

Carbon Atom	β-GalNAc	α-GalNAc	*H. influenzae* type f
C-1	98.1		104.9
C-2	56.5		54.2 (5.0)
C-3	73.8		77.5 (6.5)
C-4	70.6		69.6
C-5	77.8		76.3
C-6	63.8		63.3
—NC(O)CH$_3$	177.7		177.0
—NC(O)CH$_3$	25.0		25.2
C-1		93.8	96.9 (6.1)
C-2		53.0	50.9 (6.8)
C-3		70.4	72.8
C-4		71.3	77.0
C-5		73.2	74.0
C-6		64.0	63.6
—NC(O)CH$_3$		177.0	176.5
—NC(O)CH$_3$		24.8	24.7
—OC(O)CH$_3$			175.6
—OC(O)CH$_3$			23.0

[a] GalNAc is to be equated with 2-acetamido-2-deoxygalactose. Phosphorus to carbon scalar couplings are given in parentheses.

The position of the anomeric resonances at 105 and 97 ppm suggests the presence of both the α and β configurations. The presence of the two anomers is further reflected in the resonance positions of the two respective C-2 atoms, especially subsequent to removal of the O-acetyl groups (see later. The resonances at 63.1 and 63.6 ppm are characteristic of the carbon atoms of a hydroxymethyl group and demonstrate that the O-6 of each N-acetyl galactosyl residue is unsubstituted (that at least one such residue contains an unsubstituted 6-hydroxyl group is shown by the failure of the polymer to consume periodate).

To ascertain the site of O-acetylation, the polymer was O-deacetylated with ammonium hydroxide. Removal of the acetyl group was seen to cause *inter alia*, an approximately 2 ppm downfield shift of one of the N-acetyl galactosamine residues (50.9 to 52.7 ppm); this established the site of O-acetylation as C-3.

The presence of scalar ^{31}P coupling to both C-1 and C-2 of the α N-acetylglucosamine portion of the polymer indicated phosphate attachment at C-1. This was readily confirmed by dephosphorylation. The 105 ppm signal was not shifted by dephosphorylation while the 97 ppm resonance was

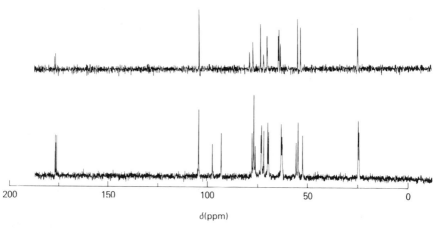

Figure 9. ^{13}C NMR spectrum of the *H. influenzae* type f capsular polysaccharide following dephosphorylation (mild acid hydrolysis followed by alkaline phosphatase catalyzed hydrolysis) (bottom tracing) and subsequent $NaBH_4$ reduction (top tracing).

affected (see Figure 9). Reaction of the hydrolyzed material with sodium borohydride resulted in the loss of the anomeric resonances generated by hydrolysis, while the 105 ppm resonance was unaffected (see Figure 9).

The structure of the type f polysaccharide follows from the foregoing observations. Commencing with the α-N-acetylgalactosamine residue, it was shown that phosphate is attached to C-1, O-3 is acetylated, and O-6 is unsubstituted. By elimination, therefore, O-4 must be the site of linkage to the second galactosamine of the repeating unit. The terminus of this linkage must be the C-1 of the β-N-acetylgalactosamine residue, as the C-1 β resonance is unperturbed by O-dephosphorylation. Additionally, as O-6 of the β-N-acetylgalactosyl residue is unsubstituted, the phosphate group must be attached to either C-3 or C-4. The observation of a ^{31}P scalar coupling (J = 5.0 Hz) to C-2 evidences attachment at C-3; such a large coupling through four bonds (as would be the case for attachment at C-4) would be unlikely, whereas it is expected for a three-bond coupling. Attachment of the phosphate at C-3 is substantiated by the observation of an ~1.2 ppm downfield shift of the C-2 resonance on O-dephosphorylation. The structure of the type f polysaccharide is therefore as depicted on page 219.

3.6. *H. influenzae*, types d and e

The *H. influenzae,* types d and e capsular polysaccharides differ from the previously discussed capsular antigens (types a, b, c, and f) in that they do not contain phosphate; in common with the types a, b, c, and f antigens, however, they are (1) simple disaccharide repeating units and (2) are acidic. The acidic nature of the types d and e polysaccharides was first revealed by

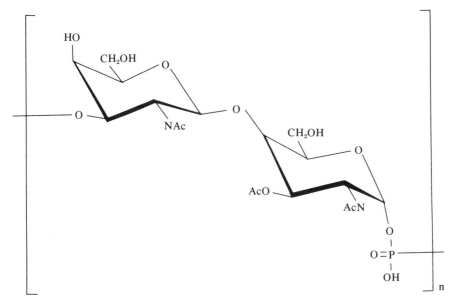

Haemophilus Influenzae type f
Polysaccharide

their anodic migration during immunoelectrophoresis; both ^{31}P NMR and chemical analysis failed to reveal phosphorus in the polymer. The simple disaccharide nature of the two polymers was best seen in their ^{13}C NMR spectra (see Figures 10 and 11).

The structure of the type e antigen is the more readily determined, and it will be discussed first. As mentioned previously, the type e polysaccharide is composed of an acidic repeating unit. The acid nature of the repeating unit is revealed by recording the ^{13}C NMR spectrum at pH values above and below the pK_a of the carboxyl group (we are now assuming that a carboxyl group is responsible for the acidic nature of the polymer). At pH 7.0, two signals, at 177.5 and 176.0 ppm, are observed in the carbonyl region of the ^{13}C NMR spectrum; upon lowering the pH to 1.0, three signals are observed in this same region, 177.8, 177.1, and 174.5. The 2–3 ppm upfield shift on lowering pH is characteristic of the behavior of carboxylic acid carbonyl carbon resonances (Wehrli and Wirthlin, 1978). The type e polysaccharide was shown to be composed of equal molar amounts of N-acetyl glucosamine and N-acetyl mannosamine uronic acid (Egan *et al.*, 1980c). No other components were detected.

The anomeric carbon resonances are seen at 103.1 and 102.4 ppm; the N-acetyl methyl group carbons are seen at 25.3 and 24.9 ppm; the C-2 carbon resonances at 57.2 and 55.4 ppm (these are the carbons bearing the N-acetyl groups). A single hydroxymethyl group carbon resonance is observed at 63.7 ppm; the observation of one and not two —CH_2OH carbon resonances is in accord with one of the sugar residues being an acid (a C-6 carboxylic acid).

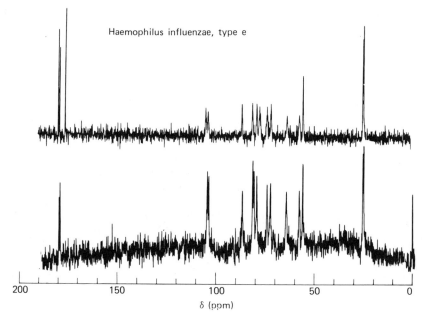

Figure 10. ^{13}C NMR spectrum of the capsular polysaccharide from *H. influenzae* type e at pH 7.0 (bottom tracing) and pH 1.0 (top tracing). Both spectra were recorded at *ca.* 80°C under conditions otherwise similar to those described in Figure 3.

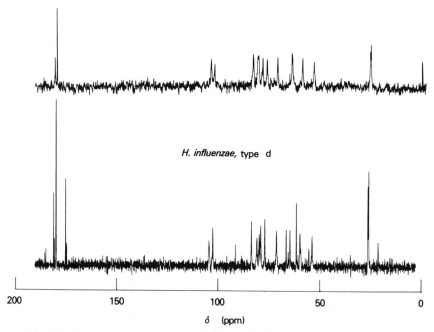

Figure 11. ^{13}C NMR spectrum of the capsular polysaccharide from *H. influenzae* type d at pH 1.0 (bottom tracing) and pH 7.0 (top tracing). Both spectra were recorded at ~80°C under conditions otherwise similar to those described in Figure 3.

The resonance at 85.7 ppm is distinct from the other resonances and corresponds to an alkylated C-3 carbon of glucose.

From its resonance position, it is readily established that glucosamine is substituted at C-1 (the resonances at 102.5 and 103.1 could not correspond to either an unsubstituted C-1, α or β, or to a substituted α configuration), O-6 is unsubstituted, and C-3 is substituted (as evidenced by its unique position). The mannosamine uronic acid is also linked at C-1, and C-6 is the site of the carboxylic acid; substitution could be either at C-3 or C-4. The resolution of the site of attachment was accomplished by periodate oxidation.

The intact type e polysaccharide does not consume periodate, whereas the N-deacetylated polymer does consumer periodate. This demonstrates that the O-3 of mannosamine is unsubstituted and that the site of attachment of the glucosamine residue to mannosamine is C-4. The ^{13}C NMR spectrum of the type e polysaccharide is in accord with this substitution pattern. ^{13}C NMR data for the type e polysaccharide and model compounds are collected in Table V. ^{13}C and ^1H NMR studies indicated the β configuration for both sugars. The structure of the type e polysaccharide is given below.

R = ala, ser, thr

Although the type d polysaccharide is composed of N-acetyl glucosamine and N-acetylmannosamine uronic acid as the sole sugars, comparison of Figures 10 and 11 show that the type d polysaccharide is considerably more complex; this complexity is due to the presence of small, appended groups. These appended groups were removable by mild hydrolysis with NaOH and identified as serine, threonine, and alanine. Following their removal, the ^{13}C NMR spectrum of the polysaccharide was recorded and, as seen in Figure 11, it is considerably simpler (see Table VI). An analysis of the spectrum indicated 1 and 4 linkages on N-acetylglucosamine and 1 and 3 linkages on N-acetylmannosamine uronic acid (Egan *et al.*, 1980c).

Since the C-5 resonance of the N-acetyl mannosamine residue of the untreated polymer did not titrate, substitution at the carboxyl carbon is indicated. The amino acids are linked to C-6 through amidic linkages; this was determined by (a) the failure to oxidize the amino acids with ninhydrin

Table V. ^{13}C NMR chemical shifts (ppm) for *H. influenzae*, type e capsular polysaccharide and model compounds

Carbon Compound	NAcGlc		*H. influenzae*[a] type e		NAcMan	
	α	β			β	α
C-1	94.8	97.4	[103.6; 102.6]		95.9	95.9
C-2	56.8	59.4	[56.4; 54.4]		56.9	56.0
C-3	73.3	76.5	85.1; 72.8		74.8	71.7
C-4	72.7	72.5	70.8; 80.2		69.4	69.6
C-5	74.1	78.4	78.4; 76.6	(79.8)	77.1	74.8
C-6	63.4	63.2	62.6; 174.5	(176.7)	63.3	63.3

[a]Chemical shift assignments must be regarded as very tentative; chemical shifts that may be interchanged are enclosed in brackets, []. The type e polysaccharide chemical shifts refer to pH 1.0; the values in parentheses, (), refer to pH 7.0.

Table VI. ^{13}C NMR chemical shifts (ppm) for *H. influenzae* type d capsular polysaccharide[a]

Carbon	Chemical shift
C-1	
C-1′	102.1 and 100.4
C-2	57.89
C-2′	52.2
C-3	75.1
C-3′	69.3
C-4	81.8
C-4′	79.1
C-5	77.1
C-5′	77.6 (79.5)
C-6	62.9
C-6′	175.1 (177.8)
N—C(O)CH$_3$	
N—C(O)CH$_3'$	177.1
NC(O)CH$_3$	
NC(O)CH$_3'$	24.85 and 25.19

[a]Assignments of resonances must be regarded as highly tentative. The values in parentheses refer to pH 7.0; all other values refer to pH 1.0. The primed carbons refer to mannosamine uronic acid; the unprimed carbons refer to glucosamine.

(a reagent specific for —NH$_2$ groupings) and (b) through direct observation of the amidic nitrogens by ^{15}N NMR. The low values of the directly bonded ^{13}C(1)—^{1}H(1) coupling constants for both sugars (160 Hz and 164 Hz) evidenced the β anomeric configurations for both sugars. The structure of the type d polymer is presented below.

R = Serine, Threonine, and alanine.

4. NEISSERIA MENINGITIDIS

4.1. Introduction

Bacteria of the genus, *Neisseria,* species *meningitidis,* are common inhabitants of the upper respiratory tract of man, the only known natural host for this organism. Morphologically, *N. meningitidis* is a small (0.8 × 0.6 μm) gram-negative cocci and has the tendency to grow in pairs and sometimes tetrads or clusters. Encapsulated strains of *N. meningitidis* are virulent and noncapsulated strains are avirulent. Meningococcal disease, most often as purulent meningitis, is fairly commonly seen in children but is also observed in adults of all ages. As was the case for *H. influenzae,* the nature of the capsule appars to be of importance.

There are eight serologically distinct meningococcal polysaccharides, designated Groups A, B, C, 29e, W 135, X, Y, and Z. Approximately 90% of meningococcal disease is caused by Groups A, B, and C and these are often associated with acute epidemics or high levels of endemic disease (Gold and Lepow, 1976; Artenstein *et al.*, 1971). In the absence of vigorous chemotherapy, the mortality associated with meningococcal meningitis is high, in excess of 85%; with prompt antibiotic treatment, the mortality can be reduced below 1%. The emergence of antibiotic-resistant strains and the central nervous system damage in "cured" individuals stress the need for prevention.

Susceptibility to meningococcal disease is associated with a lack of serum antibodies against the organism in question and protective antibodies are specific for the capsule polysaccharide (Robbins, 1978). Attempts to vaccinate adults with the Groups A and C polysaccharide have been extremely successful, both under epidemic and nonepidemic situations (Gold and Artenstein, 1971; Wahdan *et al.*, 1973; Erwa *et al.*, 1973; Peltola *et al.*, 1976, 1977). Two major problems to the elimination of Groups A, B, and C disease are extant. First, the A and C vaccines are not very effective in infants, the group with the highest attack rate; second, it has not yet been possible to produce an efficacious Group B polysaccharide vaccine (Wyle *et al.*, 1972).

4.2. *N. meningitidis,* Groups B and C

Groups B and C *N. meningitidis* organisms are highly pathogenic and possess phagocytosis inhibiting capsules composed of sialic acid (5-acetamido-3,5-dideoxy-D-glycero-D-galacto-nonulosonic acid). Although both capsules are homopolymers of sialic acid, they differ in structure; as immunogens, they also differ extensively, with the Group C polysaccharide being a good immunogen and the Group B polysaccharide being an extremely poor immunogen.

The Group C polysaccharide was first isolated by Watson and Sherp (1958) and later shown by Watson *et al.* (1958) to be a homopolymer of sialic acid. The Group B polysaccharide was isolated in pure form by Gotschlich *et al.* (1969) and characterized by Liu *et al.* (1971a) as a homopolymer of sialic acid. Additionally, these authors showed that the B and C polysaccharides differed in linkages and, moreover, that the Group C polysaccharide was O-acetylated whereas the Group B polysaccharide was not.

The detailed nature of the distinction between the B and C polysaccharides remained unclear until the ^{13}C NMR studies of Bhatacharjee *et al.* (1975). ^{13}C NMR spectra of the purified B and C polysaccharides (Figure 12) were consistent with their composition as pure sialic acid polymers, and showed that the Group C polysaccharide was O-acetylated (~1.16 equivalents of O-acetyl per sialic acid residue); the O-acetyl content was established by comparative integration of the N—C(O)$\underline{C}$H$_3$ and O—C(O)CH$_3$ methyl group signals.

The carbon spectrum of the O-deacetylated Group C polysaccharide (the polymer was O-deacetylated by treatment with 0.5 M NaOH at 37° for 3 h) was distinct from that of the Group B, indicative of a difference in linkages (see Figure 12). From comparison with model compounds, the linkage sites in both polymers could be established. ^{13}C NMR spectral data for the B and C polysaccharides as well as for model compounds are collected in Table VII.

Resonances in α- and β-methyl-N-acetylneuraminic acid were assigned by a combination of techniques: correlation with model compounds; specific chemical modifications; and single resonance ^{1}H decoupling. The linkage of the Group B polysaccharide was established as 2 → 8 on the basis of: (1) an approximately 6 ppm downfield shift of the C-2 resonance relative to 2-O-methyl sialic acid; (2) an approximately 7 ppm downfield shift of the C-8

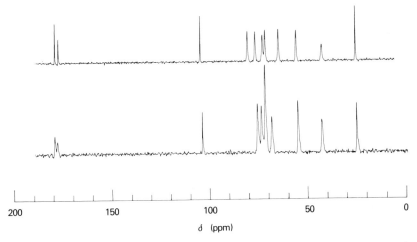

δ (ppm)

Figure 12. ^{13}C NMR spectra of the *N. meningitidis* Group B (top tracing) and O-deacetylated Group C (bottom tracing) capsular polysaccharides. Both spectra were recorded under conditions similar to those described in Figure 3. The O-deacetylated Group C polysaccharide is equivalent to the *Apicella* strain.

resonance, relative to 2-O-methyl sialic acid; (3) an upfield shift of approximately 2 ppm for the C-9 carbon. The α chirality at C-2 was assigned to the B polysaccharide by comparison with the spectra of the α and β forms of the model monomer; of particular use in this assignment were the changes in resonance positions of C-4 and C-6 on going from the β to the α forms.

The ^{13}C NMR spectrum of the O-deacetylated Group C polymer shows that it is an anomerically homogeneous polymer of sialic acid differing in linkage from the Group B polysaccharide. The downfield shift of C-9 (2.0 ppm) and upfield shift of C-8 (0.6 ppm) and the resonance position for C-2 establish the linkage as being 2 → 9. In accord with this assignment, the Group C polymer consumed 1 equivalent of periodate (the group B polymer does not consume periodate). From the resonance positions of C-4 and C-6, the α configuration was assigned to the C-2 carbon.

C-7 and C-8 were established as the sites of acetylation on the basis of the chemical shifts of C-6, C-7, C-8, and C-9, and a lack of shift for the C-3 resonance. It was expected that acetylation would produce an approximately 3 ppm downfield shift of the acetylated carbon and an approximately 3 ppm upfield shift of the vicinal carbons. C-3 and C-4 were unaltered in shift subsequent to O-deacetylation, thereby ruling out C-4 as a site of acetylation. The anomeric region of the C-polysaccharide displayed three resonances, one of which corresponded to the O-deacetylated material and accounted for ~25% of the C-2 resonances. On this basis, it was concluded that ~25% of the repeating units were not acetylated. Mass balance requires that some units be acetylated at both C-7 and C-8. Evidence for C-7 acetylation derived from the observation of an upfield shift of ~0.6 ppm for ~65%

Table VII. ^{13}C NMR chemical shifts (ppm) for *Neisseria meningitidis* Groups B and C capsular polysaccharides and related model compounds

Polysaccharides and monomers	C-1	C-2	C-3	C-4	C-5	C-6	C-7	C-8	C-9	NCO$\underline{C}$H$_3$	NCOC$\underline{H}$$_3$
Methyl-β-D-NAc neuraminic acid	175.6	102.2	42.2	69.4	54.7	72.8	71.0^a	73.4^a	66.3	177.7	24.9
Methyl-α-D-NAc neraminic acid	173.9	101.9	41.6	70.1	54.5	75.6	71.1	73.6	65.9	177.7	24.8
Group B polysaccharide	176.0	103.7	42.5	71.0^a	55.2	75.9	72.0^a	86.4	64.0	177.7	25.0
O-deacetylated Group C poly-saccharide	176.5	103.0	42.8	71.1	54.6	75.2	71.1	73.0	67.9	177.7	24.9

^{a}Assignments may be reversed.

of the C-5 signal (this is a long range effect of acetylation). This upfield shift would derive from both C-7 monosubstituted and from C-7, C-8 disubstituted residues. Mass balance ruled out the possibility that the entire upfield shifted resonance was due to disubstitution, as this would require more O-acetyl than observed, and therefore by inference, C-8 monosubstituted units must be present to some extent. Direct evidence for the presence of monoacetylated C-8 units was not obtained. The structures of the *N. meningitidis* Group B and Group C capsular polysaccharides are presented below.

A Group C meningococcal polysaccharide, differing from the "normal" in that it does not contain O-acetyl groups, has been isolated by Apicella (1974). Interestingly, it appears to be more immunogenic than the acetylated variant (Glode *et al.*, 1979).

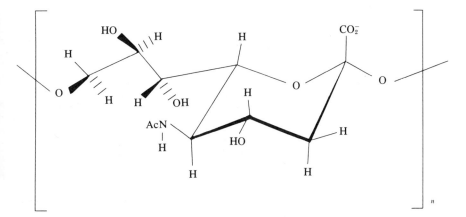

Neisseria meningitidis, Group B polysaccharide

Neisseria meningitidis, Group C polysaccharide

4.3. *N. meningitidis,* Groups Y and W135

The Y and W135 capsular polysaccharides also contain sialic acid. However, unlike the B and C polysaccharides, they are not homopolymers. In addition to sialic acid, the Y polysaccharide contains glucose and the W135 polysaccharide contains galactose. Additionally, the Y polysaccharide is O-acetylated. The Y organism was first described by Slaterus (1961) and the W135 organism by Evans *et al.* (1968); the structures of these two capsular polysaccharides were determined by Bhattacharjee *et al.* (1976).

The ¹³C NMR spectra for the Y, O-deacetylated Y, and W135 polysaccharides are presented in Figure 13; spectral data for these polysaccharides

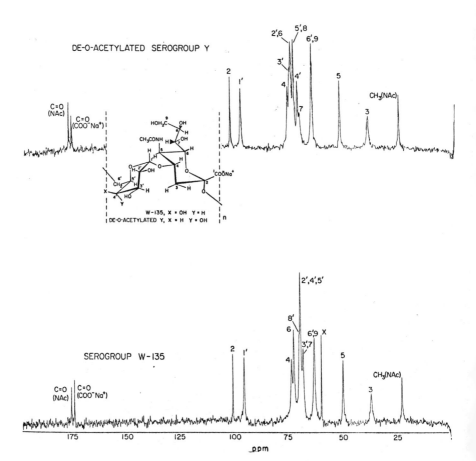

Figure 13. ¹³C NMR spectra of the capsular polysaccharides deriving from *N. meningitidis* Groups W135 and Y (top spectrum and next page). Spectra were determined at 25 MHz (reprinted from *Canadian Journal of Biochemistry*, by permission).

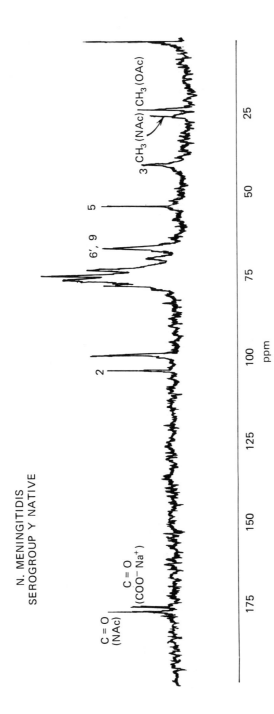

Figure 13. Continued.

as well as glucose and galactose models are collected in Table VIII. From inspection of Table VIII, it is seen that signals for the C-3, C-4, C-5, and C-6 carbon atoms of glucose and for the C-1, C-2, C-6, C-7, C-8, and C-9 carbon atoms of sialic acid in the disaccharide obtained from the O-deacetylated Group Y polysaccharide are within 0.4 ppm of their values in the monomeric sugars. The glucose C-1 resonance is shifted downfield by ~3.1 ppm relative to the monomer; in the sialic acid portion of the disaccharide, C-3 and C-5 are shifted upfield by 3.3 and 2.5 ppm, respectively, and C-4 is shifted downfield by 5.5 ppm. These shifts are characteristic of a 1 → 4 glucosyl sialic acid linkage. Furthermore, the C-1 resonance of glucose occurs at 96.4 ppm, indicative of the α configuration. From inspection of the ^{13}C resonances in the polymer, one observes that the glucose C-6 and the sialic acid C-2 resonances have shifted downfield by approximately 2 and 5 ppm, respectively, relative to their values in the disaccharide (or the individual free sugars); thus, the remaining linkage is from the C-2 carbon atom of sialic acid to the C-6 carbon atom of glucose. From the position of the C-4 and C-6 carbon atom resonances in the sialic acid portion of the polysaccharide, one concludes that sialic acid possesses the α configuration at C-2.

A similar analysis of the W135 polysaccharide shows that it differs from the Y polysaccharide with respect to glucose being replaced by galactose. W135 is not acetylated. The structures of the meningococcal Y and W135 capsular polysaccharides are presented below.

W−135, X = OH Y =H
DE-O-ACETYLATED Y, X = H Y = OH

The Y polysaccharide contains 1.3 equivalents of O-acetyl per unit of sialic acid. Although the ^{13}C spectrum was too intractable for a detailed analysis of the sites of O-acetylation, it could nonetheless be shown that C-8

Table VIII. ¹³C NMR chemical shifts (ppm) for meningococcal Group Y and W-135 polysaccharides, disaccharides, and related monomers[a]

Compound	Hexose Moiety						Sialic Acid Moiety										
	C-1	C-2	C-3	C-4	C-5	C-6	C-1	C-2	C-3	C-4	C-5	C-6	C-7	C-8	C-9	C(O)C̲H₃	C̲(O)CH₃
α-D-glucose	94.7	74.1	75.4	72.3	74.1	63.3	—	—	—	—	—	—	—	—	—	—	—
Group Y disaccharide	97.9	74.2	75.5	71.9	73.7	63.1	176.	97.5	38.2	74.9	52.3	72.9	70.9	72.9	65.9	24.9	177.0
O-deacetylated Group Y	98.0	74.9	75.7	71.8	73.7	65.3	175.8	103.	39.6	76.4	52.4	74.9	71.0	73.7	65.3	25.0	177.1
W-135 disaccharide	98.2	71.9	71.0	72.3	73.9	63.6	177.0	97.6	38.2	74.1	52.4	73.1	70.7	73.1	66.0	25.0	177.2
W-135 polymer	97.7	71.9	70.9	71.8	71.8	65.4	175.9	103.	39.4	75.9	52.3	74.9	70.8	72.5	65.4	25.2	177.2

[a]Chemical shifts for galactose may be found in Table III.

and C-9 of sialic acid were not acetylated. Thus, C-7 of sialic acid as well as C-3 and C-4 of glucose remain as possible sites for O-acetylation.

A serogroup designated BO was described by Evans *et al.* (1968). On the basis of the ^{13}C NMR spectrum, it was shown that BO was identical to Y, save possessing a slightly greater O-acetyl content; BO should therefore be classified as a Y strain.

4.4. *N. meningitidis,* Groups A and X

^{13}C NMR spectroscopy has been used to verify (Bundle *et al.*, 1974) the structure of the group A meningococcal polysaccharide, originally shown (Liu *et al.*, 1971b) to be a homopolymer of partially O-acetylated 2-acetamido-2-deoxymannosyl phosphate. The ^{13}C NMR spectra of the Group A polysaccharide and the O-deacetylated Group A polysaccharides are shown in Figure 14.

The presence of phosphate is readily ascertained from the ^{31}P couplings to C-1, C-2, C-5′, and C-6′. The scalar couplings to these carbons suffice to establish the linkage in the polymer as 1 → 6.

The polymer contains ~0.74 equivalents of O-acetyl per mannosamine unit. The site of O-acetylation is readily established as C-3 on the basis of the observation of an approximately 2.4 ppm downfield shift at C-2, an approximately 2.4 ppm downfield shift at C-4, and an approximately 3.5 ppm upfield shift at C-3 occurring upon O-deacetylation at C-3 (see Figure 14).

Comparison of the C-1 resonance position of the group A polymer with that from α and β N-acetylglucosamine-1-phosphates establishes the β configuration at C-1 for the Group A polymer. The structure of the Group A polysaccharide is presented below.

NEISSERIA MENINGITIDIS, GROUP A

The Serogroup X polysaccharide is compositionally a homopolymer of 2-acetamido-2-deoxyglucose linked 1 → 4 by a phosphoric diester group (Bundle *et al.*, 1973, 1974). The ^{13}C NMR spectrum of the Group X polysac-

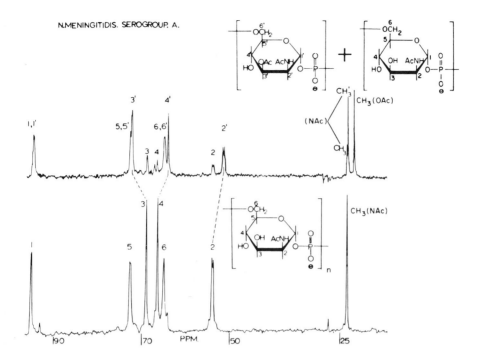

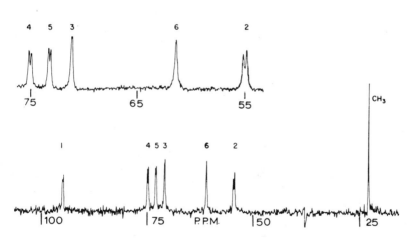

Figure 14. ¹³C NMR spectra of the *N. meningitidis* Groups A and X capsular polysaccharides. Spectra were recorded at 25 MHz (reproduced from *Journal Biological Chemistry*, by permission).

Table IX. ^{13}C NMR chemical shifts (ppm) for the meningococcal Groups A and X capsular polysaccharides[a]

Compound	C-1	C-2	C-3	C-4	C-5	C-6	—NC(O)$\underline{C}H_3$	—N$\underline{C}$(O)CH$_3$
Group X	96.8	56.4	72.7	76.7	74.8	62.9	24.9	177.2
O-deacetylated Group A	98.0	55.9	71.3	68.7	75.1	67.2	24.9	177.4

[a]Chemical shift values for α and β GlcNAc-4-phosphate may be found in Table III.

charide is shown in Figure 14. The resonance position of C-6 (61.3 ppm) indicates that it is unsubstituted. ^{31}P scalar couplings to C-1, C-2, C-4, and C-5 indicate a 1 → 4 linkage. Comparison of chemical shifts for model compounds with that of the polymer substantiate the proposed linkage and, moreover, the α configuration at C-1. ^{13}C NMR spectral data for the group A and Group X polysaccharides as well as for model compounds are assembled in Table IX. The structure of the meningococcal Serogroup X polysaccharide is shown below.

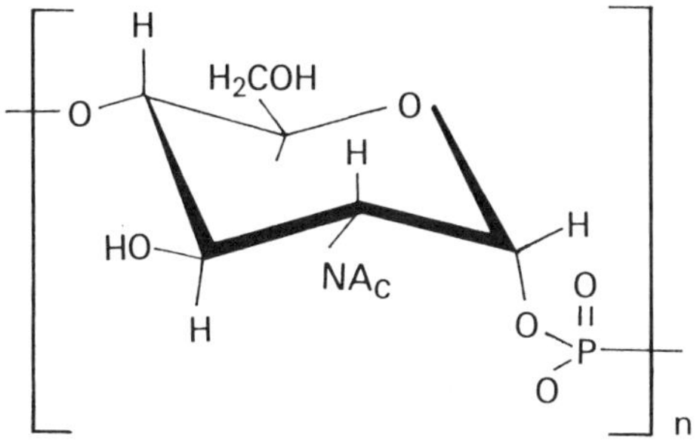

NEISSERIA MENINGITIDIS, GROUP X

4.5. *N. meningitidis,* Group 29e

The Group 29e meningococcal organism was first recognized and isolated by Evans *et al.* (1968). The capsular polysaccharide from this organism was shown by Bhattacharjee and Jennings (1974) to be composed of equal molar amounts of D-galactosamine and 3-deoxy-D-manno-octulosonic acid (KDO); O-acetyl groups were also found to be present. Although KDO is a common component of lipopolysaccharides, this was the first report of its isolation from a bacterial capsule.

The structure of the O-deacetylated 29e polysaccharide was elucidated (Bhattacharjee *et al.*, 1978) by comparison of its ^{13}C NMR spectrum with model compounds (see Figure 15 and Table X). In this fashion, it was established that the KDO residue was linked at C-7 and C-1 and the galactosamine residue at C-1 and C-3; both sugars were found to possess the α configuration. The structure of the meningococcal Group 29e polysaccharide is given in Figure 15.

Consistent with this structure, the polymer was found to consume 1 equivalent of periodate (no formaldehyde was released), releasing unoxidized 2-acetamido-2-deoxygalactose and erythritol. Methylation analysis resulted in the production of a neutral sugar (following re-N-acetylation), identified as 4,6-di-O-methyl galactosamine by both gas chromotography and mass spectroscopy.

The originally isolated polymer contained 0.70 equivalents of O-acetyl per

Table X. ^{13}C NMR chemical shifts (ppm) for meningococcal Group 29-e capsular polysaccharide and related compounds

Compound	C-1	C-2	C-3	C-4	C-5	C-6	C-7	C-8	—OCH$_3$ (methyl glycoside)
O-deacetylated Group 29-e	176.1	105.5	37.7	70.0	68.4	74.9	80.5	65.4	
Methyl-β-3-deoxy-manno-octulosonic acid (Na salt)	176.4	104.0	37.1	70.2	68.1	76.2	71.9	66.8	54.5
Methyl-α-3-deoxy-manno-octulosonic acid (Na salt)	178.1	104.1	36.8	69.0	68.7	74.1	72.1	65.8	53.5

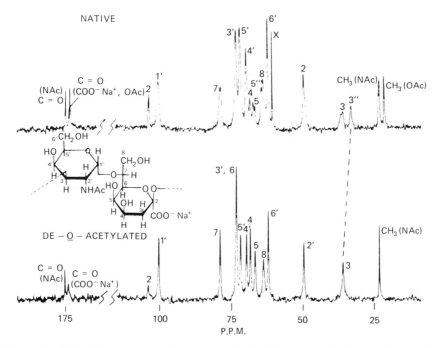

Figure 15. ¹³C NMR spectrum and structure of the *N. meningitidis* Group 29e capsular polysaccharide. Spectrum was recorded at 25 MHz (reproduced from *Biochemistry*, by permission).

equivalent of galactosamine. Using the previously mentioned acetylation chemical shift changes, the sites of O-acetylation could be ascertained. It was found that ~30% of the KDO residues were not O-acetylated, 40% were acetylated at C-4, and the remaining units were acetylated at C-5.

On the basis of a sensitivity of the chemical shift of the C-8 resonance of KDO (as the sodium salt) to anomeric configuration at C-1 of KDO, a hydrogen bond between the carboxylate and C-8 was proposed for the α configuration. A similar correlation was also noted with the sialic acid polymers. On this basis, it was thought that KDO and sialic acid as found in the meningococcal polysaccharides, have similar steric (spatial) arrangements.

4.6. *N. meningitidis*, Group Z

The Group Z meningococcal capsular polysaccharide was the final antigen of this series characterized. Its structure has been elucidated at the National Research Council in Canada (Jennings, 1979) and confirmed at the Bureau of Biologics (Frasch *et al.*, 1979). Compositionally, the polysaccharide is com-

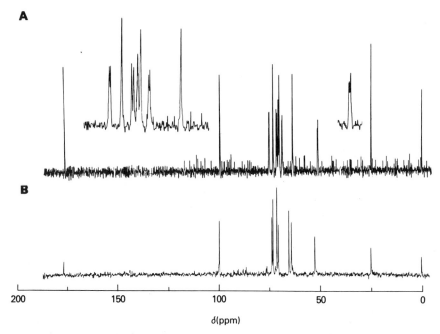

Figure 16. ¹³C NMR spectrum of the *N. meningitidis* Group Z capsular polysaccharide acquired under conditions similar to those described in Figure 3 (A); the dephosphorylated Group Z polysaccharide is shown in B. The inset tracing in A is an expansion of the spectral region between approximately 50 and 80 ppm.

posed of equal molar amounts of glycerol, N-acetyl galactosamine, and phosphate.

From the ¹³C NMR spectrum in conjunction with the observation that the material does not consume periodate, the structure is readily determined. The ¹³C NMR spectrum is shown in Figure 16. Nine resonances, several of which show discernible ³¹P scalar couplings, are observed in the 40–110 ppm chemical shift range (see Table XI). The resonance at 100.2 ppm is assignable to the galactosamine C-1 carbon atom; following dephosphorylation, this resonance is unaltered (see Figure 16) and is therefore directly attached to glycerol. Its chemical shift indicates the α configuration.

As only a single —CH₂OH group resonance is observed (at 64.0 ppm), and this must derive from the galactosamine residue by virtue of the nonconsumption of periodate, the anomeric carbon of galactosamine must be linked to the C-1 carbon of glycerol. Phosphate is linked to the C-3 carbon of glycerol and either the C-3 or C-4 carbon of galactosamine. The presence of an observable ³¹P scalar coupling to C-2 evidences attachment at C-3. Consistent with this conclusion, the C-2 resonance underwent an approximately 2 ppm downfield shift on dephosphorylation. The structure of the *N. menin-*

Table XI. ^{13}C NMR Chemical shifts (ppm) and ^{31}P–^{13}C Coupling Constants (Hz) for the *N. meningitidis* Group Z capsular polysaccharide[a]

Compound	C-1	C-2	C-3	C-4	C-5	C-6	C-1'	C-2'	C-3'
α-2-acetamido-2-deoxy-D-galactose	93.8	53.0	70.4	71.3	73.2	64.0			
N. meningitidis, Z	99.8	51.5 (5.3)	75.5 (6.1)	71.0	73.6	63.9	70.5	71.7 (8.0)	69.1 (5.4)
Dephosphorylated *N. meningitidis*, Z	100.0	52.6	71.3	71.3	73.2	64.1	70.5	73.8	65.4
β-2-acetamido-2-deoxy-D-galactose	98.1	56.5	73.8	70.6	77.8	63.8			

[a]The primed carbons refer to glycerol and the unprimed carbons to N-acetyl galactosamine.

gitidis Group Z capsular polysaccharide is as indicated. The chirality associated with the glycerol residue has not yet been established.

5. CROSS-REACTING *ESCHERICHIA COLI*

5.1. Introduction

The ability to resist invasive *H. influenzae* and *N. meningitidis* disease has been correlated with the presence of serum antibodies (IgG) directed against the polysaccharide capsule (Fothergill and Wright, 1933; Norden, 1972; Schneerson *et al.*, 1971; Goldschneider *et al.*, 1973). The inverse relationship between incidence of disease and antibody titre is shown in Figure 17. Although the nature of this protective effect has not been clearly defined, a fair body of evidence suggests that antibody-mediated and complement-enhanced phagocytosis is the protective mechanism (Alper *et al.*, 1970; Roberts, 1970). The antibodies may be acquired passively during intrauterine life (presumably the origin of the low incidence of disease shown during the first 6 months of life, see Figure 17) or actively synthesized during development as the result of active immunization, asymptomatic carriage of the homologous organism, or an encounter with a cross-reacting species. It is this latter mechanism, namely, encounter with a cross-reacting organism that will now be addressed.

Historically, antibodies to organisms such as *Salmonella typhi* and *Vibrio cholerae* were found to occur in laboratory aninals that had had little opportunity to encounter these organisms (Lovell, 1934; Mackie and Finkelstein, 1932). Similarly, for example, antibodies to *N. meningitidis* Group A are found in the majority of children and young adults in the United States despite the fact that this organism is only rarely encountered in the U.S. as a disease or nasopharyngeal isolate (Goldschneider *et al.*, 1973). The simplest rationalization for the presence of antibodies directed against nonencoun-

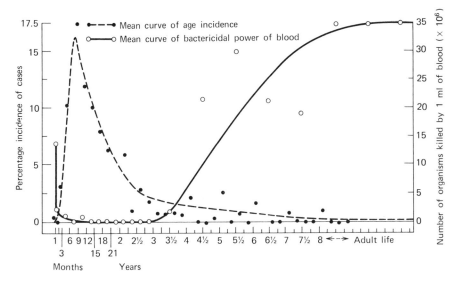

Figure 17. Relation between the age incidence of *H. influenzae* caused meningitis and bactericidal antibody titers in blood (reproduced from *Journal of Immunology*, by permission).

tered organisms is that the antibodies were elicited in response to encountered organisms and that the thus formed antibodies are reactive with the antigenic structure of the pathogen in question (Michael and Rosen, 1963; Myerowitz *et al.*, 1973). *E. coli* strains have been found that cross-react with *H. influenzae* type b and *N. meningitidis* Groups A, B, and C (Robbins *et al.*, 1976; Robbins, unpublished observations).

Additional interest in *E. coli* exists because certain strains are themselves the etiologic agent for invasive disease. Two prominent *E. coli* diseases are neonatal meningitis and childhood urinary tract infections. The *E. coli* bacteria responsible for these diseases are encapsulated. The K1 capsular type is the most commonly associated with neonatal meningitis (Robbins *et al.*, 1974); there also appears to be a capsular specificity associated with urinary tract infections (Kaijser *et al.*, 1977), but we will not have occasion to discuss these capsule types.

The remainder of this Section will concern the *E. coli* capsules of the organisms that are cross-reactive with *H. influenzae,* type b (K100), cross-reactive with *N. meningitidis* Group B (K1), and *N. meningitidis* Group C (K92).

5.2. *E. coli* K100

Compositionally, the K100 capsule is identical to that derived from *H. influenzae*, type b and is thus composed of ribose, ribitol, and phosphate in a molar ratio of 1:1:1. The ^{13}C NMR spectra of the K100 and type b polysac-

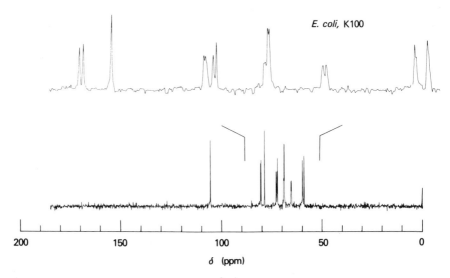

Figure 18. ^{13}C NMR spectrum of the capsular polysaccharide from *E. coli* strain Easter (K100; O75; H5) acquired under conditions similar to those described in Figure 3. The inset shows an expansion of the region between approximately 50 and 85 ppm.

charides each display ten resonances; the positions of the resonances deriving from each polysaccharide, however, differ, evidencing a distinction in linkages. The ^{13}C spectrum of the K100 polymer revealed one anomeric carbon resonance, two hydroxymethyl carbon resonances, one phosphorylated hydroxymethyl carbon resonance, and six secondary hydroxyl group and glycosyloxylated carbon resonances (the K100 spectrum is shown in Figure 18; spectral data are assembled in Table XII). The K100 structure has been determined by Egan *et al.* (1980d).

Dephosphorylation (mild hydrolysis with base followed by treatment with alkaline phosphatase) resulted in a ten-carbon disaccharide, whose ^{13}C NMR spectrum (Figure 19) revealed, *inter alia,* three hydroxymethyl group carbon resonances and an unaltered anomeric carbon resonance. The intact polysaccharide consumed 1 equivalent of periodate, liberating neither formaldehyde nor formic acid. Sugar analysis of the periodate-treated polysaccharide showed that it was the ribitol residue that was oxidized. The ^{13}C NMR spectrum of the periodate-treated material was relatively complex and, interestingly, showed no carbonyl absorptions (Figure 20). The most ready explanation for these findings is that the "aldehydic" carbonyls exist as a mixture of hemiacetals and acetal hydrates. After reduction with sodium borohydride, the ^{13}C NMR spectrum was recorded and is shown in Figure 21. The spectrum consisted of ten resonances, five of which displayed scalar couplings to phosphorus. Four hydroxymethyl group carbon resonances were observed, one of which was scalar coupled to phosphorus; the anomeric carbon resonance of ribose was unaltered.

Table XII. ^{13}C NMR chemical shifts (ppm) and $^{31}P–^{13}C$ Coupling Constants (Hz) *E. coli* K100 and *H. influenzae*, type b capsular polysaccharides[a]

Carbon Atom	*H. influenzae* type b	*E. coli* K100
Ribose		
C-1	109.4	108.75
C-2	77.0[b] (4.9)	76.89[b] (2.4)
C-3	76.5[b] (3.1)	76.12[b] (5.5)
C-4	84.7 (6.7)	84.40 (6.1)
C-5	65.1	63.80[b]
Ribitol		
C-1	71.4	63.03[b]
C-2	74.2[b]	82.50
C-3	72.8[b]	72.56
C-4	73.6 (7.9)	72.95 (7.3)
C-5	69.5 (6.1)	69.69 (5.5)

[a] Values in parentheses represent ^{31}P coupling constants to carbon.
[b] Assignments may be reversed.

The foregoing spectral and chemical findings are sufficient to piece together the structure of the K100 polysaccharide. The periodate data indicates that either C-2 or C-3 of ribose is substituted since ribose is not consumed by the oxidation. Thus, one of the hydroxymethyl group carbons observed in the intact polysaccharide must derive from ribose (ribose C-5); the other hydroxymethyl group carbon that was observed would therefore correspond to ribitol C-1. The dephosphorylated product revealed three hydroxymethyl group carbons, this third —CH$_2$OH group must therefore derive from ribitol C-5, the erstwhile phosphoric diester ribitol terminus. Several lines of evidence show that the anomeric carbon of ribose is linked to C-2 of ribitol. The consumption of 1 equivalent of periodate rules out C-4 attachment (attachment at C-4 would require the consumption of 2 equivalents of periodate and the production of formic acid and formaldehyde). C-3 attachment, while consistent with the consumption of 1 equivalent of periodate, is not consistent with the finding that no formaldehyde was generated by the oxidation. Attachment at C-2 is, however, in accord with all the periodate data. Moreover, the finding of a ten-carbon saccharide subsequent of periodate treatment and borohydride reduction is consistent only with C-2 attachment; C-3 and C-4 attachments would generate, respectively, nine- and eight-carbon polymers. (The low molecular weight of the periodate-oxidized, borohydride-reduced material was established by gel-permeation chromatography on Sephadex G-25.) The site of phosphate

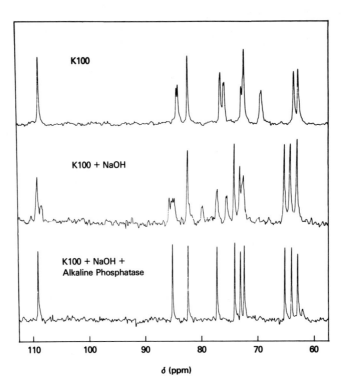

Figure 19. ^{13}C NMR spectra of the *E. coli* K100 capsular polysaccharide (top) and following mild hydrolysis with NaOH (middle) and subsequent treatment with alkaline phosphatase (bottom). Spectra were recorded under conditions similar to those described in Figure 3.

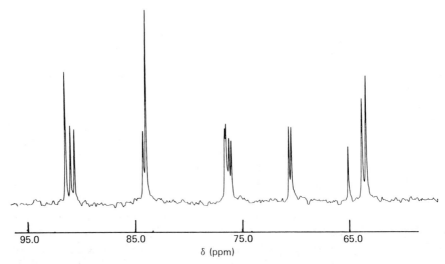

Figure 20. ^{13}C NMR spectrum of the *E. coli* K100 capsular polysaccharide following oxidation with sodium periodate; only the region between approximately 60 and 95 ppm is displayed. The spectrum was acquired under conditions similar to those described in Figure 3.

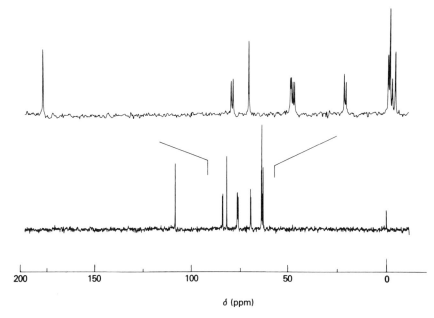

Figure 21. ¹³C NMR spectrum of the *E. coli* K100 capsular polysaccharide following periodate oxidation and sodium borohydride reduction; the spectrum was acquired under conditions similar to those described in Figure 3.

attachment to ribose must be either C-2 or C-3; the observation of a ^{31}P scalar coupling to C-4 and the lack of a scalar coupling to C-1, serves to establish C-3 as the site of phosphate attachment.

The β configuration for the C-1 of ribose is established from its ^{13}C chemical shift (the α and β C-1 resonances of the two methyl ribofuranosides occur at 105 and 110 ppm, respectively). The structure of the K100 polysaccharide is as indicated on page 246.

It still remains, however, to determine whether the C-1 of ribose is linked to the pro-**R** or pro-**S** carbon of ribitol; we are currently pursuing this aspect of the problem.

5.3. *E. coli* K92 and K1

The *E. coli* K92 capsular polysaccharide is a pure polymer of sialic acid. In this regard it is similar to the *N. meningitidis* Groups B and C capsular polysaccharides. The similarity between the Group C meningococcal polysaccharide and the K92 polysaccharide is further attested to by their cross-reactivity; K92 is, however, not cross-reactive with the Group B meningococcal polysaccharide. Unlike the Group C but similar to the Group B polysaccharide, the K92 polysaccharide is susceptible to neuraminidase catalyzed hydrolysis; however, unlike the Group B polysaccharide, both

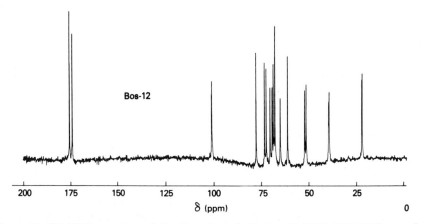

the K92 and Group C polysaccharides are susceptible to acid-catalyzed methanolysis. The structural basis for the meningococcal Group B and Group C like features of the K92 capsule has been supplied (Egan *et al.*, 1977).

The ^{13}C NMR spectrum of the K92 polysaccharide is shown in Figure 22. From the appearance of the spectrum (20 major resonances are discernible) it can be readily established that the K92 polysaccharide is neither a simple α-2 → 8 nor α-2 → 9 linked homopolymer of sialic acid (as are the menin-

Figure 22. ^{13}C NMR spectrum of the *E. coli* strain Boston 12 (K92; O16; NM) capsular polysaccharide. The spectrum was recorded at a resonance frequency of 67.89 MHz using a 15,151.5 Hz spectral window, 90° pulse, 2.5 s pulse repetition rate using a polysaccharide sample of 100 mg dissolved in 2.0 ml of H_2O at pH 7.0 in a 10 mm NMR tube and at a temperature of ~25°C.

gococcal Group B and Group C polysaccharides, respectively). A possible explanation for the number of resonances observed for the K92 polysaccharide, consistent with its chemical and immunological properties, is that the polymer contains alternating α-2 $\rightarrow$ 8 and α-2 $\rightarrow$ 9 linkages. A peak-by-peak comparison of the ^{13}C spectrum of the K92 polysaccharide with those for the Group B and O-deacetylated Group C polysaccharides supports this interpretation. Thus, the spectrum of the K92 polymer is nearly identical to the composite spectrum derived by combining the individual Group B and Group C spectra. The small displacements in chemical shift that are noted (less than about 0.5 ppm) for several resonances are readily understood in terms of environmental changes accompanying heteropolymer formation; that is, a strict retention of chemical shifts is not expected.

Three K92 strains were investigated, Bos-12 (O16; K92; NM), N-67 (O13; K92; H4), and MT 411 (O23; K92; H4). The ^{13}C spectrum for all three were identical; they thus share the same structure, depicted below.

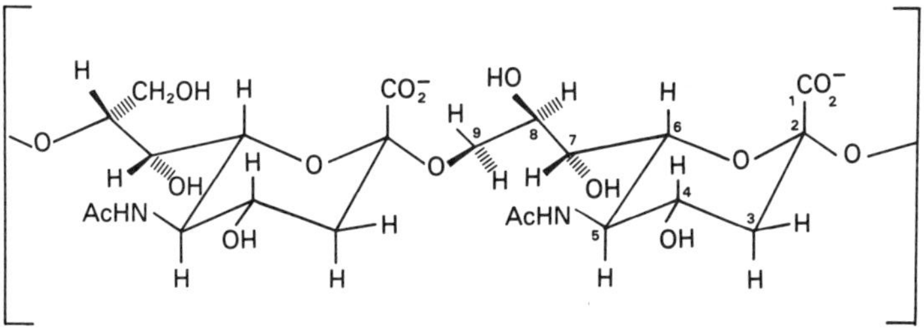

The heteropolymer structure for the K92 capsular antigen satisfactorily explains its cross-reactivity with the Group C meningococcal capsular antigen, its susceptibility to neuraminidase catalyzed hydrolysis, and its susceptibility to acid catalyzed methanolysis. It does not, however, satisfactorily explain the lack of cross reaction with the Group B meningococcal derived polysaccharide; a more detailed investigation as regards, *inter alia*, the determination of end groups and higher order structure, is in order.

E. coli K1 is cross-reactive with the Group B meningococci (K1 does not cross-react with Group C). Inspection of the ^{13}C NMR spectrum of the K1 polysaccharide reveals that it is identical to that derived from *N. meningitidis* Group B. The reason for the cross reaction is thus the trivial one of polysaccharide identity.

Among K1 strains, an interesting variation in form exists, characterized by weak and strong halo formations when the K1 organisms are grown on agar plates impregnated with Group B meningococcal antiserum from horses. The form variation was characterized at the Bureau of Biologics on a molecular level (Ørskov *et al.*, 1979) as deriving from the presence (faint halos) or absence (strong halos) of O-acetyl groups. The O-acetyl groups were readily seen in the ^{13}C NMR spectra of the capsular polysaccharides from the various strains (see Figure 23). A more detailed analysis of the

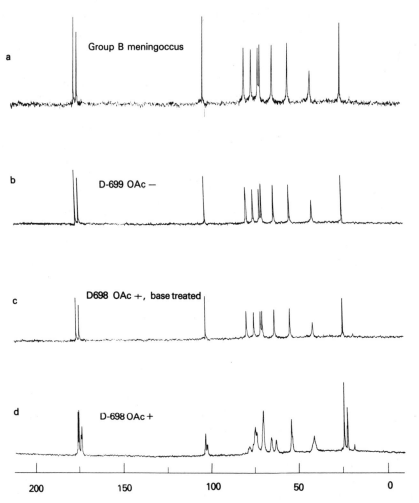

Figure 23. ^{13}C NMR spectra of the *E. coli* K1 O-acetyl variants acquired under conditions similar to those described in Figure 22. The spectrum of the meningococcal Group B capsular polysaccharide (a) is shown for comparison with the O-acetyl negative (b), O-acetyl positive (d) and O-acetyl positive variant subsequent to base catalyzed hydrolysis of the O-acetyl group (c).

redistribution of resonances on adding acetylation indicated that the sites of acetylation were C-7 and C-9. The extent of acetylation was determined by comparative integration of the N-O-acetyl methyl group signals.

The variation in O-acetyl content is not without consequence. It renders identification of K1 disease isolates difficult, as horse anti-B meningococcal serum is often used in diagnosis. The relationship between pathogenicity and O-acetyl content has not been established, although work in this area is currently underway. Additionally, it has been noted that the O-acetyl positive variant is more immunogenic, at least in rabbits; the K1 O-acetyl

positive variant is thus a possible candidate for a vaccine against Group B
meningococcal disease.

6. MICRODYNAMIC BEHAVIOR OF CAPSULAR POLYSACCHARIDES

Among the physical parameters that are thought to influence the im-
munogenicity of a polysaccharide antigen are size and rigidity (Abramoff,
1970); these parameters might also relate to bacterial virulence. Size is a
complex parameter and its assessment involves knowing the individual
polymer chain length, the polydispersity in length, the degree of chain
aggregation. For purposes of vaccine licensure (Wong *et al.*, 1977), size is
characterized simply by the point of elution of polysaccharides on a gel-
chromatograph column. It is clearly necessary to establish more rigorously
the variables of size, both chemically and physically.

Rigidity, defined in terms of segmental mobility, may be probed by ^{13}C
NMR relaxation times (T_1, T_2, $T_{1\rho}$, and the relaxation-related parameter, the
nuclear Overhauser effect. The use of magnetic resonance methods to probe
molecular motion is thoroughly discussed by London in Chapter 1 of this
book and the reader is referred to that Chapter. Relaxation studies have been
begun in this laboratory with the meningococcal Groups B and C polysaccharides. In addition to their availability, we have chosen to work with these
particular polysaccharides because of their differing immunogenicities (despite their close structural similarity) and because they are simultaneously
being investigated by laser light scattering techniques (Tsunashima *et al.*,
1978).

On the basis of the light scattering studies, it has been concluded that the
meningococcal Group C polysaccharide has a weight-average molecular
weight of ~5×10^5 daltons and a Stokes' radius of ~250 Å. In addition,
analysis indicated that the polymer behaved like a spherical random coil. ^{13}C
NMR studies, described below, support this random-coil model, but indicate
a considerably smaller Stokes' radius.

Relaxation data were obtained for the Group B and Group C polysaccharides at ^{13}C NMR resonance frequencies of 25.05 and 67.89 MHz (see
London, Chapter 1, in this volume, regarding the need to determine relaxation data at two differing resonance frequencies). Representative inversion-
recovery spectra are shown in Figure 24 and a plot of $\ln(M_t - M_0)$ versus
time for several differing carbon atoms is shown in Figure 25. The linearity
of the plot reflects behavior that can be analyzed in terms of a single
exponential relaxation; that is, we did not see evidence for statistically
distinct aggregates (this does not, however, rule out such a possibility).
Excluding methyl group and nonprotonated carbon atoms, NT_1 values for all
the carbons in the Group C polysaccharide and all the carbons except C-9 in
the Group B polysaccharide were equal see Table XIII; the relaxation times
for the two polymers were relatively similar.

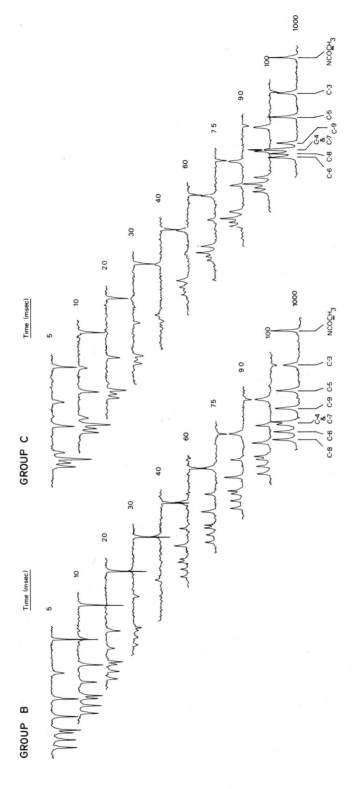

Figure 24. ^{13}C NMR spectra (25 MHz) of the meningococcal Groups B and C capsular polysaccharides obtained under the inversion-recovery pulse sequence (180–τ–90), the τ values are indicated in milliseconds; the spectra were determined at 25°C.

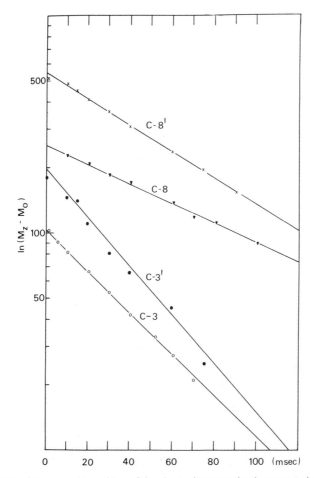

Figure 25. Plots of the natural logarithm of the change in magnetization versus time for the C-3 and C-8 carbon atoms of the meningococcal Group B (primed carbons) and Group C, Apicella strain (unprimed carbon) corresponding to the spectra displayed in Figure 24.

The Stokes–Einstein–Debye equation (1) relates the rotational correlation time, τ_c, to the radius of a spherical particle, a, at a given absolute temperature, T, in a medium of defined viscosity, η; k is the Boltzmann constant.

$$\tau_c = \frac{4\pi a^3 \eta}{3\,kT} \tag{1}$$

A spherical particle of radius 250 Å would have an associated rotational correlation time of $\sim 10^{-5}$ s. The dipole–dipole ^{13}C spin-lattice relaxation time, T_1, for a C–H vector undergoing isotropic rotational reorientation with $\tau_c \sim 10^{-5}$ s, would be in the hundreds of seconds range at the frequencies studied. As is seen in Table XIII, this is clearly not in line with the observed

Table XIII. Relaxation parameters for the *Neisseria meningitidis* Group C polysaccharide; 37 °C; pH 7.0; 200 mg/ml

Carbon	25 MHz		68 MHz	
	observed	calculated	observed	calculated
C-3	.070	.074	.160	.157
C-6	.078	.084	.176	.175
C-9	.080	.082	.170	.169

Calculated overall correlation time: 8.9×10^{-9} s
Calculated internal correlation time: 3.0×10^{-10} s

data. To fit the relaxation data, a model for the motion incorporating both a much smaller particle size and considerable internal mobility was required. This is detailed below.

An attempt was first made to fit the relaxation time to a model consisting of simple isotropic rotational reorientation with no associated internal rotation. If the T_1 data were fit at 25 MHz, the observed and theoretically predicted values at 67.89 MHz were widely divergent and, of course, vice versa; see Figure 26. A more complicated model was mandated. (Models based on anisotropic rotational reorientation with no internal reorientation did not account for the observed field dependence of T_1). It was thus necessary to consider internal (segmental) motions. There are many models from which to choose and, in the absence of additional information, it is usually difficult to prefer one over the other. As a point of departure, however, we chose a model incorporating overall isotropic reorientation and isotropic internal reorientation. With this simple model, we were able to account for the field dependence of T_1 for the Group C polysaccharide. The overall reorientation time, τ_c, was 9×10^{-9} s and an internal reorientation correlation time (τ_G) was 3×10^{-10}. However, at this point one would hesitate to attach too much physical significance to these correlation times. Employing the same model with the Group B relaxation data resulted in a poorer (but nonetheless acceptable) fit.

Alternate models of rotational reorientation, both overall and segmental, are now being pursued, to define better the limits of the possible models that can be employed. Although it may not prove possible to uniquely define the rotational reorientational process in the Groups B and C polysaccharides, it is nonetheless clear from the NMR data heretofore gathered that *considerable and comparable flexibility attends the two polymers*. It therefore does not seem tenable that the difference in observed antigenicity stems from differences in mobility of the individual chains.

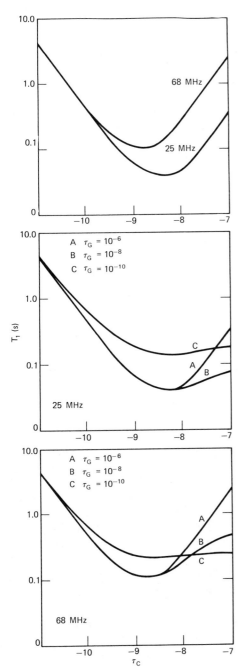

Figure 26. Theoretical plots of the spin-lattice relaxation time for a C–H vector (1.09 Å) undergoing isotropic rotational reorientation as a function of correlation time (τ_c) and resonance frequency (top) and including internal isotropic reorientation (τ_G) (middle and bottom).

7. FUTURE PROSPECTS

Structural determinations of polysaccharide capsules are merely the preamble to the story of the host–pathogen interaction. The very difficult task—defining virulence in terms of more fundamental chemical and physical principles—lies ahead.

At the outset of this Review, it was noted that capsules were necessary for bacterial virulence (but, however, that not all encapsulated bacteria were virulent). It was also emphasized that the mechanism whereby the capsule protects the bacterium was not at all understood. Presumably, it is then the investigator's role to pursue the relationship between capsule *structure* (in its widest possible definition) and pathogen virulence. Relatively few bacteria cause disease in uncompromised human hosts; 14 or so *S. pneumoniae* serotypes (of 83 known serotypes), 3 *N. meningitidis* serotypes (of 8 known serotypes), 1 *H. influenzae* serotype (of six known serotypes), 8 *E. coli* serotypes (out of approximately 105 known types), and similarly with several other bacterial species. Most bacteria are not pathogenic. Although this review has focused on two bacterial systems, *H. influenzae* and *N. meningitidis*, one might choose any of the above bacterial systems as a paradigm for defining pathogenicity, each having its own virtues and defects. For example, the pneumococci could be pursued and one might look for overlapping trends and patterns in the capsules; the drawback is that the system is exceedingly complex. The *H. influenzae*, on the other hand, are very simple. However, this very same simplicity precludes the recognition of patterns among the virulents and hinders that among the nonvirulents. The meningococci are in between.

For the *H. influenzae* system, there are many investigational avenues to pursue. Namely to fully characterize the polysaccharides in terms of chain length, ability to aggregate, polydispersity, and end groups and the influence of these parameters on antigenicity/immunogenicity and virulence. It is necessary to characterize the polysaccharides, *as they exist on the bacterium;* here solid-state NMR techniques will prove of great value. It is also necessary to characterize the interactions of polysaccharides with their surroundings; for example, how hydrophilic are they? How does hydrophilicity relate to ability to avoid phagocytosis? How do the polysaccharides interact with metal ions? And how does this interaction influence pathogenicity? And again we must distinguish the polysaccharide as invading bacterium and as isolated polysaccharide (vaccine). We must examine how the bacterial polysaccharide capsules interact with the phagocytic cell, the complement mediated immune system (both the classical and nonclassical systems), and the noncomplement mediated phagocytic system involving heparin and an unknown opsonin (Molnar *et al.*, 1979).

Indeed, there are many problems to be answered and, very likely many questions and problems yet to be defined. However, it is certain that magnetic resonance, which has proven to be so informative in studies of human physiology, and that has opened the way for the study of bacterial virulence and vaccine prophylaxis, will help in further defining the etiology of bacterial disease.

ACKNOWLEDGMENTS

It is a pleasure to mention and acknowledge a number of individuals at the Bureau of Biologics who have participated directly in the structural studies mentioned herein, provided criticism and insight, provided direction, and provided skilled technical help; in particular, Bob Tsui, Rachel Schneerson, Peggy Climenson, Bob Boykins, Darrell Liu, and John Robbins. It is also a pleasure to acknowledge the scientific acumen and open spirit of Harry Jennings, Lennart Kenne, and Bengt Lindberg.

REFERENCES

P. Abramhoff (1970), in *Biology of the Immune Response*, P. Abramoff and M. LaVia, Eds., McGraw-Hill, New York, NY, pp. 13–35.

Alexander, H. E. (1943), *Am. J. Dis. Child.* **66**, 160.

Alper, C. A., N. Abramson, R. B. Johnston, Jr., J. H. Jandl, and F. S. Rosen (1970), *New Engl. J. Med.* **282**, 351.

Anderson, P. W., R. B. Johnston, and D. H. Smith (1972) *J. Clin. Invest.* **51**, 31.

Apicella, M. A. (1974), *J. Infect. Dis.* **129**, 147.

Artenstein, M. S., H. Schneider, and M. D. Tingley (1971), *Bull. W. H. O.* **45**, 275.

Bellanti, J. A., and D. H. Dayton, Eds. (1975), *The Phagocytic Cell in Host Resistance*, Raven Press, New York, NY.

Bhattacharjee, A. K., and H. J. Jennings (1974), *Biochem. Biophys. Res. Commun.* **61**, 489.

Bhattacharjee, A. K., H. J. Jennings, C. P. Kenney, A. Martin, and I. C. P. Smith (1975), *J. Biol. Chem.* **250**, 1926.

Bhattacharjee, A. K., H. J. Jennings, C. P. Kenney, A. Martin, and I. C. P. Smith (1976), *Can. J. Biochem.* **54**, 1.

Bhattacharjee, A. K., H. J. Jennings, and C. P. Kenney (1978), *Biochemistry* **17**, 645.

Bodenhausen, G., R. Freeman, and G. A. Morris (1976), *J. Magn. Reson.* **23**, 171.

Boykins, R., and T.-Y. Liu (1980), *J. Biochem. Biophys. Methods* **2**, 71.

Branefors-Helander, P., B. Classon, L. Kenne, and B. Lindberg (1979), *Carbohydr. Res.*, accepted for publication.

Branefors-Helander, P., C. Erbing, L. Kenne, and B. Lindberg (1976), *Acta Chem. Scand., Ser. B.* **30**, 276.

Branefors-Helander, P., C. Erbing, L. Kenne, and B. Lindberg (1977), *Carbohydr. Res.* **56**, 117.

Branefors-Helander, P., L. Kenne, and L. Lindqvist (1979), *Carbohydr. Res.*, accepted for publication.

Bundle, D. R., H. J. Jennings, and I. C. P. Smith (1973), *Can. J. Chem.* **51**, 3812.

Bundle, D. R., I. C. P. Smith, and H. J. Jennings (1974), *J. Biol. Chem.* **249**, 2275.

Cozzone, P. J., and O. Jardetzky (1976), *Biochemistry* **15**, 4853.

Cooper, N. R., and D. C. Morrison (1978), *J. Immunol.* **120**, 1862.

Crisel, R. M., R. S. Baker, and D. E. Dorman (1975), *J. Biol. Chem.* **250**, 4926.

Davis, B. D., R. Dulbecco, H. N. Eisen, H. S. Ginsberg, W. B. Wood, and M. MacCarty, Eds. (1973), *Microbiology*, 2nd ed., Harper & Row, New York, NY.

Egan, W., T.-Y. Liu, D. Dorow, J. S. Cohen, J. D. Robbins, E. C. Gotschlich, and J. B. Robbins (1977), *Biochemistry* **16**, 3687.

Egan, W., F.-P. Tsui, P. A. Climenson, and R. Schneerson (1980a), *Carbohydr. Res.*, **80**, 305.

Egan, W., F.-P. Tsui, and R. Schneerson (1980b), *Carbohydr. Res.*, **79**, 271.

Egan, W., F.-P. Tsui, and R. Schneerson (1980c), *Carbohydr. Res.*, submitted.

Egan, W., F.-P. Tsui, and R. Schneerson (1980d), *J. Biol. Chem.*, in press.

Elwell, L. P., J. DeGraaf, D. Seibert, S. Falkow (1975), *Infect. Immunity.* **12**, 404.

Erwa, H. H., M. A. Haseeb, A. A. Idris, L. Lapeyssonnie, W. Sanborn, and J. Sippel (1973), in *Table Ronde sur l'immunoprophylaxie de la meningite cerebro-spinale*, LeMas D'Artigny, Edition Fondation Merieux, p. 66. Marcy L'Etoile, France.

Evans, J. R., M. S. Artenstein, and D. H. Hunter (1968), *J. Bacteriol.* **95**, 1300.

Fearon, D. T., M. R. Daha, J. M. Weiler, and K. F. Austen (1976), *Transplant. Rev.* **32**, 12.

Fothergill, L. D., and J. Wright (1933), *J. Immunol.* **24**, 273.

Francis, Jr., T., and W. S. Tillett (1930), *J. Exp. Med.* **52**, 573.

Frasch, C. E., D. E. Craven, G. E. Hoff, C.-M. Tsai, B. A. Fraser, and W. Egan (1979), unpublished results.

Fraser, B. A., F.-P. Tsui, and W. Egan (1979), *Carbohydr. Res.* **73**, 59.

Fraser, D., G. C. Geil, and R. Feldman (1974), *Am. J. Epidemiol.*, **100**, 29.

Freeman and Morris (1979), *Bull. Magn. Res.*, **1**, 1.

Friebolin, H., N. Frank, G. Kerlich, and F. Siefert (1976), *Makromol. Chem.* **177**, 845.

Garegg, P. J., B. Lindberg, and B. Samuelsson (1977), *Carbohydr. Res.* **58**, 219.

Glode, M. P., E. Lewin, A. Sutton, C. T. Le, E. C. Gotschlich, and J. B. Robbins (1979), *J. Infect. Dis.* **139**, 52.

Gold, R., and M. S. Artenstein (1971), *Bull. W. H. O.* **45**, 279.

Gold, R., and M. L. Lepow (1976), *Adv. Pediatr.* **23**, 71.

Goldschneider, I., M. L. Lepow, E. C. Gotschlich, F. T. Mauck, F. Bachl, and M. Randolph (1973), *J. Infect. Dis.* **128**, 769.

Gotschlich, E. C., T.-Y. Liu, and M. S. Artenstein (1969), *J. Exp. Med.* **129**, 1349.

Hudson, R. F. (1965), *Structure and Mechanism in Organo-Phosphorus Chemistry*, Academic Press, New York, NY.

Jennings, H. J., and I. C. P. Smith (1978), *Meth. Enzymol.* **50**, 39.

Jennings, H. J. (1979), unpublished results.

Kaijser, B., U. Jodal, L. A. Hanson, G. Lidin-Janson, and J. B. Robbins (1977), *Lancet* **1**, 663.

Lindberg, B. (1972), *Meth. Enzymol. XXVIII*, 178.

Lindberg, B., and L. Kenne (1979), unpublished results.

Liu, T.-Y., E. C. Gotschlich, F. T. Dunne, and E. K. Jonssen (1971a), *J. Biol. Chem.* **246**, 4703.

Liu, T.-Y., E. C. Gotschlich, E. K. Jonssen, and J. R. Wysocki (1971b), *J. Biol. Chem.* **246**, 2849.

Lovell, R. (1934), *J. Comp. Pathol.* **47**, 107.

MacCleod, C. M., R. G. Hodges, M. Heidelberger, and W. G. Bernhardt (1945), *J. Exp. Med.* **88**, 369.

Mackie, T. J., and M. H. Finkelstein (1932), *J. Hygiene* **32**, 1.

Michael, J. G., and F. S. Rosen (1963), *J. Exp. Med.* **118**, 619.

Mims, C. A. (1977), *The Pathogenesis of Infectious Diseases*, Academic Press, New York, NY.

Molnar, J., F. B. Selder, M. Z. Lai, G. E. Siefring, R. B. Credo, and L. Lorand (1979), *Biochemistry* **18**, 3909.

Müller-Eberhardt, H. J. (1975), *Ann. Rev. Biochem.* **44**, 697.

Myerowitz, R. L., Z. T. Handzel, R. Schneerson, and J. B. Robbins (1973), *Infect. Immun.* **7**, 137.

Norden, C. W. (1974), *J. Infect. Dis.* **130**, 489.

Nunez, H. A., T. E. Walker, R. Fuentes, J. O'Connor, A. Serianni, and R. Barker (1977), *J. Supramol. Struc.* **6**, 535.

Ørskov, F., I. Ørskov, A. Sutton, R. Schneerson, W. Lin, W. Egan, G. E. Hoff, and J. B. Robbins (1979), *J. Exp. Med.* **149**, 669.

Parke, J. C., R. Schneerson, and J. B. Robbins (1972), *J. Pediatr.* **81**, 765.

Peltola, H., P. H. Makela, O. Elo, O. Pettay, O-V. Renkonen, and A. Sivonen (1976), *J. Infect. Dis. Scand.* **8**, 169.

Peltola, H., P. H. Makela, H. Kaythy, H. Jousimies, E. Herva, K. Hallstrom, A. Sivonen, O-V. Renkonen, O. Pettay, V. Karanko, P. Ahvonen, and S. Sarna (1977), *New Engl. J. Med.* **297**, 686.

Perlin, A. S. (1976), *Int. Rev. Sci., Org. Chem., Ser. 2*, **7**, 1.

Pittman, M. (1931), *J. Exp. Med.*, **53**, 471.

Robbins, J. B. (1978), *Immunochemistry* **15**, 839.

Robbins, J. B., G. H. McCracken, E. C. Gotschlich, F. Ørskov, I. Ørskov, and L. A. Hanson (1974), *New Engl. J. Med.* **290**, 1216.

Robbins, J. B., R. Schneerson, J. C. Parke, Jr., T.-Y. Liu, Z. T. Handzel, I. Ørskov, and F. Ørskov (1976), in *The Role of Immunological Factors in Infectious, Allergic, and Autoimmune Processes*, R. F. Beers, Jr., and E. G. Bassett, Eds., Raven Press, New York, NY, pp. 103–120.

Roberts, R. B. (1970), *J. Exp. Med.* **131**, 499.

Schneerson, R., L. P. Rodrigues, J. C. Parke, and J. B. Robbins (1971), *J. Immunol.* **107**, 1081.

Schreiber, R. D., M. K. Pangburn, P. H. Lesavre, and H. J. Muller-Eberhardt (1978), *Proc. Natl. Acad. Sci. U.S.A.* **75**, 3948.

Sell, H. W., R. E. Merrill, E. O. Doyne, and E. P. Zimsky (1972), *Pediatrics* **49**, 206.

Sharon, N. (1975), *Complex Carbohydrates: Their Chemistry, Biosynthesis, and Functions*, Addison-Wesley, Reading, MA.

Shaw, D. (1976), *Fourier Transform N. M. R. Spectroscopy*, Elsevier, Philadelphia, PA.

Slaterus, K. W. (1961), *Antonie van Leeuwenhoek J. Microbiol. Serol.* **27**, 304.

Smith, D. H., and J. B. Robbins (1974), *Prev. Med.* **3**, 445.

Tsunashima, T., K. Moro, B. Chu, and T.-Y. Liu (1978), *Biopolymers* **17**, 251.

Wahdan, M. H., F. Rizk, A. M. El-Akkad, A. A. El-Ghoroury, R. Hablas, N. I. Girgis, A. Amer, W. Boctar, J. E. Sippel, E. C. Gotschlich, R. Trian, W. R. Sanborn, and B. Cvetanovic (1973), *Bull. W. H. O.* **48**, 667.

Ward, J. I., T. F. Tsai, G. Filice, and D. W. Fraser (1978), *J. Infect. Dis.* **138**, 421.

Watson, R. G., and H. W. Scherp (1958), *J. Immunol.* **81**, 331.

Watson, R. G., G. V. Marinetti, and H. W. Scherp (1958), *J. Immunol.* **81**, 337.

Wehrli, F. W., and T. Wirthlin (1978), *Interpretation of Carbon-13 NMR Spectra*, Heyden, Philadelphia, PA.

Winkelstein, J. A., and A. Tomasz (1977), *J. Immunol.* **118**, 451.

Winkelstein, J. A., and A. Tomasz (1978), *J. Immunol.* **120**, 174.

Wong, K. H., O. Barrera, A. Sutton, J. May, D. Hochstein, J. D. Robbins, J. B. Robbins, P. D. Parkman, and E. B. Seligman, Jr. (1977), *J. Biol. Stand.* **5**, 197.

Wyle, F. A., M. S. Artenstein, B. L. Brandt, E. C. Tramont, D. L. Kasper, P. L. Altieri, S. L. Berman, J. P. Lowenthal (1972), *J. Infect. Dis.* **126**, 514.

Zinsser, H. (1964), *Microbiology*, 13th ed. Appleton-Century-Crofts, New York, NY, pp. 41–51.

Six

Nucleic Acid Structure, Conformation, and Interaction

Martin P. Schweizer

Department of Medicinal Chemistry
The University of Utah
Salt Lake City, UT 84112

1. INTRODUCTORY REMARKS

In this account I will attempt to provide an overview of the important contributions that nuclear magnetic resonance has made toward deciphering solution structure, conformation, and interaction of nucleic acids and their components. The treatment is not necessarily comprehensive; rather, literature has been cited which I think best illustrates key features of the various topics. The chapter is divided into three sections: (1) Oligonucleotides; (2) Transfer RNA; and (3) High Molecular Weight Nucleic Acids and their Complexes. In Section 1, I cover the conformation of single-stranded oligonucleotides and the conformation of monomers within the oligomer. Also, I examine duplexes of short complementary oligonucleotides. In Section 2, I highlight areas of investigation on transfer RNA which have not or are not being covered in the many other reviews emerging on this important class of nucleic acid; for example, ^{13}C-enrichment studies and the influence of modified nucleosides. Finally, in Section 3, I present the exciting current focus of ^{31}P and ^{1}H NMR on polynucleotides, DNA, and their complexes such as nucleosomes and ribosomes.

2. OLIGONUCLEOTIDES

2.1. Conformational Details of Single Strands

The use of ^{1}H, ^{13}C, and ^{31}P-NMR has been particularly fruitful in the past few years in the elucidation of the solution structure and conformation of the monomeric units of nucleic acids. Progress in this area has been thoroughly reviewed (Ts'o, 1974a; Davies, 1978). How do the conformational properties change as a result of incorporation into dimers, trimers, etc. In other words, are the individual conformations dependent upon chain length? How appropriate are the monomers and dimers as models for oligo- and polynucleotides?

Data bearing on these questions has recently appeared in the literature. Before we discuss these findings, let us consider in a general way the various torsional angles which govern the conformation of segments in nucleic acids.

In Figure 1 is depicted the triribonucleoside diphosphate unit —ApYpA—,[1] with the various torsional angles designated by Greek letters according to the convention due to Sundaralingam (1969). This trimer sequence is found in the anticodon sequence of phenylalanine specific transfer RNA (tRNA^{Phe}) of torula yeast. The involvement of the tricylic guanine derivative, Y (Wybutosine), in the structure and conformation of an oligonucleotide from the anticodon loop of this tRNA will be discussed in Section 3. The furanose ring is flexible and the time-average conformation may be described by an

---

[1]*Abbreviations Used:* The common ribonucleosides are adenosine (A), cytidine (C), guanosine (G), uridine (U), ribothymidine (rT). Deoxyribonucleosides are designated by the prefix d-; for example, d-A is 2'-deoxyadenosine. 5' monophosphates are—pN whereas 2', 3' phosphates are Np—. The common 3'—5' internucleotide bond for a typical triribonucleoside diphosphate is illustrated thusly; NpNpN. PMR = Proton magnetic resonance. NOE = nuclear Overhauser enhancement; ^{31}P-{^{1}H} and ^{1}H-{^{1}H}, respectively, denote observation of ^{31}P and ^{1}H while irradiating ^{1}H. DSS = 2, 2-dimethyl-2-silapentane sulfonate.

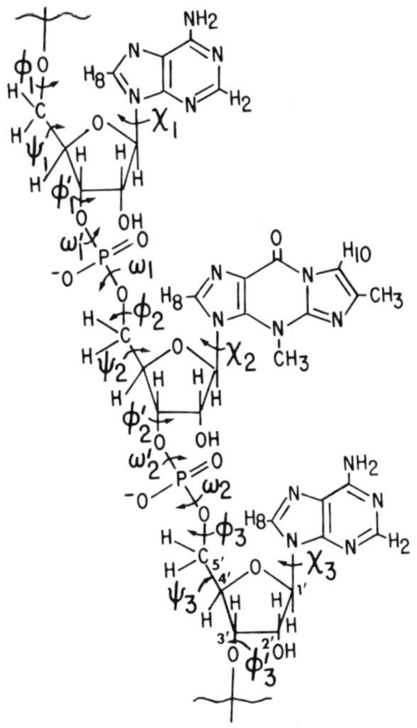

Figure 1. Torsion angles for backbone and glycosidic rotations in the trinucleotide segment —ApYpA— (see text).

equilibrium population distribution involving various puckered forms, for example, 2'-endo, 3'-endo, 2'-exo, 3'-exo, etc., where "endo" refers to displacements out of the plane in the same direction as the $C_{4'}$—$C_{5'}$ bond and "exo" is in the opposite direction.

What has NMR been able to tell us about the preferred domains of the various torsional angles involved in the spatial relationships between base and sugar rings (χ), sugar–phosphate backbone (ϕ_i, ψ_i, ψ'_i, ϕ'_i, ω'_i, ω_i) and furanose pucker?

The early 100 MHz PMR results have been reviewed by Ts'o (1974b). Basically, consideration of dimerization shifts [chemical shifts of base, 1' proton, and 2' proton between mixtures of constituent monomers and dinucleotide monophosphates (Ts'o et al., 1969; Kondo et al., 1970; Fang et al., 1971; Schweizer et al., 1968)], the specific effects of phosphate-bound paramagnetic Mn(II) on base protons (Chan and Nelson, 1969), the effects of the phosphate anion on base protons of 5'-nucleotide units (Schweizer et al., 1968; Danyluk and Hruska, 1968) and computed versus experimental ORD and CD profiles (Johnson and Tinoco, Jr., 1969) all lead to the conclusion that the individual monomer units in the dimers are preferentially in the anti glycosidic conformation (Figure 2a). The sugar–phosphate backbone adopts a right-handed helical turn with the bases stacked. The sugar pucker in highly stacked ribodimers is 3'-endo (^3E), whereas in deoxyribo dimers the out-of-plane preference is 2'-endo (^2E).

With the advent of high-field spectrometers, spectral simulation programs,

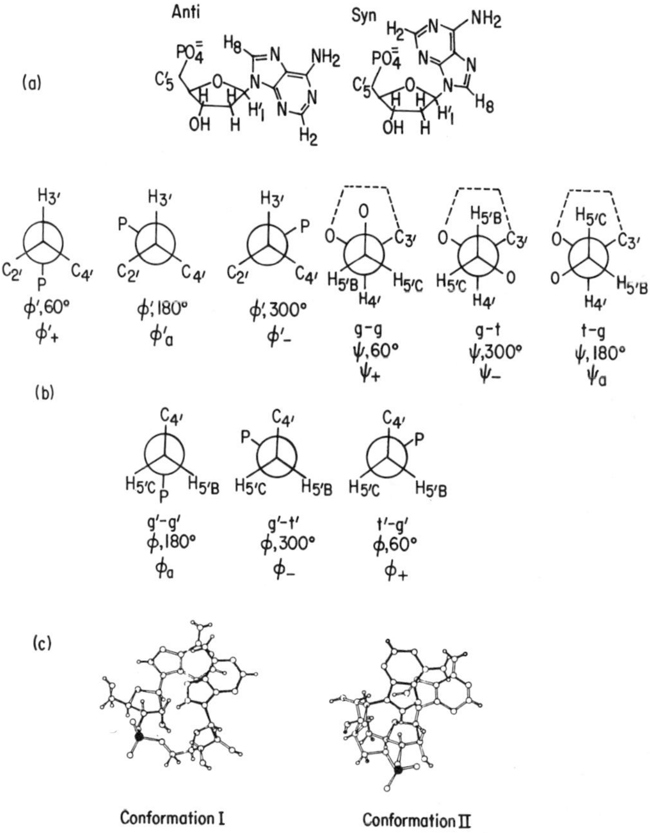

Figure 2. (a) *Anti* and *syn* conformers of 5′-AMP. (b) Newman projections for rotamers about $C_{3'}$–$O_{3'}$, $C_{4'}$–$C_{5'}$ and $C_{5'}$–$O_{5'}$ bonds. (c) ApA in two base stacked unformations: I, *anti*, 3′-endo, g^-, $\omega'(330°)$, $\omega(320°)$, $g'g'$, gg, 3′-endo, *anti*; II, *anti*, 3′-endo, g^-, $\omega'(80°)$, $\omega(50°)$, $g'g'$, gg, 3′-endo, *anti*.

and specific deuteration techniques, it has been possible to separate and assign all proton resonances in dimers and trimers. Sarma and Danyluk and coworkers, utilizing these techniques, have measured proton chemical shifts and have related experimentally determined $^3J_{\text{H-H}}$, $^3J_{\text{P-H}}$ coupling constants with the corresponding torsional angles via the well-known Karplus relationships.[2] Based on these data, they have refined and embellished the earlier models of the structure and conformations of dimers and trimers (Lee *et al.*, 1976; Ezra *et al.*, 1977; Cheng and Sarma, 1977a; Kondo *et al.*, 1975; Evans and Sarma, 1976; Cheng *et al.*, 1978; Kondo and Danyluk, 1976).

[2]M. Karplus (1959), first described the relationship between vicinal H–H coupling constants and θ, the dihedral angle between the interacting spins. The expression is $J = A \cos^2 \theta - B \cos \theta + C$ where the constants are associated with the nature of the substituents on the atoms of the rotating bond.

Table I. Effect of dimerization on individual nucleotide conformational properties in ApA

Conformational Unit	Monomer	Dimer
C—M (glycosidic bond)	$\chi_{CN} > 45°$	$\chi_{CN} \to 0°$
$C_{4'}$—$C_{5'}$	gg (74%)	gg (85%)
$C_{5'}$—$O_{5'}$	$g'g'$(75%)	$g'g'$(86%)
$C_{3'}$—$O_{3'}$	$g^+ \leftrightarrow g^-$	g^-
P—$O_{5'}$	free rotation	(320° and 80°)
P—$O_{3'}$	torsional angle	(330° and 50°)
	$= \pm 36°$ and/or	
	free rotation	
ribose ring	2'-endo, 3'-endo	2'-endo, 3'-endo
	(65:35)	(40:60)

The work from these laboratories as well as that of Chan (Kroon *et al.*, 1974) and Altona (Altona *et al.*, 1976) have shown that increasing the chain length of ribose derivatives, particularly for highly stacked purine-containing systems, has a marked effect on the conformational properties of the individual nucleotide residues, whereas in nonstacked ribo derivatives and in the deoxy series, the monomeric properties are sustained with oligomerization.

2.1.1. Ribose Systems. The data in Table I illustrates what happens to the mononucleotide preferred conformation when the stacked homologous ribo dinucleoside monophosphate, ApA is formed (see Figure 2*b* for Newman projections about the various flexible backbone bonds and the designations for the various conformational states). The relative amounts of the conformers were determined from vicinal and four bond $H_{4'}$–P coupling-constant measurements and the following empirical relationships (Wood *et al.*, 1973):

$$C_{4'}\text{—}C_{5'} \text{ bond } \%gg \sim \frac{13 - \Sigma}{10} \tag{1}$$

$$C_{5'}\text{—}O_{5'} \text{ bond } \%g'g' \sim \frac{23 - \Sigma'}{18} \tag{2}$$

where

$$\Sigma = {}^3J_{H_{4'}-H_{5'}} + {}^3J_{H_{4'}-H_{5''}}$$

and

$$\Sigma' = {}^3J_{P-H_{5'}} + {}^3J_{P-H_{5''}}.$$

In later works (Lee *et al.*, 1976) the equations have been slightly changed. The % ^3E or ^2E are estimated from $^3J_{1'-2'}$ and $^3J_{3'-4'}$ coupling via the empirical sum $J_{1'-2'} + J_{3'-4'} = 9.5$ Hz based upon measurements of many furanose systems. Intramolecular stacking of the hydrophobic adenine bases (as can be seen from dimerization shifts) leads to a shift in the *anti, syn* glycosyl equilibria towards a more highly *anti* orientation. The out-of-plane pucker of the furanose ring shifts from 2'-endo (^2E) to 3'-endo (^3E) which accommodates the overlap of base rings by decreasing the steric hindrance of the 2'-OH group. Accompanying the glycosyl and ribose puckering alterations are shifts in the backbone towards a higher degree of g g and g' g' character (see Figure 2b) about $C_{4'}$—$C_{5'}$ and $C_{5'}$—$O_{5'}$ bonds and a greater fraction of g^- about the $C_{3'}$—$O_{3'}$ bond, for example, 90° displacement of the 3' phosphates. Free rotation occurs about the P—$O_{3'}$ and P—$O_{5'}$ bonds; in the base-stacked dimer the ω', ω torsions can either be centered around 325° or 50–80° depending upon the stacked form, that is, right-handed helical stack or loop (Figure 2c). The dimer must pass through some unfolded form(s) as an intermediate between these two stacked species.

This type of behavior described for ApA does not occur in nonstacking systems; for example, adenosine monomers, dimers, trimers, poly(A) at high temperature, or the flexible uridine series (Evans and Sarma, 1976).

When one proceeds to larger stacked structures such as ApApA and poly (A) (Evans and Sarma, 1976) the preference for ^3E ribose pucker is even greater (67–78%) than in ApA, presumably again due to the driving force of the hydrophobic adenine rings to stack. Information on the backbone conformation in stacked ApApA and poly(A) could not be obtained at 270 MHz due to spectral overlap. Perhaps higher fields (e.g., 600 MHz) will be of help here. It would be of interest to monitor the backbone in larger, single-stranded oligonucleotides in order to discern subtle changes at individual residues, for example, "end" effects and possible loopout conformations for poorly stacking bases.

Even without the ability to monitor the backbone conformation in larger molecules, perturbations in base stacking from what one would expect from an extended helical dimer array have been noted by Kondo and Danyluk (1976) and Kroon *et al.*, (1974) for oligoriboadenylates, mainly by monitoring base and $H_{1'}$ protons. In the former study, Kondo and Danyluk compared the change in chemical shifts of H_2 protons of the adenine ring in proceeding from ApA to ApApA. The H_2 proton, particularly on an adenine of the 3'-attached adenosine in ApA experiences a large degree of ring current diamagnetic shielding from the neighboring adenine ring of the 5'-attached adenosine when the dimer is in the preferred g^-g^- right-handed stack. One would suppose that in the stacked trimer, the H_2 of the central adenine would be most shielded due to diamagnetic neighboring adenines on both sides. Thus one would expect the H_2 shift trend in the trimer to be —pA < Ap < pAp but one finds that the progression is —pA < pAp < Ap. This indicates that the adenines of Ap and pAp interact more strongly than the pAp + pA pair. Even though the ApApA stack appears to be somewhat irregular with respect to interaction of adenines, the percentage ^3E furanose pucker as monitored by $J_{1'-2'}$ is greatest for the central residue, which is indicative of a high degree of stacked character for this internal segment.

Kroon *et al.* (1974) reach the same conclusion as Kondo and Danyluk with respect to the pattern of base stacking in ApApA and postulate a time average stacked array with the adenine rings displaced (by rotating ω', ω bonds) from the "normal" expected extended dimer stack to reflect stronger Ap— + —pAp— interaction at H_2 than —pAp— + —pA. In the tetramer and pentamer, the interior units, at least as far as the H_2 shielding criterion is involved, appear to be no more stacked, and are perhaps less stacked, than the ApAp—terminal dimer segment. In the pentamer H_2 of one interior base seems to be more strongly shielded by nearest neighbor adenines. These data point out that the local conformational features definitely vary with oligonucleotide chain length.

So far we have dealt with homogeneous adenine ribonucleotide systems. What is the situation with respect to oligonucleotides containing pyrimidine and other purine bases? Lee *et al.* (1976) and Ezra *et al.* (1977) have extensively applied high-field PMR measurements of chemical shifts, vicinal and four-bond coupling constants, specific deuteration techniques, and computer simulation to arrive at proposed solution conformations of various "mixed" dinucleoside monophosphates as well as dipurine (Pu) and dipyrimidine (Py) dimers. The distillate of these studies is that the conformationally flexible molecules can partition amongst several extended and folded forms. As discussed above for ApA, shifts in furanose ring pucker toward ^3E and decreases in χ (in the *anti* range) accompany stacking tendencies, with Pu—Py dimers being generally higher in this regard than other classes of dimers, the trend progressing Pu—Py > Pu—Pu > Py—Py > Py—Pu. In terms of the backbone torsion angles, $C_{4'}$—$C_{5'}$ is gg (70–90%), $C_{5'}$—$O_{5'}$ is $g'g'$ (70–90%). The torsion about the $C_{3'}$—$O_{3'}$ bond seems to be coupled to the furanose pucker, such that preference for ^3E results in a g^- orientation ($H_{3'}$—$C_{3'}$—$O_{3'}$—P = 34–38°). The percent right-handed stack form is highest in Pu—Py dimers.

Using methods alternative to comparative chemical shifts and vicinal coupling constants, other workers have arrived at somewhat different conclusions than presented above on the preferred glycosyl conformation of dinucleoside monophosphates (that each nucleotide unit is mainly in the preferred *anti* domain of χ). Chachaty *et al.* (1977) studied the 250 MHz PMR of ribo dimers GpU and UpG. They measured spin-lattice relaxation times, T_1, for the various protons using standard inversion-recovery methods (180°–τ–90° sequence). Interproton distances r_{ij} were computed from X-ray data, modulated by ribose dynamic factors such as $^2E \leftrightarrow {}^3E$ equilibria and time-average conformation of the exocyclic methyl carbinol. [They obtained this conformational data from vicinal H–H coupling-constant measurements. It is curious to note that for the guanosine portion of GpU, two key couplings for determining sugar-ring pucker, $J_{1'-2'}$ and $J_{3'-4'}$, differed from those of Ezra *et al.* (1977) by 20%]. A function of the correlation time, τ_c, for isotropic reorientation of the molecule in question (actually either 3'- or 5'-mononucleotide segment) was calculated from the T_1's and calculated r_{ij} using the relationship

$$T_{1i}^{-1} = \frac{3}{10} \gamma_H^4 + \hbar^2 f(\tau_c) \Sigma_i \langle r_{ij}^{-6} \rangle$$

Minima in the plots of an "agreement" factor $\Delta f(\tau_c)/f(\tau_c)$ versus the glycosidic torsional angle show that both *anti* and *syn* conformational states are populated. When this ratio is plotted as a function of the mole fraction of *syn* conformer for both G and U residues, G is found to be predominantly *syn* in both GpU and UpG (75–80%), whereas U is mainly *anti* in GpU (85%) and is more evenly distributed between the two conformational states in UpG (40% *anti*, 60% *syn*). From this study it may be predicted that the two dimers with such high fractions of *syn* glycosyl conformers are primarily not in *anti–anti* base stacked arrangements. Ezra *et al.* (1977) estimate the amount of stacked forms of GpU and UpG are 27 and 10%, respectively.

Hart (1978) has published interesting ^{31}P {^1H} and ^1H {^1H} NOE data on ApA which have been interpreted in terms of a significant proportion of *syn* glycosyl conformer at 25°C. Statistically significant enhancements were obtained for H_8 protons of the 3'- and 5'-adenylyl units upon irradiation of the respective $H_{1'}$ protons, which argues for the *syn* range of χ so that these protons are close to one another. That the *anti* conformer is significant, in agreement with the chemical-shift and coupling-constant data, comes from the simultaneous irradiation of $H_{2'}$ and $H_{3'}$, leading to enhancements in H_8. This was so for both nucleotide segments. Interestly enough, proton–proton interaction between the nucleotide residues was shown by the enhancement of H_8 on Ap— when $H_{1'}$ of —pA was irradiated. Also, when $H_{2'}$ plus $H_{3'}$ of Ap— was irradiated, the area of H_8 of —pA was affected positively. These data confirm the compact folded structure for the dimer. Evidence for close contact of phosphorus with $H_{3'}$ of —pA was forthcoming from the large enhancement of phosphorus when $H_{2'}$ and $H_{3'}$ of —pA were irradiated.

It is of interest to note what effect methyl substitution on the ribose 2'-oxygen has upon the oligonucleotide conformation since these substituted derivatives occur naturally in transfer RNA, eukaryotic messenger RNA, and ribosomal RNA. Data from Sarma's lab (Cheng and Sarma, 1977b; Singh *et al.* 1976) has shown that 2'-O-methylation of the 3'-nucleotide unit of a strongly stacking dimer such as ApA does inhibit the base–base interaction, based upon smaller dimerization shifts and less shift difference between $H_{5'}$ and $H_{5''}$ of the —pA unit. For a pyrimidine system such as C_mpC, the 2'-O-methylation has a less noticeable effect on base–base interaction (Cheng and Sarma, 1977b). Alderfer *et al.* (1974) first reported that 2'-O-methyl groups hindered base stacking as monitored by PMR. Since stacking and furanose pucker shifts to ^3E preference seem to be coupled, the destacking in A_mpA leads to a decrease in percentage of ^3E conformer; χ_1 and χ_2 also increase with the destacking. There is steric interference between the

2'-O-methyl group and the 3'-phosphate such that the conformation about the $C_{3'}$—$O_{3'}$ bond is in the g^+ rather than g^- domain (Figure 2); this finding comes mainly from $^4J_{H_{2'}-P}$ which is 1.2 Hz in A_mpA and only 0.3 Hz in ApA. The g^+ state is apparently coupled to ^2E furanose pucker, so from the drop in ^3E preference mentioned above, the equilibrium $^3Eg^- \leftrightarrow {}^2Eg^+$ is shifted to the right for A_mpA. Undoubtedly the 2'-O-methyl–backbone-phosphate interaction results in an alteration of the average conformation about P—$O_{3'}$ and P—$O_{5'}$.

In conjunction with the smaller effect of 2'-O-methylation on stacking in CpC versus ApA there are concomitantly lesser effects on other conformational features of the pyrimidine system (Cheng and Sarma, 1977b). In the Cp group there is a small shift towards a higher proportion of ^3E pucker and an associated decrease in χ. Again, there is some small perturbation to the ω and ω' torsion as in A_mpA. Conformer populations about $C_{1'}$—$C_{5'}$ and $C_{5'}$—O_5, are unaffected in C_mpC, whereas in A_mpA the latter shifts slightly towards higher $g'g'$.

2.1.2. Deoxyribose Systems. It may be seen from the foregoing discussion that in single-stranded ribo furanose oligonucleotides, the preferred conformational properties markedly change as chain length increases, particularly in systems in which there is a high degree of base stacking. The conformational details of each nucleotidyl residue is dependent upon the nature of neighboring bases, 2'-O-methylation of the sugar, and whether or not the unit is in exterior or interior location in the chain. Thus small single-stranded oligoribonucleotides are not really good models for polyribonucleotides.

What about deoxyribonucleotide systems? The NMR evidence suggests that the preferred conformational features of the monomer are preserved with chain lengthening through dideoxyribonucleoside monophosphates (Ts'o, 1974b; Cheng and Sarma, 1977a) and higher oligomers (Cheng et al. 1978; Altona et al. 1976). At the dimer level, on the basis of dimerization shift and coupling-constant data (Cheng and Sarma, 1977a), the preferred solution conformations include the following properties:

χ_{CN}, anti, $\chi_1 < \chi_2$

^2E pucker, irrespective of nature of base and stacking propensity

ω', $\omega = g^-g^-$ (except d-TpT and d-TpC which are tg^-)

ψ_1, ψ_2, ($C_{4'}$—$C_{5'}$) = gg (60°)

ϕ_2 ($C_{5'}$—$O_{5'}$ of 5' unit) = g'g'(180°)

$\phi_{1'}$($C_{3'}$—$O_{3'}$ of 3' unit) ~200°

These parameters characterize a predominantly *anti, anti* right-handed stacked dimer with 2'-endo pucker in both deoxyribose rings. Raising the temperature to affect destacking has its largest effect on the P—$O_{3'}$ where ω'

goes from g^- to t. The extent of stacking appears to be greater than in ribose systems, perhaps due to the lack of steric interference of the 2'-OH (Fang *et al.*, 1971).

Cheng *et al.* (1978) have examined the 270 MHz PMR of the deoxy trinucleoside diphosphates d-TpTpC and d-TpTpT and found that these trimers retain the conformational features of the constituent dimers.

Another deoxy trimer, d-TpTpA, has been studied by Altona *et al.* (1976) using PMR at 360 MHz. Using vicinal coupling and chemical shifts they find that the time-average pucker for the three sugar rings is 62–65% biased toward ^2E, which is essentially the same in deoxy dimers and monomers. The fraction of ^2E conformer increases somewhat for the interior Tp from dimer to trimer, 63–72%. The backbone conformation maintains the characteristics of the component dimer, with the values of the various torsions very close to those found by Cheng and Sarma (1977a). Again, the glycosyl conformations are predominantly *anti* and the most abundant trimer conformer is one in which the bases are stacked and the backbone assumes a right-handed helical twist.

Adding the 5'-adenyl unit to d-TpT results in significant (0.08–0.09 ppm) upfield shifts of H_6, $H_{1'}$, and the 2'-protons of the 3'-terminal thymidine. These long-distance shielding effects are thus of next-nearest-neighbor origin as first proposed by Kan *et al.* (1973). Rather than an effect propagated within the framework of the major stacked right-handed helical conformer, Cheng and Sarma (1977a) believe these upfield shifts arise from a minor conformer in which the two sets of phosphate torsions, ω_1', ω_1 and ω_2', ω_2 are g^+g^+ and the central Tp is looped out of the stack, allowing interaction between the terminal bases in a somewhat compressed trimer. This suggestion does not seem to be entirely satisfactory either, since the $H_{5'}$, $H_{5''}$ of the terminal 3'-Tp residue should also be shielded by the adenine ring, yet these protons are only shifted upfield approximately 0.04–0.05 ppm.

Due to the high degree of complexity at the trinucleotide level (see Figure 3 for d-TpTpA) it is probably not feasible to completely analyze the spectra of larger chains, even at 360 MHz. We will have to await the next generation spectrometers at 600 MHz or greater for these experiments.

2.2. Oligonucleotide Duplexes

The formation of double-stranded complexes by complementary oligonucleotides in H_2O solution may be followed by PMR monitoring of the hydrogens involved in base-pairing interactions, both ring imino protons (−9 to −15 ppm from DSS) and exocyclic amino protons (−6 to −9 ppm). For large nucleic acids such as transfer RNA, resonances due to base pairing —NH protons are clearly present at room temperature and persist with temperature increase until the order–disorder transition point, T_m, has been reached, as first shown by Kearns *et al.* (1971). Peaks of imino protons may broaden and disappear altogether at temperatures below the optical T_m of the com-

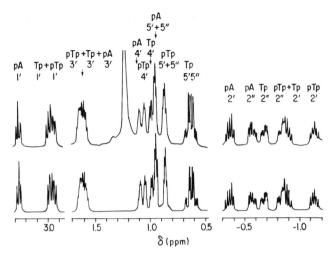

Figure 3. 360 MHz PMR spectrum of the sugar protons of d-TpTpA, 65°C, pD 7.4; upper plot is experimental, lower is computer simulated (from Altona *et al.*, 1976). Permission from Federation of European Biochemical Societies.

plex. Amino protons involved in hydrogen bonds exchange out at somewhat higher temperatures (Crothers *et al.* 1973). It is usually necessary to examine small complementary oligonucleotide duplexes at low temperatures in order to "see" the hydrogen-bond interactions.

The exchange processes governing the line broadening can be generalized by the following:

$$\text{helix} \xrightleftharpoons[K_{\text{closed}}]{K_{\text{open}}} 2 \text{ coil} \xrightarrow{K_{\text{ex}}} H_2O$$

where a proton involved in a hydrogen bond in a duplex can reside on either strand as the helix opens with a rate K_{open} and then can exchange with solvent. Under many situations the rate of exchange is determined by the rate of helix opening so that $K_{\text{ex}} \gg K_{\text{open}}$.

For dinucleotides, complementary hydrogen bonding with amino, but not imino protons has been observed with d-pGpC in H_2O at 2.2°C (Krugh and Young, 1975). The noncomplementary d-pGpT did not give evidence of hydrogen-bonding interaction.

The duplex structure of several complementary deoxyoligonucleotides has been probed by Patel and coworkers by PMR at 270 and 360 MHz and ^{31}P at 145.7 MHz. These include G–C tetra- and hexanucleotides (Patel, 1976, 1977, 1979a), the mixed hexanucleotide d-ApTpGpCpApT (Patel, 1974), and the octanucleotide d-GpGpApApTpTpCpC (Patel and Canuel, 1979). In practical terms, one must have at least a tetranucleotide in order to observe imino proton hydrogen bonding at temperatures above 0°C. The various complementary G–C containing oligomers examined included d-GpCpGpC, dCpCpGpG, dGpGpCpC, d-CpGpCpG. In these systems double-strand helix formation could be ascertained by the broad resonances at about -13 ppm in H_2O solution below $\sim 10°C$. Of these guanine imino resonances, those assigned to terminal base pairs were most readily exchanged with solvent due to incipient helix opening as temperature was raised. This behavior may be termed "fraying." Cytosine amino protons involved in hydrogen bonding resonate around -8 to -9 ppm, while amino protons not engaged in hydrogen bonding but exposed to solvent produce signals at -6.5 to -7.5 ppm. These two-proton peaks, corresponding to terminal and internal residues, finally exchange out about 30–35°C. Rates of amino group rotation intermediate on the NMR time scale result in the inability to observe guanine 2-amino protons until about 10°C below the melting temperature ($T_m \sim 50°C$).

Chemical shift versus temperature plots for these oligonucleotides show large upfield changes with lowered temperature for base (particularly H_5 of cytidine) and $H_{1'}$ protons, indicative of base stacking as duplexes are formed. This is particularly true for internal bases, whereas terminal bases are somewhat less dramatically effected. The upfield shifts of terminal base protons continue after internal proton shifts have leveled off, which may be due to end-to-end stacking of the duplexes. Other sugar protons beside $H_{1'}$ were little affected.

Using T_1 measurements, it has been possible to show that the predominant glycosic conformations are *anti*. This is done by noting that the ratio $T_1(H_8)/T_1(H_{1'})$ which Akasaka (1974) showed was about 0.5 for *anti* conformations and about 1.5 for *syn* orientations, equals 0.6. Other conformational features which can be discerned include the predominate pucker for the sugar ring, ^2E, from $J_{1'-2''}$ and $J_{1'-2''}$. No other vicinal couplings could easily be determined in these molecules due to the complex spectra. However, the downfield shift of ^{31}P resonances with temperature increase as the duplexes give rise to coils is indicative of an alteration in ω', ω about $P—O_{3'}$, and $P—O_{5'}$ bonds from gg (stacked) to gt (unstacked) (Gorenstein, 1975; Gorenstein and Kar, 1975).

So, as might be predicted from the earlier discussions, the main conformational features of higher-order deoxyribose oligonucleotides, insofar as they may be determined, are essentially the same as found in monomers, dimers, and trimers. The small duplex deoxynucleotide helices are most closely related to the B form of DNA. This fact can be determined by (1) taking the coordinates derived from X-ray analysis for the various base-overlap geometries, A, B, and D (Arnott et al., 1973; Arnott and Hukins, 1973; Arnott et al., 1974), (2) using these distances in calculating ring-current effects at given protons with shielding contours as determined by Giessner-Prettre et al. (1976), and (3) comparing these computed nearest-neighbor-ring-current induced shifts with those experimentally found upon duplex formation.

In a study of a larger complimentary axially symmetric deoxyoligonu-cleotide, d-GpGpApApTpTpCpC, Patel and Canuel (1979) have shown that this species has similar properties compared with the shorter G–C oligo-mers, albeit a lower T_m, ~35°C (0.1 M NaCl) due to the lowered stability of the A–T pairs. Two hydrogen-bonded thymidine —N_3 proton peaks are downfield (−13.75 and −13.85 ppm) from the guanosine —N_1H resonances (−12.75 and −12.95 ppm). The lower field G–N_1H peak exchanges out ~14°C below the T_m and is assigned to the terminal G–C base pairs. The other imino proton peaks broaden out at the duplex-strand transition.

Comparison of upfield shifts on nonexchangeable base proton resonances upon duplex formation with those calculated from ring-current shielding effects show the B-DNA geometry gives the closest fit. These comparisons are somewhat crude and the resonances from internal bases have not been assigned.

The ^{31}P shifts as a function of temperature seem to monitor duplex ↔ strand equilibria with the added interesting overtone that each resolved peak (5 out of 7) shows a different curve, which seems to indicate differential ω',ω rotations. This behavior was also seen for d(CpG)$_3$ (Patel, 1976).

As an example of a minimal ribose helical duplex, r-CpCpGpG was studied by PMR at 220 MHz by Arter et al. (1974). As might be expected, this symmetric molecule displayed two resonances in the guanine–N_1H region, −12.4 to −13.2 ppm from DSS. The lower field resonance disappeared at 17°C, whereas the peak at −12.4 ppm was exchanged out at 26°C. So, in a similar manner as in the deoxy series, the more rapidly exchanging reso-nance would be assigned to terminal base pairs undergoing early helical opening (fraying) as temperature is increased. The calculations of Arter et al., (1974) show that the positions of hydrogen-bonding imino proton peaks are most nearly comparable with chemical shifts calculated using A-RNA (11-fold helices) crystal coordinates (Arnott et al., 1969). The nonexchange-able G_{H_8} protons are upfield shifted to a larger extent upon duplex formation than would be expected on the basis of forming a helix with A-RNA geometry. This may be due to end-to-end stacking of duplexes, particularly on the guanine face. No information was presented on the furanose ring pucker or backbone conformation of this tetranucleotide duplex.

Ts'o and coworkers (Borer et al., 1975; Kan et al., 1975) studied the self-complementary hexaribonucleotide ApApGpCpUpU. The 100 and 220 MHz PMR data obtained with this molecule were similar in many aspects of duplex structure as reported by Arter et al. (1974) on r-CpCpGpG:

1. The terminal base pair imino protons are most sensitive to thermal per-turbations and their behavior can be due to fraying of the helix ends.
2. The fit between experimental coil → duplex induced shifts and those calculated from ring-current shielding curves and A-RNA (11-fold helices) or A'RNA (12-fold helices) geometry was better than for B-DNA geometry.

3. The predicted shifts for the terminal adenine base protons was less than experimental, probably due to duplex end-to-end stacking.

Additionally, these workers showed that increase in temperature leads to a corresponding increase in $^3J_{1'-2'}$. At high temperature ($\sim$80°C), the hexamer values are close to those for the monomers themselves; at low temperature, the couplings become quite small ($\sim$1.5 Hz). This observation is consistent with the earlier discussions on the $^2E \rightarrow {}^3E$ shift in the furanose pucker equilibrium which occurs when ribose mononucleotides are oligomerized, particularly in base-stacked systems.

Davanloo et al. (1979a) have employed ^{31}P relaxation and chemical-shift measurements in studies of deoxyribooligonucleotides, including various dinucleotide diphosphates and the octamers d-$(pA)_8$ and d-$(pA)_3$ pGpC$(pT)_3$. The pH-sensitive terminal phosphate resonance was found $\sim$4 ppm downfield from the internucleotide ^{31}P resonance. Only the latter was temperature sensitive, reflecting rotation about P—$O_{3'}$ and P—$O_{5'}$ bonds (Gorenstein, 1975; Gorenstein and Kar, 1975) as well as changes in ring-current effects of the bases. The chemical shift of the average internucleotide phosphate in the octamer d$(pA)_8$ was different than in d$(pA)_2$, probably reflecting differences in the time average ω',ω values and adenine ring-current effects. In sequence isomers, d-pGpT versus d-pTpG, the phosphodiester chemical-shift difference was 0.6 ppm, (again undoubtedly due to differences in ω',ω and base overlap geometry).

These T_1 and NOE measurements are particularly useful since they provide an assessment of the motional dynamics of the oligonucleotide backbone. For purely dipolar interactions between ^{31}P and nearby protons (and in the region of motional narrowing due to rapid molecular motion), NOE (max) = 2.24 ($n + 1$). The dipolar T_{1D}, is related to NOE (obs) by

$$T_{1D} = T_{1(obs)} \frac{NOE\ (max)}{NOE\ (obs)}$$

and

$$\frac{1}{T_{1(obs)}} = \frac{1}{T_{1D}} + \frac{1}{T_{1,CSA}} + \frac{1}{T_{1,SR}} + \frac{1}{T_{1P}}$$

where CSA, SR, and P are respectively the relaxation contributions from mechanisms due to shift anisotropy, spin rotation, and paramagnets. The NOE measurements of Davanloo et al. (1979a) were 30–80% of the dipolar maximum, which illustrates that dipole–dipole interactions are the chief contributors to ^{31}P relaxation in these oligonucleotides. Assuming overall isotropic tumbling of the molecules, one can extract, having measured $1/T_1$, an overall isotropic correlation time, τ_R, from a plot of calculated $1/T_{1D}$ versus τ_R, in which the $1/T_{1D}$ values were derived from a relaxation model in which ^{31}P in the internucleotide bond interacts with the two adjacent $C_{5'}$

protons and the $C_{3'}$ proton. For the dinucleotides, one obtains $\tau_R = r \times 10^{-10}$ s at 16°C and for the octamers, $\tau_R = r \times 10^{-9}$ s at 18°C.

The question arises as to the possible modulation of this overall motion by internal motion along the ribose–phosphate backbone ($\Delta\omega'$, $\Delta\omega$) as established by ^{31}P shifts as a function of temperature. Using a model configuration whereby such an internal motion would consist of rotation about the P—O bond and the vectors between P and $C_{5'}$ and $C_{3'}$ protons would have an angle of 60° with respect to the axis of rotation, Davanloo et al. (1979a) found that in order for internal phosphate flexing to be observed in these small molecules, the correlation time for such motion would necessarily have to be 10^{-10} or faster, which is not observed. These internal motions may be on the same scale as the overall isotropic tumbling (10^{-9} s) and thus not discernible. As we shall mention in Section 4, for large polynucleotides with slower overall tumbling rates, the internal backbone motion on this time scale must be reckoned with as a contributor to overall ^{31}P relaxation rates.

3. TRANSFER RNA

Over the past few years, transfer RNA as a class of important biopolymers has attracted a large following of magnetic resonance practitioners who have employed PMR, ^{13}C, ^{19}F, and ^{31}P techniques to probe its solution structure and elucidate dynamic properties. Numerous excellent reviews have appeared (or will soon be published): by Sigler (1975), Kearns (1976), Bolton and Kearns (1978), Patel (1978a), Redfield (personal communication), and Reid and Hurd (1977; personal communication). These works have primarily stressed the exciting PMR data for low-field exchangeable imino and amino protons and the nonexchangeable high-field protons. In the present account I will focus primarily on other areas, including specifically ^{13}C-labeled tRNA and the function of unusual nucleosides, but initially will deal with some of the latest interesting findings from PMR investigations of hydrogens involved in secondary and tertiary structural interactions and dynamics.

3.1. PMR Investigations

3.1.1. Solution Structure. Figure 4 displays the three-dimensional diagram for crystalline yeast tRNA^{Phe} determined using X-ray diffraction techniques (Robertus et al., 1974; Kim et al., 1974; Rich and RajBhandary, 1976). Hydrogen bonds between base pairs in helical regions are denoted by open bars, tertiary hydrogen-bond interactions are signified by solid bars, and short rods represent non-hydrogen-bonded residues. Numerous NMR investigations have been directed towards finding an answer to the questions as to the similarity of the tertiary structure of tRNAs in solution compared with that in the solid state. If one takes atomic coordinates from the refined structure (Ladner et al., 1975; Quigley and Rich, 1976; Sussman and Kim, 1976) of yeast tRNA^{Phe}, uses ring-current shielding values (Giessner-Prettre et al., 1976; Arter and Schmidt, 1976) and intrinsic chemical shifts for

3-D structure of yeast tRNA^{Phe}

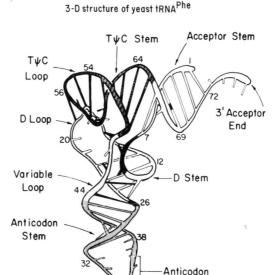

Figure 4. Three-dimensional structure of yeast tRNA^{Phe} as determined by Robertus *et al.* (1974); Kim *et al.* (1974).

isolated base pairs, spectra can be computed which agree quite well with experimental results (Robillard *et al.*, 1976, 1977). Some discrepancies between simulated and experimental spectra have been obtained (Geerdes and Hilbers, 1977; Kan and Ts'o, 1977). In these instances different sets of ring-current shielding values and/or intrinsic chemical shifts were used. By and large the calculated and experimental spectra are close enough to instill some degree of confidence that on the whole, crystal and solution structures are quite similar.

Work is proceeding on the assignment of secondary and tertiary —NH and —NH$_2$ resonances for a variety of tRNAs in the -9 to -15 ppm region. There are some 20 plus secondary H— bonds in addition to 7 or more tertiary interactions which makes for a difficult but necessary task if the solution structure of tRNA is to be known. As an example of this activity, Hurd and Reid (1979) have recently assigned the imino resonance due to the m^{7}G46—G22—C13 triple tertiary interaction in several tRNAs. The existence of such an interaction was postulated from the crystal structure (Ladner *et al.*, 1975; Quigley and Rich, 1976; Sussman and Kim, 1976). Figure 5 illustrates the effect of mild acid catalyzed removal of m^{7}G on the low-field PMR spectrum of *E. Coli* tRNA$_I^{Val}$. The peak labeled ''F'' at -13.4 ppm disappears. This procedure was repeated for other m^{7}G-containing tRNAs with similar results, establishing the identity of this tertiary imino hydrogen bond. Other perturbations are also manifested in the spectrum due to m^{7}G removal. These perturbations could be due to minor changes in the secondary and tertiary structure in the region of the excision.

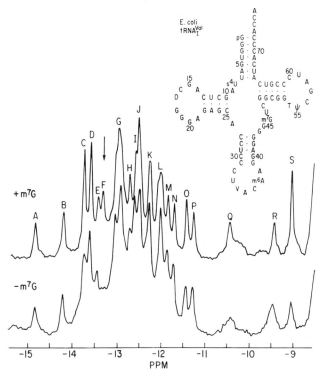

Figure 5. Low-field PMR spectrum of *E. coli* tRNA$_1^{Val}$ before and after m^{7}G46 removal; 45°C in 10 m*M* cacodylate, pH 7.0, 100 m*M* NaCl, 15 m*M* MgCl$_2$ [reprinted with permission from Hurd and Reid (1979), *Biochemistry*, **18**, 4017–4024, copyright by the American Chemical Society].

Johnston and Redfield (1978, 1979) have developed a clever selective NOE procedure for assigning specific proton resonances in tRNA spectra. This technique is illustrated in Figure 6, from the work of Hurd and Reid (1979) for *E. coli* tRNA$_1^{Val}$ at 35°C. The upper spectrum in both portions of the figure is that obtained normally without preirradiation. Below this spectrum in the top half of the figure is a difference spectrum obtained by subtracting a spectrum in which the line at −11.95 ppm was preirradiated with a weak 100 ms pulse from a "control" spectrum in which the preirradiation pulse was set in a "blank" region, for example, −11.6 ppm. The same process is repeated in the lower half of the figure, except the preirradiation pulse was set at −11.35 ppm. It is clear that the only lines affected are the two at −11.95 and −11.35 ppm. The phenomenon observed in this experiment is the transfer of magnetization from the irradiated proton spins to nearby protons. Since this saturation transfer is strongly distance dependent (approximately $1/r^6$), only protons separated by about 3.5 Å will undergo the interactions. The NOE occurs between protons in this low-field chemical shift region, where imino protons resonate. The only imino protons which would be within 3.5 Å are those involved in the G50—U69 wobble base pair in the acceptor stem.

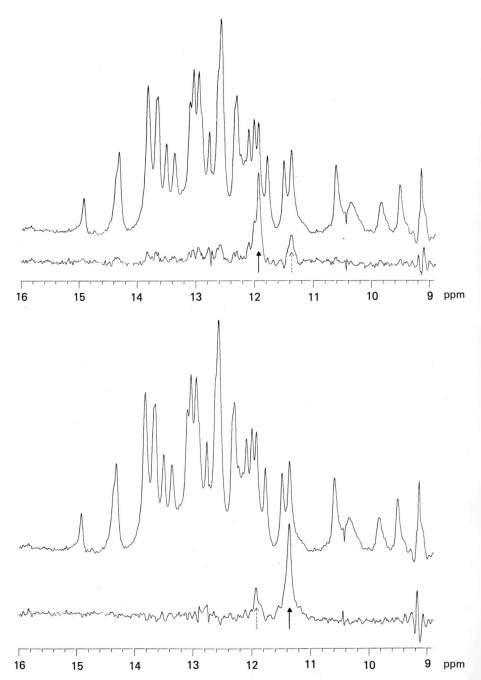

Figure 6. Saturation transfer experiment with tRNA$_i^{Val}$. Identification of G–U base pair (see text).

U69   G50

G–U base pairs in *E. coli* tRNA[fMet] and yeast tRNA[Phe] have been similarly assigned (Johnston and Redfield, 1979). Thus one observes in solution an interaction postulated in the yeast tRNA[Phe] crystal (Ladner *et al.*, 1975).

Johnston and Redfield (1978, 1979) have observed NOEs between other closely spaced protons. For instance, in yeast tRNA[Phe], irradiation of a resonance in the AU imino proton region at −14.35 ppm results in a 30% NOE of a peak at −7.73 ppm which is in the chemical-shift region where aromatic carbon-bound hydrogens resonate. This interaction may arise from either of two reverse Hoogsteen base pairs:

U8   A14

m¹A58   T54

Studies such as these in addition to others involving base substitution, omission, and chemical modification have led to a fairly complete assignment of tertiary imino proton resonances (Reid *et al.*, 1979).

3.1.2. Conformation and Dynamics.   Having the necessary fairly complete assignment scheme for imino protons involved in tertiary hydrogen bonds, one is able to obtain information on conformational features and dynamic properties via the temperature dependence of the exchange of these protons with solvent. In this way a comprehensive picture of the sequential melting of various hydrogen-bonded regions may be obtained, with or without the presence of magnesium ion, known to stabilize tRNA tertiary structures.

Redfield and coworkers (Johnston and Redfield, 1979; Johnston *et al.*, 1979) have extensively studied the exchange kinetics of hydrogen-bonded protons of yeast $tRNA^{Phe}$ at 270 MHz. As discussed above for oligonucleotide duplexes, the lifetimes of these hydrogen-bonded protons in various states can be discussed in terms of interconversion between several states.

$$hH \xrightleftharpoons[k \text{ closed}]{k \text{ open}} hC \xrightarrow{k_{ex}} H_2O$$

where H denotes the helical or hydrogen-bonded state, and C = the coil or open state from which exchange with solvent can occur, governed by k_{ex}. The rate of solvent exchange, R, is

$$R = \frac{k_{open} \cdot k_{ex}}{k_{closed} + k_{ex}}$$

At low temperatures (<25°C), R in the range of about 0.005 to 10 s^{-1} can be measured by real-time solvent exchange techniques as described by Johnston *et al.*, (1979). These procedures involve rapid passage of H_2O solutions of tRNA through Sephadex-G25 gel-filtration columns equilibrated with D_2O at 0°C followed by collection and Fourier Transform NMR measurement at the temperature of interest.

At higher temperature, where the exchange rates are above 16 s^{-1}, R can be obtained as the rate of recovery of magnetization in a saturation recovery experiment of the type similar to that described above for NOE experiments, with the modification that a delay time, τ_T, is used between the preirradiation and observation pulses. The intensity of an irradiated line with a given delay τ_T can be written

$$A_T = A_\infty [1 - e(-\tau_T R)]$$

By plotting the log of the line intensity, A_T versus τ_T, one can obtain R, the recovery rate, which is the rate of helix opening provided $k_{ex} \gg k_{open}$.

Using these techniques, Johnston and Redfield (1979) found that in yeast $tRNA^{Phe}$ in the absence of Mg(II), the following "melting" sequence occurs, as noted by the exchange rates: (1) 23 to 36°C, partial melting (exchange with solvent) of tertiary structure, including T54—m^{1}A58, U8—A14, m^{7}G46—

G22, G15—C48 and some acceptor-stem secondary resonances such as G4—U69, A5—U68, A67—U6 and G1—C72, A31—ψ39 in the anticodon stem; (2) 36 to 42°C, complete loss of tertiary structure and acceptor-stem hydrogen bonds; 3° hydrogen bonds are characterized by rates 60 to 120 s^{-1}, while 2° hydrogen bonds have rates 28 to 45 s^{-1}; (3) between 42 and 51°C dihydrouridine and anticodon stems melt; at 51°C, there is an estimate that 70% of the TψC stem is still intact.

With Mg(II) stabilization (15 mM), no melting occurs to 47°C; at 47°C m^{7}G46—G22, G4—U69 and A31—ψ39 show increased exchange; from 50 to 61°C the A31—ψ39, m^{7}G46—G22, and G15—C48 resonances disappear and the acceptor stem partially melts. Generally A–U base pairs showed faster exchange kinetics than did the G–C base pairs, indicating they open more readily upon thermal perturbation.

Reid and Hurd (1977) have interpreted their temperature-melt data on *E. coli* tRNA^{Phe} to mean that the last helical region to open was the acceptor stem. This helix has six G–C pairs and one A–U pair at position 7. This high G–C content confers added stability. By studying *E. coli* tRNA^{Phe} at 58°C where only the acceptor stem helix is still closed, Reid and Hurd (personal communication) have used saturation-recovery techniques as described above to measure lifetimes of individual base pairs in the acceptor stem. An example is shown in Figure 7 for the G3—C70 base pair. The recovery rate or exchange rate is 25 s^{-1}.

3.2. ^{13}C NMR Investigations

An important alternative technique to observation of protons in tRNA is the use of ^{13}C NMR, particularly utilizing site specific ^{13}C enrichment. This technique overcomes the inherently poor sensitivity and peak overlap of natural-abundance approaches and allows one to capitalize on the advantages of ^{13}C NMR including broad chemical shift range, and relative ease in relating relaxation data to dynamic features of the molecule. Komoroski and Allerhand (1972, 1974) first studied tRNA by ^{13}C NMR. Incorporation of ^{13}C-enriched methyl groups into tRNA of a methionine auxotroph of *E. coli* was accomplished by Agris *et al.*, (1975). These workers have recently reported on a followup study on unfractionated tRNA from this auxotroph (C6) in which the assignments of the ^{13}C-enriched methyl groups were made (Tompson *et al.*, 1979). These assignments were based upon comparison of chemical shifts with the modified nucleosides of known structure and from peak intensities in relation to their prevalence in tRNA.

Hamill *et al.*, (1976, 1980) have also made use of a nutritional auxotroph in the incorporation of 90% [4-^{13}C]uracil into tRNA of *Salmonella typhimurium* JL-1055, a uracil-requiring strain. Most of the unusual uracil nucleosides found in tRNA involve some modification at the 5 position so enrichment at C$_4$ provides potential opportunity to monitor readily the behavior of these modified compounds as well as uridine itself in various

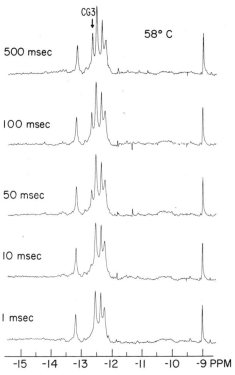

Figure 7. Saturation recovery experiment for the GC$_3$ base pair of the acceptor stems of *E. coli* tRNA^{Phe}, 58°C, 5 m*M* phosphate, pH 7, 100 m*M* NaCl. (See text; Reid and Hurd, personal communication.)

locations in the polymer backbone. In addition, C$_4$ is a quaternary carbon with a relatively long T_1 (~1.5–3 s), so the resonance is narrow.

Figure 8 displays the 25 MHz ^{13}C spectrum for the enriched unfractionated *S. typhimurium* JL-1055 tRNA (150 mg/ml H$_2$O), taken at 37°C. The chemical shift scale is related to TMS. Mass spectrometric analysis showed that the ^{13}C isotopic enrichment in uridine and pseudouridine was 45–46%. C$_4$ carbons of 4-thiouridine (s^{4}U) and dihydrouridine (D) were readily assigned by comparison with model nucleosides. This is the first reported observation of s^{4}U in tRNA by ^{13}C NMR. Pseudouridine (ψ) and ribothymidine (rT) C$_4$ resonances fell in the chemical shift band of the unmodified uridines.

Note the narrow dihydrouridine line, which may be due to faster motion of the tRNA segment containing this residue compared with the overall dynamics of the tRNA. Note also the 4–5 ppm range in chemical shifts for the uridines, which is a sampling of the divergent molecular environments for these residues at various locations in the polymer. As temperature is increased, the uridine band narrows between 45–60° as secondary and ter-

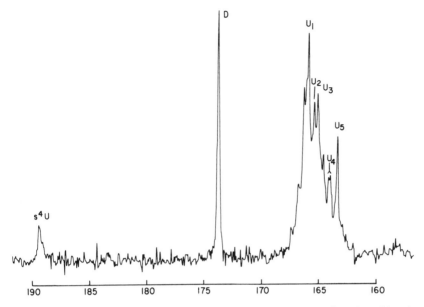

Figure 8. 25 MHz ^{13}C NMR spectrum of [4-^{13}C]uracil-labeled bulk tRNA from *S. typhimurium*, JL-1055, 140 mg/ml, 40 mM MgCl$_2$, 2.2 mM EDTA, 2 mM dithiothreitol, 1% dioxane. Chemical shifts were measured from dioxane and converted to TMS by adding 66.3 ppm (Hamill *et al.*, 1980).

tiary structure melts out. The T_1 and NOE for C$_4$ of D increase, indicating D and T loops are no longer interacting via tertiary hydrogen bonds. The narrowing of the uridine band is complete at 82°C. At this temperature a small resonance 1 ppm upfield of the main uridine band at 165.2 ppm was assigned to ribothymidine.

A model to account for the ^{13}C relaxation data is proposed which assesses contributions from proton dipolar, ^{14}N dipolar and chemical shift anisotropy mechanisms. Using this model the rotational correlation time for overall motion of the tRNA, τ_R, calculated from the T_1 data, was 3×10^{-8} s. Relaxation of unmodified uridines at low temperature ($<45°$C) is dominated by the overall tRNA motion. However, the data on dihydrouridine is indicative of more rapid motion than experienced by the overall tumbling, suggesting that the D-containing segment is more flexible than the tRNA molecular framework.

In order to more readily correlate ^{13}C-NMR parameters with molecular structure and dynamics, and to relate these properties to biological function, we have begun a program to chromatographically separate and purify individual isoaccepting tRNA species from ^{13}C-labeled *E. coli* bulk tRNA (Schweizer *et al.*, 1980). The uracil-requiring *E. coli* strain SO-187 was grown with 90% [4-^{13}C]uracil and the bulk tRNA isolated. Mass spectrometric analysis showed that the uridines were 75% enriched in ^{13}C at C_4 whereas cytidine was only 2–3% labeled. tRNA$_\text{I}^\text{Val}$ was purified by sequential chromatography on columns of BD cellulose, DEAE-Sephadex A50, and Sepharose 4B as suggested by Reid *et al.*, (1977). The cloverleaf structure of *E. coli* tRNA$_\text{I}^\text{Val}$ with labeled uridines and their structures are shown in Figure 9. In Figure 10a the 25 MHz ^{13}C spectrum of a 23 mg/ml D$_2$O solution of tRNA$_\text{I}^\text{Val}$ (~1 mM) is displayed. The solution contains 30 mM K$_2$HPO$_4$ (pD = 7.3), 15 mM MgCl$_2$, 150 mM NaCl, 1.5 mM EDTA, 1.5 mM Na$_2$S$_2$O$_3$, and 0.03% NaN$_3$. 16,304 transients were accumulated and transformed. 4-thiouridine (1) and dihydrouridine (2) are readily apparent. The resolution of the uridine band (3–10) is much better than in Figure 9, as is to be expected since the overlap of 50–60 individual polymers is no longer present. By comparison with authentic material, peak #11 is tentatively assigned to V34 in the anticodon loop, which is uridine-5-oxyacetic acid (Murao *et al.*, 1970). Figure 10b shows the labeled uridines in *E. coli* tRNA$_\text{I}^\text{Val}$ superimposed on the yeast tRNA$^\text{Phe}$ backbone.

We have used paramagnetic metal-ion binding in efforts to assign the resonances arising from common uridine residues in addition to ribothymidine and pseudouridine (#3–10). Cobalt [Co(II)] and manganese [Mn(II)] supposedly bind to the hinge or bend region as determined crystallographically (Jack *et al.*, 1977; Holbrook *et al.*, 1977; Quigley *et al.*, 1978). Hurd *et al.*, (1979) have used these paramagnetic effects as an aid in assignment of tertiary imino proton resonances. Figure 10c displays the three-dimensional structure of tRNA$_\text{I}^\text{Val}$ with the Mn(II) and Co(II) sites as found in yeast tRNA$^\text{Phe}$ (Jack *et al.*, 1977).

The Mn(II) and Co(II) effects are shown in Figures 11 and 12. The initial binding sites appear to be different for the two paramagnetic metals, as can be noted by the dihydrouridine peak which is affected before s^{4}U upon Mn(II) addition. With Co(II), s^{4}U goes first. These data are not completely analyzed, but it is noteworthy that Mn(II) appears to bind closer to D than Co(II), which appears to bind closer to s^{4}U, in agreement with the X-ray data on tRNA$^\text{Phe}$ of yeast. Both titrations suggest that the metals bind in the bend region and that residues as V34, U33, U25 in the anticodon would be expected to be among those least affected. A preliminary assignment scheme is presented in Table II, based upon the metal-ion effects and the attendant distances between metal and uridine residues.

3.3. Function of Modified Nucleosides

Many exotic nucleosides have been found in transfer RNA (McCloskey and Nishimura, 1977). Their functional roles in the biochemistry of tRNA have been somewhat difficult to pin down. The highly substituted purine nucleosides located at the 3′ end of the anticodon triplet sequence have been found to be necessary for proper codon–anticodon interaction, but the mechanism whereby their influence is exerted remains somewhat obscure

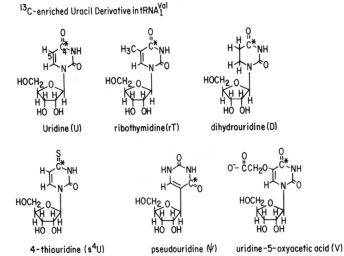

E.coli B and K 12 tRNA$_I^{Val}$ Cloverleaf structure

^{13}C-enriched Uracil Derivative in tRNA$_I^{Val}$

Uridine (U) ribothymidine (rT) dihydrouridine (D)

4-thiouridine (s^{4}U) pseudouridine (ψ) uridine-5-oxyacetic acid (V)

Figure 9. Cloverleaf structure of *E. coli* tRNA$_I^{Val}$ with labeled uridine residues underlined and molecular structure of these labeled uridines.

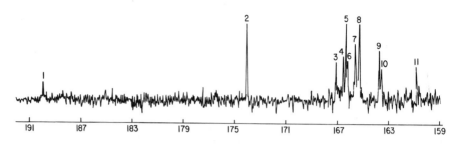

(a)   25 MHz ^{13}C-spectrum of 4-^{13}C-Uracil labeled tRNA$^{Val}_I$ from *E.Coli* SO-187

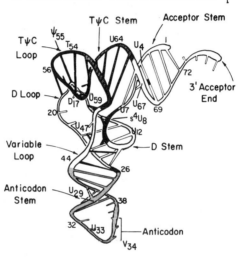

(b) 3-D framework with ^{13}C-enriched Uracils as in tRNA$^{Val}_I$

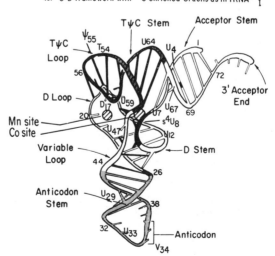

(c) 3-D framework with ^{13}C-enriched Uracils as in tRNA$^{Val}_I$

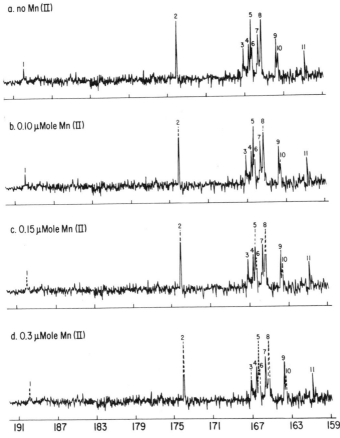

Figure 11. Titration of [4-¹³C]uracil labeled tRNA$_I^{Val}$ with Mn(II).

and complex. Recently an interesting facet of the potential role for a hydrophilic anticodon adjacent modified nucleoside, N-[9-(β-D-ribofuranosyl) purin-6-yl-carbamoyl] threonine, t⁶A, was revealed. Biochemical data (Miller et al., 1976) indicating that this residue may be a site for magnesium binding in tRNA was confirmed by carbon-13 NMR data on the nucleoside itself, using manganese as a paramagnetic analog (Schweizer and Hamill, 1978). Metal binding primarily to the carboxyl and secondarily to the base (N₇) and probably ureido carbonyl is indicated, since the carboxyl carbon intensity dropped first, followed by C₅ and NHC̲(O)NH. Potentiometric

Figure 10. (a) 25 MHz ¹³C NMR spectrum of [4-¹³C]uracil-labeled tRNA$_I^{Val}$ from E. coli SO-187. (See text for sample conditions.) (b) Three-dimensional framework of yeast tRNA^{Phe} with labeled uracils in E. coli tRNA$_I^{Val}$ superimposed. (c) Same as in (b), with Mn(II) and CO(II) sites as determined crystallographically (Jack et al., 1977).

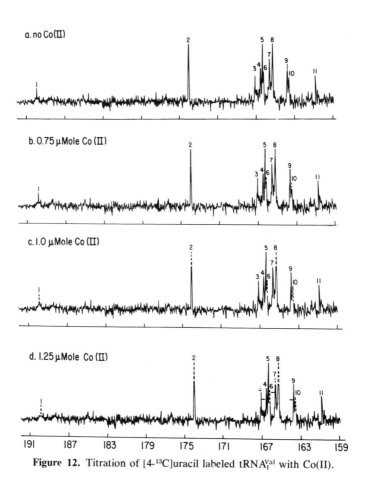

Figure 12. Titration of [4-^{13}C]uracil labeled tRNA$_I^{Val}$ with Co(II).

Table II. Assignment of resonances in 25 MHz spectrum of [4-^{13}C] uracil-enriched *E. coli* SO-187 tRNA$_I^{Val a}$

Peak Number (Chemical Shift)[b]	Assignment	Nucleoside Standard (Chemical Shift)[b]
1 (189.2)	4-Thiouridine	4-Thiouridine,s^{4}U (190.9)
2 (173.4)	dihydrouridine	dihydrouridine,D (173.5)
5* (165.7), 6 (165.6), 8* (164.8) 10 (163.1)	U$_7$, U$_{12}$, U$_{47}$, U$_{59}$	Uridine, U (165.8)
3 (166.6), 5* (165.7), 8* (164.8) 9 (163.3)	ψ, rT, U$_{64}$, U$_{67}$	Pseudouridine, ψ (165.0) Ribothymidine, rT (166.0)
4 (166.0), 7 (165.1)	U$_{29}$, U$_{33}$, U$_4$	
11 (160.4)	Uridine-5-oxyacetic acid	Uridine-5-oxyacetic acid. (V) (161.4)

[a]Figure 10 peaks 5, 7 and 8 contain two carbons. Data taken from Schweizer *et al.* (1980).

[b]Measured in ppm from internal dioxane, converted to TMS by adding 66.3 ppm. An asterisk indicates an assignment of only a part of a peak.

$$\begin{array}{ccc} O & CO_2^- \\ \parallel & \mid \\ NHCNHCHCHCHCH_3 \\ & \mid \\ & OH \end{array}$$

t⁶A

titrations have further confirmed the formation of metal–t⁶A complexes using both Mg(II) and Mn(II), where the formation constants at neutral pH were 3×10^5 and 10^6 M^{-1}, respectively (Reddy *et al.*, 1979). Monophosphates and oligonucleotides containing t⁶A will be examined next. These data are in contrast to those reported by Watts and Tinoco (1978) on the dimer Upt⁶A. Using different techniques, CD and UV absorption, they could find no evidence of metal binding. The reasons for these discrepancies are unresolved.

Watts and Tinoco (1978) also employed 360 MHz PMR in studies of various dinucleoside monophosphates containing modified nucleosides as found adjacent to anticodon segments in various tRNAs. Compared with ApA, dimers, such as Api⁶A, Apms²i⁶A, and ApϵA, in which the 5′-adenylyl unit had substitutions (of N^6-(Δ^2-isopentenyl); N^6-(Δ^2-isopentenyl) and 2-methylthio; and a third ring [1, N^6 ethenoadenine], ϵA, a model of the tricyclic wybutosines found next to anticodons in eukaryotic tRNA[Phe]), were less stable in stacked conformations. Dhingra *et al.*, (1978) have reported ApϵA had a more destacked conformer, g^+t, about the P—O₃′, P—O₅′ bonds than did ApA.

We have employed 360 MHz PMR to study the conformation of a hexanucleotide containing wybutosine, Y_t, isolated from the anticodon loop of torula yeast tRNA[Phe], GmpApApY$_t$pApψ (Dea, *et al.*, 1978). GmpApA is the anticodon triplet. Our chemical-shift versus temperature data definitely show that the Y_t nucleoside is a locus for temperature-induced destacking (Y_t protons show large shifts to low field with temperature increase) in this molecule. These unusual hydrophobic nucleosides may serve as a hinge point for the polynucleotide backbone so that the anticodon segment can be manipulated into proper position for complexation with the codon triplet.

Davanloo *et al.*, (1979b) have used 270 MHz PMR to study the influence of the nucleoside at position 54 upon the thermal stability of tRNA[fMet] isolated from three different organisms. Ribothymidine is at position 54 in wild type *E. coli*. In mutant Trm⁻, uridine occupies this position and in the thermophile, *Thermus thermophilus* HB8, 2-thioribothymidine. By monitoring the high-field regions as a function of temperature, Davanloo, *et al.* (1979b) found that the order–disorder transition temperature increased in the order of tRNAs containing U54 < T54 < s²T54. Presumably this added stability is due to increases in tertiary interactions between dihydrouridine and TψC loops. This interpretation is somewhat complicated by several base pair changes in the thermophile which may also affect the thermal stability.

Another modified nucleoside which is usually found in tRNA is pseudouridine [5-(β-D-ribofuranosyl)uracil, denoted ψ], usually located at position 55. In some tRNAs additional ψ residues are present, for example at position 39, in those tRNAs which participate in regulating biosynthesis of the corresponding amino acid. Hurd and Reid (1977) propose, based upon 360 MHz PMR studies, that ψ39 exists in the *syn* glycosidic conformation, forming Watson–Crick hydrogen bonds to A31 via the imino proton at N_1 rather than with N_3H as would occur with the nucleoside in the *anti* conformation. In the *syn* conformation, N_3H would be available to interact in the major groove. Actually it seems plausible that N_1H could so interact when ψ is in the normal *anti* conformation. The primary reason for postulating the abnormal *syn* conformation and the —N_1H interaction with A31 was in fact that this imino proton occurs at ~0.4 ppm higher field than —N_3H of uridine which could, in most part, explain why the A—ψ31 hydrogen-bonded resonance in *E. coli* tRNAs such as tRNA[His], tRNA[Lys], tRNA[Phe], etc. occurs 0.6 ppm to higher field than A–U resonances.

3.4. ³¹P NMR Studies

Gueron and Shulman (1975) recorded 109 MHz ³¹P spectra of *E. coli* tRNA[Glu] and yeast tRNA[Phe] which displayed several small resonances both upfield and downfield from the main phosphodiester peak. These resonances shifted when the tRNA structure was manipulated—for example, by changing Mg(II) concentration—leading to the suggestion that the small peaks may be due to diester moieties in environments strongly affected by tertiary interactions of the polymer. Hayashi *et al.* (1977) showed that separate resonances could be seen for coil and helical states in yeast tRNA[Phe]. Recently, Salemink *et al.* (1979) have published a very nice study of yeast tRNA[Phe] in which structural modifications allowed several of the shifted ³¹P resonances to be assigned to particular loci, for example, the anticodon loop. Low-field PMR of the hydrogen-bonded imino and amino protons served to calibrate the extent of structural perturbation resulting from a particular operation. That the small shifted resonances are due to unique phosphodiester environments as a result of tertiary interactions was demonstrated by cyanoethylation of ψ, which disrupts association between D- and TψC loops. Pancreatic RNase, in addition to producing phosphomonoesters by hydrolyzing the acceptor—ACCA terminus, caused disruption in the anticodon loop structure (^{31}P resonances disappeared) as a result of endonucleolytic cleavage at U33. These same resonances did not change upon RNase T$_1$ treatment, which cannot attack the anticodon loop of this tRNA. These enzymes did cause two other peaks to disappear, which are undoubtedly from other looped-out regions. Removal of the hydrophobic Y base which is known to result in an altered anticodon-loop conformation, resulted in perturbation of the ^{31}P peaks which were affected by the pancreatic RNase treatment.

4. STRUCTURE, CONFORMATION, AND DYNAMICS OF HIGH-MOLECULAR-WEIGHT NUCLEIC ACIDS AND THEIR COMPLEXES

In this Section, I will present the essence of recent reports primarily using PMR and ^{31}P NMR to investigate the structure, conformation, and interaction of high-molecular-weight single-stranded and duplex polynucleotides, nucleic acids alone, and in functional complexes such as nucleosomes and ribosomes. These applications represent an exciting new era in the utilization of NMR for investigations of structure and dynamics of nucleic acids, particularly their "backbones," in biologically important entities. Kearns (1977) has reviewed some of the earlier work in these areas.

4.1. Polynucleotides and Nucleic Acids

As was discussed in Section 2, ribose and deoxyribose self-complementary oligonucleotides capable of forming double-stranded complexes displayed base-overlap geometries in solution akin to those found in the solid state, for example, A-RNA or A'-RNA for ribose derivatives and B-DNA for deoxy compounds. The preferred 2'-endo conformation of the sugar ring found at the mononucleotide level was maintained for the deoxy derivatives, whereas for the ribose compounds, the 2'-endo, 3'-endo equilibrium shifted toward the 3'-endo form in the oligonucleotides. Patel, who has used high-field (360 MHz) instrumentation to examine many oligonucleotides by PMR and ^{31}P NMR has also investigated polynucleotide systems. In particular, he has compared, in detail, the helical properties of two representative synthetic RNA and DNA systems, poly (IC) and poly d(IC) (Patel, 1978b). The duplex states for the two polymers were readily determined by observing the exchangeable base-paired inosine—N$_1$H at -15 ppm and cytosine amino protons at -6.4 and -7.6 ppm. The transition temperatures in the duplex $\rightarrow$ strand conversions were 50.0 and 53.5°C, respectively. Going through this transition, the change in chemical shifts for inosine H$_8$ and H$_2$ protons were larger for the RNA model whereas H$_6$ and H$_{1'}$ of cytidine experience greatest shifts in melting of the DNA model. These changes reflect the difference in base overlap geometries in the two types of helices, A-RNA and B-DNA.

For the RNA system the 2'-endo, 3'-endo mix in preferred ribose pucker, which shifted toward 3'-endo upon oligomerization of mononucleotides, becomes mainly 3'-endo when the coil undergoes duplex formation. On the other hand, the poly d(IC) sugar pucker is primarily 2'-endo in both single strand and duplex, as is also true at the mononucleotide level.

Temperature-induced transitions prior to, and after the duplex → coil melt occur for both helix types. Before the double-strand complexes break up, there are apparently changes in the average glycosidic and backbone conformations, as reflected in the chemical-shift changes for base and $H_{1'}$ protons. After the melt, destacking of single strands continues, again manifested in the low-field shifts for base protons as temperature is increased.

High salt (4.0 M NaCl) caused shift changes chiefly at $I-H_8$ and $C-H_5$ in poly d(IC) which may arise from torsional angle changes in the polymer backbone. Patel et $al.$ (1979) have more thoroughly investigated the salt-induced backbone conformational changes in two smaller polymers of (dGdC), having chain lengths 16 and 20–30. In this study, one of the most striking effects of high salt was reflected in the 145.7 MHz ^{31}P spectra for the 16-mer. Instead of the single broad resonance seen at low salt (0.2 M NaCl), two peaks are observed in 4 M NaCl—one at the low-salt position; the other about 1.5 ppm to lower field. This result may be interpreted in terms of an alternating conformation of the backbone such that the ω',ω torsion angles of the dC(3'—5')dG diester linkage have different average values than the dG(3'—5')dC bridge.

Not only are the average phosphate-bond conformations different in this alternating B-DNA high-salt form, but there is evidence from the 360 MHz PMR data that there is a corresponding change of glycosidic conformation at every other monomer unit. In particular, one $H_{1'}$ resonance shifts about 0.5 ppm downfield in high salt. In addition, changes occur in the overlap of adjacent bases as evidenced by the ~0.3 ppm downfield shift of the $G-N_1$ proton and ~0.15 ppm upfield shift at $C-H_5$ in 4 M salt.

With respect to the backbone conformation of another synthetic DNA, poly d(AT) of >300 nucleotides in length, Patel (1979b) and Patel and Canuel (1976) did not observe salt-induced nonequivalence, but the 145.7 MHz ^{31}P resonances in 1 M NaCl is asymmetric, suggesting at least two dissimilar environments for the phosphates in the duplex. The nonequivalence was enhanced by intercalation of the mutagen proflavin, resulting in two partially resolved peaks, one being downshifted from the position in the absence of the drug.

That an alternating backbone conformation does exist in poly d(AT) has recently been reported by Shindo et $al.$ (1979) for the duplex with a homogeneous chain length of 145 nucleotides from micrococcal nuclease digestion of chromatin composed of nucleosome core histones and high-

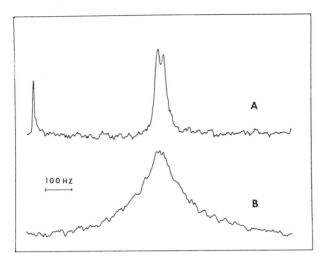

Figure 13. Proton decoupled 109.3 MHz ^{31}P NMR spectra of: (a) 145 base pair duplex poly [(dA)–(dT)] and (b) 2500 base pair duplex poly [(dA)–(dT)], 1 mg/ml, 30°C, 19,000 and 28,000 transients. Conditions used: 10 mM Tris pH 8.0, 1 mM EDTA (Shindo *et al.*, 1979a).

molecular-weight poly d(AT). The 109.3 MHz ^{31}P spectra of high-molecular-weight and 145 base-pair duplex poly d(AT) is shown in Figure 13. The 24 Hz separation for the smaller system persists with temperature increase until the duplex → coil transition, with $T_m = 38°C$. The random coil produces one resonance. An alternating backbone conformation for such an alternating sequence polynucleotide has been observed in crystals of the tetramer, (dAdT)$_2$ (Viswamitra *et al.*, 1978). The sugar pucker in the solid also had alternating characteristics; this feature may also exist in solution for the 145 base-pair polymer, since one of the two ^{31}P resonances was significantly broader in the proton-coupled spectrum (Shindo *et al.*, 1979).

Several other recent papers have utilized ^{31}P chemical shifts to assess nucleic acid backbone conformation, the influence of the base sequence, and heterogeneity on local diester and overall backbone conformations. For example, separate resonances for single strand and duplex helical states have been seen for salmon sperm DNA (Mariam and Wilson, 1979; Yamada *et al.*, 1978a), and calf thymus DNA (Mariam and Wilson, 1979). The coil states in the DNA samples were produced by alkaline pH and high temperature. Yamada *et al.* (1978a) also examined apurinic acid in which most of the purine bases have been hydrolyzed off leaving essentially a single-stranded polymer which has rapid segmental backbone motion similar to poly U. The 145.7 MHz ^{31}P spectrum contained several sharp peaks, presumably due to heterogeneity in the pyrimidine base composition. This base sequence and composition is important in polynucleotide ^{31}P chemical shifts. One may ask whether each sequence permutation has a certain unique average diester rotational conformation characterized by ω',ω angles. This was the case at the dinucleotide level and apparently is also present in the polymers.

The local characteristics of the conformation of individual phosphodiester links are not so readily discernible via discrete chemical shifts in rigid, random-sequence nucleic acids or homogeneous polynucleotides. In these cases ^{31}P relaxation techniques have been fruitfully applied to ferret out information about the backbone conformation and dynamics. At 40.6 MHz the ^{31}P spectrum for poly (G) consists of one broad resonance with a linewidth of about 100 Hz (Yamada et al., 1978b). Values of the spin–spin relaxation time, T_2, measured by the Hahn spin echo technique and calculated from the linewidth at half peak height differed by a factor of 10. Since the T_2 values should be equivalent in a homogeneous situation, the ^{31}P resonance is apparently being inhomogenously broadened by the chemical-shift distribution which exists as a result of a distribution of rotational conformers at local phosphodiester bonds.

Thorough ^{31}P chemical-shift and relaxation studies have recently been accomplished with precise-chain-length, homogeneous-sequence DNA (140–145 base pairs) from chicken erythrocyte nucleosomes (Shindo, 1980; Shindo et al., 1980), and calf thymus nucleosomes (Klevan et al., 1979) as well as from high-molecular-weight calf thymus DNA (Bolton and James, 1979a). In the work on 145 base-pair DNA from chicken erythrocyte nucleosomes, Shindo (1980) and Shindo et al., (1980) studied the frequency dependence of ^{31}P shifts, T_1 values, half-height linewidths ($\Delta v_{1/2}$) and ^{31}P$\{^1$H$\}$NOE. Reasonable fits of the relaxation data were obtained for an isotropically rotating model with a calculated overall correlation time τ_R of 6.5×10^{-7} s. Chemical-shift dispersion was found to account for roughly one-half the observed linewidth of 103 Hz at 109 MHz and 19°C. Of the balance of the linewidth, due to dipolar and chemical-shift anisotropy contributions to relaxation effects, the latter mechanism was found to contribute 27% at 24.3 MHz, but rose to 50% at 40 MHz and 80% at 109 MHz. The overall motional correlation time obtained experimentally was 3 times shorter than calculated, which indicates that local motion at the individual diester bonds contributes. If a distribution of correlation times is used to account for the local backbone motion, good agreement of calculated and experimental T_1 values are obtained. The local motion may involve torsional twisting about the long axis of the polymer.

Somewhat different conclusions were put forward by Klevan et al. (1979) for their 140 base-pair DNA from calf thymus nucleosomes. This is probably a result of differences in treatment of the relaxation data, because they considered that the ^{31}P–^{1}H dipolar mechanism dominates the relaxation process at their single observing frequency of 36.4 MHz. Shindo et al. (1980) showed that CSA may contribute ~40% at this frequency. At any rate, it was found that internal backbone motion with a correlation time of $2–4 \times 10^{-10}$ s was a dominant contributor to the ^{31}P relaxation. The overall isotropic tumbling motion is about 1 μs, based on electric birefringence data (Hogan et al., 1978). These auhors did not specify a model for this internal motion but it must involve some concerted flexing of P—$O_{3'}$, P—$O_{5'}$, $C_{4'}$—$C_{5'}$, and perhaps sugar puckering changes to rapidly move $H_{3'}$, $H_{5'}$, $H_{5''}$, and $H_{2'}$ protons with respect to the phosphorus atom.

Bolton and James (1979a) have utilized rotating frame off-resonance T_1's ($T_{1\rho}^{off}$) in addition to T_1, ^{31}P-$\{^1H\}$NOE, and linewidth measurements to chart the backbone motions in poly (A), poly (IC) and calf thymus DNA. They fit the experimental data with a dual internal plus overall motional model. In this model, the rapid internal motion with 0.3–0.5 ns correlation times was essentially constant for the three systems (despite environmental perturbation) whereas the slower overall motions with correlation times of 10^{-5} to 10^{-7} s (attributed to bending of the polynucleotide backbone) were dependent upon temperature and salt concentration.

These workers have also investigated the motion in DNA using ^{13}C relaxation (Bolton and James, 1979b). Only the ribose carbons were observed. The aromatic carbons of the bases, which are involved in interstrand hydrogen bonding interactions, were not seen, probably due to a combination of long T_1's because of slow reorientation on a nanosecond time scale, and the fast pulse rate. T_1 and NOE values of the ribose carbons show a gradation of slow to faster motion in the order 1', 3', 4' < 2' < 5'. $C_{2'}$ may have less steric constraints to movement, perhaps occasioned by movement associated with change in sugar pucker. Thus the phosphodiester and to a lesser extent the ribose moieties in high-molecular-weight DNA are quite mobile.

Early and Kearns (1979) have reached parallel conclusions in their 300 MHz investigation of the low-field imino and aromatic protons of duplex salmon sperm and calf thymus DNA of various chain lengths, that is, rapid internal motion exists along the backbone. These conclusions were reached by comparing experimentally observed dipolar broadening of the linewidths of exchangeable imino and nonexchangeable aromatic protons with predictions using rigid rod models. Deviations using the model were seen with helical duplex fragments 200 base pairs and larger in that the observed linewidths were smaller than predicted; correlation times for aromatic protons were about one-tenth as large as for imino protons, which indicates rapid motion of ribose protons due to backbone flexibility, resulting in decreased line broadening. The estimated internal correlation times of about 5×10^{-7} s for imino protons and 5×10^{-8} s for the aromatic protons are considerably slower than nanoseconds for the ribose carbons and fractions of nanoseconds for the phosphate determined by Bolton and James (1979b) for high-molecular-weight thymus DNA. However, even the slower motion exhibited by the imino protons is 10–20 times faster than the overall bending motion of the molecule as a whole ($\tau_0 = 0.5$–1×10^{-6} s).

4.2. Nucleic Acids in Macromolecular Complexes

4.2.1. Nucleosomes. Recently several reports have employed PMR of exchangeable protons (Feigon and Kearns, 1979) and ^{31}P relaxation and shifts (Shindo *et al.*, 1980; Klevan *et al.*, 1979; Kallenbach *et al.*, 1978; Cotter and

Lilley, 1977; Simpson and Shindo, 1979) to obtain information on the structure and dynamics of DNA in nucleosome core particles from calf thymus and chicken erythrocytes, which may be compared with DNA alone.

None of the ^{31}P studies indicate that there is significant asymmetry of the resonance from the backbone phosphates to support the concept of a "kinking" model for DNA in the nucleosome core particle as proposed by Crick and Klug (1975) and Sobell et al. (1976). Rather, this data lends credence to the postulate of Finch et al. (1977), who proposed a smoothly curving DNA surrounding the histone protein octamer core. However, the motion of the DNA in the nucleosome is constricted compared to the naked DNA, as shown in Figure 14, presumably as a result of interaction with the histone core proteins.

The ^{31}P chemical shifts and relaxation data at three frequencies for nucleosomes and nucleosomal DNA are displayed in Table III, taken from Shindo (1980) and Shindo et al. (1980). That chemical-shift anisotropic contributions to the relaxation should be reckoned with is indicated by the strong frequency dependence of T_1 and line width ($W_{1/2}$). These data have been analyzed in terms of a model for DNA in nucleosomes in which isotropic motion exists, characterized by a distribution

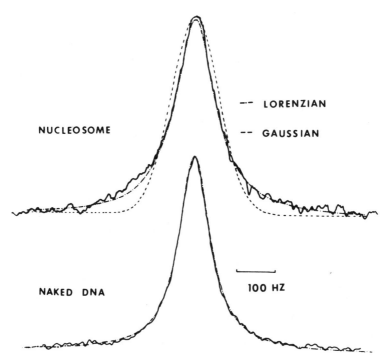

Figure 14. Proton decoupled 109.3 MHz ^{31}P spectra of nucleosomal DNA and naked DNA with least squares fitted curves; 1.5 mg/ml DNA, 5 mM Tris buffer, pH 8.0, 0.1 mM EDTA, 512 scans (Shindo et al., 1979b).

Table III. Experimental ³¹P chemical shifts and relaxation parameters plus calculated[a] relaxation parameters for DNA in nucleosomes and DNA alone

	f_0 (MHz)	Experimental[b]				Calculated[b]				
		Chemical Shift[c]	T_1^d	Linewidth (Hz)	NOE(η)	T_1^d	$W^{1/2}*^e$	$W^{1/2}$	NOE(η)	Average Correlation Time, τ
Nucleosome DNA	109.3	−4.49	3.34	130	0.22	3.8	0.77	134	0.06	3×10^{-8}
	40.3		3.00	45		3.2		42	0.12	
	24.3		2.70	25	0.50	2.2		25	0.18	
Nucleosome + 6 M urea	109.3	−4.39	2.3	205	0.20	2.3	0.70	205	0.16	1.5×10^{-8}
	24.3		2.2	35		2.0		34	0.33	
DNA Alone	109.3	−4.35	3.43	103	0.30	4.4	0.50	101	0.02–0.03	
	40.3			28	0.28	2.8		28	0.02–0.06	
	24.3		2.15	18	0.18	1.6		17	0.03–0.13	

[a] Calculations were based upon a model with isotropic motion having a distribution of correlation times with distribution parameters, $P =$ 20 for nucleosomes and $P = 12$ for nucleosomes in 6 M urea.
[b] Data taken from Shindo (1979), and Shindo et al. (1979b).
[c] Measured up field from internal trimethyl phosphate.
[d] Seconds; experimental errors less than 10%.
[e] Contribution to line width due to chemical shift dispersion in ppm.

of correlation times to account for the heterogeneity in motional rates. The calculated parameters are shown in Table III, and may be compared with the experimental values. The distribution of chemical shifts due to variation of phosphate environments in the DNA is 0.77 ppm, which is broader than the dispersion for DNA alone. An overall motional correlation time of 1.7×10^{-7} s can be calculated from the Stokes–Einstein equation assuming a radius of sphere = 55 Å. This number is about six times larger than the average correlation time of 3×10^{-8} s determined from the ^{31}P relaxation data. The same ordering of hydrodynamic and NMR overall motional correlation were seen for DNA alone. Thus the DNA must possess rapid local motion both in solution and in the histone complex. Klevan *et al.* (1979) have analyzed their ^{31}P in nucleosomes in terms of a two-correlation-time model consisting of overall tumbling of 3×10^{-7} s (from birefringence) and rapid segmental motion dominating the relaxation with a correlation time 2–4×10^{-10} s.

Shindo *et al.* (1980) perturbed the nucleosome with both urea and temperature. The correlation times in Table III show that the DNA becomes more mobile with urea addition whereas the motion is more heterogeneous as monitored by the linewidth increase. These effects upon the DNA may be due to urea disruption of the histone core complex from 3.5–5.5 M urea (Martinson and True, 1979). In the ^{31}P NMR, the urea effects are biphasic, as contrasted to the CD results (Olins *et al.* 1977), but the major NMR transition coincides with the CD data.

Temperature-induced effects are also biphasic, with transitions at about 60 and 72°C, similar to the optical melts. Apparently part of the DNA denatures at 60°; the presence of two populations of DNA species can be seen at 62° in the asymmetrical ^{31}P peak shown in Figure 15a. Two distinct premelt phosphate environments in the polynucleotide chain can readily be seen between 45 and 53°C upon examination of semisynthetic nucleosomes prepared by reconstituting histone core particles with duplex poly d(AT), shown in Figure 15b (Simpson and Shindo, 1979). Chemical-shift dispersion results in a linewidth increase from duplex 145 base-pair poly d(AT) and DNA of 43 to 103 Hz. Between these polymers and their complexes with histone core proteins are linewidth increases of 70 and 30 Hz, respectively, undoubtedly because of further heterogeneity in the phosphate environments due to differences in interactions with histones and/or a spectrum of different conformations.

Thermal perturbation of the hybrid nucleosomes apparently leads to loosening of some contacts between histones and duplex poly d(AT) as indicated by emergence of a small downfield shoulder at 40°C. Between 40 and 50°C these histone-free polynucleotide segments melt leading to the

Figure 15. (*a*) Proton decoupled 109.3 MHz ^{31}P spectrum of nucleosome particles in the premelting temperature region. (*b*) Temperature variation of ^{31}P spectra of semisynthetic nucleosome containing duplex poly (dAdT)·poly (dAdT) (Shindo *et al.*, 1979, 1980).

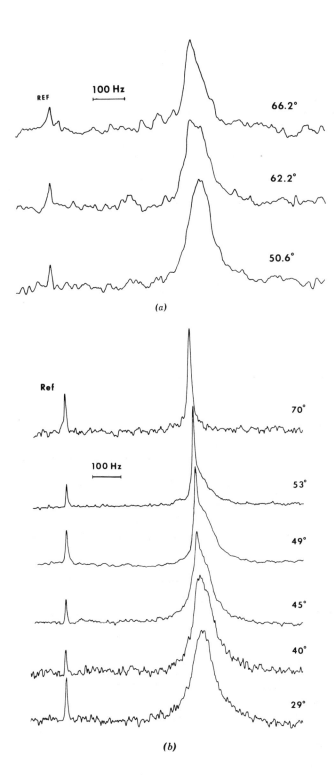

REF

100 Hz

66.2°

62.2°

50.6°

(a)

Ref

100 Hz

70°

53°

49°

45°

40°

29°

(b)

sharp downfield component of the asymmetric ^{31}P peak as seen in Figure 15b, About 100 base pairs of duplex still contact histones. Above 65–75°C, the whole structure breaks up including helix-to-coil transitions in the polynucleotide.

Summarizing the ^{31}P data on the nucleosomes, one may state that the duplex DNA molecule, although more constrained than when free in solution, still displays considerable internal motion. The interactions of DNA with histone cores leads to increased heterogeneity in the phosphate environment. Individual thermal perturbations involving both DNA and proteins can be followed in the case of semisynthetic particles containing duplex poly d(AT). Feigon and Kearns (1979) have shown that the imino protons involved in base pairing in nucleosomal DNA are immobilized to about the same extent as in DNA in solution. Thus complexing with core histone proteins cannot be followed by monitoring this rigid base paired portion of the duplex DNA molecule.

4.2.2. Ribosomes.   Tritton and Armitage (1978) have studied RNA in the *E. coli* ribosome using ^{31}P NMR shifts and relaxation at a single frequency, 36.4 MHz. The relaxation data is given in Table IV. The NOEs are significant, thus dipolar relaxation is important in the relaxation mechanisms. However, chemical-shift anisotropy (CSA) may also be of equal importance here. Shindo (1979) demonstrated a CSA contribution of at least 50% at 40 MHz for high-molecular-weight DNA. Thus it is important to assess the frequency dependence of the relaxation parameters. This mechanism (CSA) was suggested both as a possible reason to explain the T_1 of the 50S subunit being shorter than for the 70S ribosome, and also as an explanation for the lower NOE for the 70S ribosome in D_2O.

In any event, the RNA molecule can be shown to be more restricted in the ribosome than when free in solution by inspection of the linewidth data, where one sees that the linewidth increases from 55 Hz for 16S + 23S RNA,

Table IV.   36.4 MHz ^{31}P relaxation data for RNA in *E. coli* ribosomes at 23°C[a]

	NOE ($\eta + 1$)	T_1 (s)	Linewidth (Hz)
30S Subunit	1.21	1.7	169[b]
50S Subunit	1.18	1.3	
70S Ribosome	1.32	2.0	208
70S Ribosome in D_2O	1.08		
RNA (16S + 23S)	1.47[c]	2.0	55

[a]Data taken from Tritton and Armitage (1978). Spectra run in Tris buffer, pH 7.5, with 10 mM Mg (II).
[b]Linewidth in a mixture of 30S + 50S subunits upon dissociating intact 70S particle by lowering Mg (II) to 1 mM.
[c]16S RNA alone.

to 169 Hz in the 50S subunit, to 208 Hz in the intact ribosome (70S particle). One would expect a contribution of chemical-shift heterogeneity in the ribosome spectra as found for DNA in nucleosomes.

The chemical-shift data shows several sharp low-field spikes, in addition to the main phosphodiester peak, which are due to monophosphates. Thus some degradation of the RNA has occurred, which suggests these preparations have some ribonuclease activity present.

The fact that NOEs are observed provides an indication of fairly rapid motion in the RNA chain. If one uses a motional model adapted for DNA in nucleosomes (Klevan *et al.*, 1979) with overall tumbling and more rapid internal motion, the RNA backbone has segmental motion with a correlation time, τ_i, of about 10^{-9} s, whereas the overall tumbling motion, τ_0, is 10^{-6} s.

ACKNOWLEDGMENTS

The author would like to thank the following for making unpublished material available: Drs. Heisaburo Shindo, Jack S. Cohen, Dinshaw Patel, Al Redfield, Brian Reid, Ralph Hurd, David Kearns, Stephen Danyluk, W. David Hamill, Jr., and David M. Grant.

REFERENCES

Agris, P. F., F. G. Fujiwara, C. F. Schmidt, and R. N. Loeppky (1975), *Nucl. Acids Res.* **2**, 1503.

Akasaka, K. (1974), *Biopolymers* **13**, 2273.

Alderfer, J. L., I. Tazawa, S. Tazawa, and P. O. P. Ts'o (1974), *Biochemistry* **13**, 1615.

Altona, C., J. H. Van Boom, and A. G. Haasnoot (1976), *Eur. J. Biochem.* **71**, 557.

Arnott, S., S. D. Dover, and A. J. Wonacott (1969), *Acta Cryst.* **B25**, 2192.

Arnott, S., and D. W. L. Hukins (1973), *J. Mol. Biol.* **81**, 93.

Arnott, S., D. W. L. Hukins, S. D. Dover, W. Fuller, and A. R. Hodgson (1973), *J. Mol. Biol.* **81**, 107.

Arnott, S., R. Chandrasekaran, D. W. L. Hukins, P. J. C. Smith and L. Watts (1974), *J. Mol. Biol.* **88**, 523.

Arter, D. B., G. C. Walker, O. C. Uhlenbeck, and P. G. Schmidt (1974), *Biochem. Biophys. Res. Comm.* **61**, 1089.

Arter, D. B., and P. G. Schmidt (1976), *Nucl. Acids Res.* **3**, 1437.

Bolton, P. H., and D. R. Kearns (1978), in *Biological Magnetic Resonance*, L. J. Berliner, and J. Reuben, Eds., Vol. 1, Plenum Press, New York, NY, pp. 91–137.

Bolton, P. H., and T. L. James (1979a), *J. Am. Chem. Soc.*, **102**, 25.

Bolton, P. H., and T. L. James (1979b), in *Proc. Intl. Symp. Magnetic Resonance in Chemistry, Biology and Physics*, Argonne National Laboratory, USERDA June 24–28, 1979.

Borer, P. N., L. S. Kan, and P. O. P. Ts'o (1975), *Biochemistry* **14**, 4847.

Chachaty, C., T. Yokono, T-D. Son, and W. Guschlbauer (1977), *Biophys. Chem.* **6**, 151.

Chan, S. I., and J. H. Nelson (1969), *J. Am. Chem. Soc.* **91**, 168.

Cheng, D. M., and R. H. Sarma (1977a), *J. Am. Chem. Soc.* **99**, 7333.

Cheng, D. M., and R. H. Sarma (1977b), *Biopolymers* **16**, 1687.

Cheng, D. M., M. M. Dhingra, and R. H. Sarma (1978), *Nucl. Acids Res.* **5**, 4399.

Cotter, R. I., and D. M. J. Lilley (1977), *FEBS Letters* **82**, 63.

Crick, F. H. C., and A. Klug (1975), *Nature* **255**, 530.

Crothers, D. M., C. W. Hilbers, and R. G. Shulman (1973), *Proc. Natl. Acad. Sci. U.S.A.* **70**, 2899.

Danyluk, S. S., and F. E. Hruska (1968), *Biochemistry* **7**, 1038.

Davanloo, P., I. M. Armitage, and D. M. Crothers (1979a), *Biopolymers* **18**, 663.

Davanloo, P., M. Sprinzl, K. Watanabe, M. Albani, and H. Kersten (1979b), *Nucl. Acids Res.* **6**, 1571.

Davies, D. B. (1978), in *Progress in Nuclear Magnetic Resonance Spectroscopy*, J. W. Emsley, J. Feeney, and L. H. Sutcliffe, Eds., Vol. 12, Part 3, Pergamon, Oxford, England, pp. 135–225.

Dea, P., M. Alta, S. Patt, and M. P. Schweizer (1978), *Nucl. Acids Res.* **5**, 307.

Dhingra, M. M., R. H. Sarma, C. Giessner-Prettre, and B. Pullman (1978), *Biochemistry* **17**, 5815.

Early, T. A., and D. R. Kearns (1979), *Proc. Natl. Acad. Sci. U.S.A.*, **76**, 4165.

Evans, F. E., and R. H. Sarma (1976), *Nature* **263**, 567.

Ezra, F. S., C. H. Lee, N. S. Kondo, S. S. Danyluk, and R. H. Sarma (1977), *Biochemistry* **16**, 1977.

Fang, K. N., N. S. Kondo, P. S. Miller, and P. O. P. Ts'o (1971), *J. Am. Chem. Soc.* **93**, 6647.

Feigon, J., and D. R. Kearns (1979), *Nucl. Acids Res.* **6**, 2327.

Finch, J. T., L. C. Lutter, D. Rhodes, R. S. Brown, B. Rushton, M. Levitt, and A. Klug (1977), *Nature* **269**, 29.

Geerdes, H. A. M., and C. W. Hilbers (1977), *Nucl. Acids Res.* **4**, 207.

Giessner-Prettre, C., B. Pullman, P. N. Borer, L. S. Kan, and P. O. P. Ts'o (1976), *Biopolymers* **15**, 2277.

Gorenstein, D. G. (1975), *J. Am. Chem. Soc.* **97**, 898.

Gorenstein, D. G., and D. Kar (1975), *Biochem. Biophys. Res. Commun.* **65**, 1073.

Gueron, M., and R. G. Schulman (1975), *Proc. Natl. Acad. Sci. U.S.A.* **72**, 3482.

Hamill, Jr., W. D., D. M. Grant, W. J. Horton, R. Lundquist, and S. Dickman (1976), *J. Am. Chem. Soc.* **98**, 1276.

Hamill, Jr., W. D., W. J. Horton, and D. M. Grant (1980), *J. Am. Chem. Soc.*, in press.

Hart, P. (1978), *Biophys. J.* **24**, 833.

Hayashi, F., K. Akasaka, and H. Hatano (1977), *Biopolymers* **16**, 655.

Hogan, M., N. Dattagupta, and D. M. Crothers (1978), *Proc. Natl. Acad. Sci. U.S.A.* **75**, 195.

Holbrook, S. R., J. L. Sussman, R. W. Warrant, G. M. Church, and S. H. Kim (1977), *Nucl. Acids Res.* **4**, 2811.

Hurd, R. E., and B. R. Reid (1977), *Nucl. Acids Res.* **4**, 2747.

Hurd, R. E., and B. R. Reid (1979), *Biochemistry*, **18**, 4017.

Hurd, R. E., E. Azhderian, and B. R. Reid (1979), *Biochemistry*, **18**, 4012.

Jack, A., J. E. Ladner, D. Rhodes, R. S. Brown, and A. Klug (1977), *J. Mol. Biol.* **111**, 315.

Johnson, Jr., W. C., and I. Tinoco, Jr. (1969), *Biopolymers* **8**, 715.

Johnston, P. D., and A. G. Redfield (1978), *Nucl. Acids Res.* **5**, 3913.

Johnston, P. D., and A. G. Redfield (1979), in *Transfer RNA, Part 1, Structure, Properties and Recognition*, P. Schimmel, D. Soll, and J. Abelson, Eds., Cold Spring Harbor Laboratory Press, Cold Spring Harbor, NY, pp. 181–206.

Johnston, P. D., N. Figueroa, and A. G. Redfield (1979), *Proc. Natl. Acad. Sci. U.S.A.*, **76**, 3130.

Kallenbach, N. R., D. W. Appleby, and C. H. Bradley (1978), *Nature* **272**, 134.

Kan, L. S., J. C. Barrett, and P. O. P. Ts'o (1973), *Biopolymers* **12**, 2409.

Kan, L. S., P. N. Borer, and P. O. P. Ts'o (1975), *Biochemistry* **14**, 4847.

Kan, L. S., and P. O. P. Ts'o (1977), *Nucl. Acids Res.* **4**, 1633.

Karplus, M. (1959), *J. Chem. Phys.* **30**, 11.

Kearns, D. R., D. J. Patel, and R. G. Shulman (1971), *Nature* **229**, 338.

Kearns, D. R. (1976), *Prog. Nucl. Acid Res. Mol. Biol.* **18**, 91.

Kearns, D. R. (1977), *Ann. Rev. Biophys. Bioeng.* **6**, 477.

Kim, S. H., J. L. Sussman, F. L. Suddath, G. J. Quigley, A. McPherson, H. J. Wang, N. C. Seeman, and A. Rich (1974) *Proc. Natl. Acad. Sci. U.S.A.* **71**, 4970.

Klevan, L., I. M. Armitage, and D. M. Crothers (1979), *Nucl. Acids Res.* **6**, 1607.

Komoroski, R. A., and A. Allerhand (1972), *Proc. Natl. Acad. Sci. U.S.A.* **69**, 1804.

Komoroski, R. A., and A. Allerhand (1974), *Biochemistry* **13**, 369.

Kondo, N. S., H. M. Holmes, L. M. Stempel, and P. O. P. Ts'o (1970), *Biochemistry* **9**, 3479.

Kondo, N. S., F. S. Ezra, and S. S. Danyluk (1975), *FEBS Letters* **53**, 213.

Kondo, N. S., and S. S. Danyluk (1976), *Biochemistry* **15**, 756.

Kroon, P. A., G. P. Kreishman, J. H. Nelson, and S. I. Chan (1974), *Biopolymers* **13**, 2571.

Krugh, T. R., and M. A. Young (1975), *Biochem. Biophys. Res. Commun.* **62**, 1025.

Ladner, J. E., A. Jack, J. D. Robertus, R. S. Brown, D. Rhodes, B. F. C. Clark, and A. Klug (1975), *Proc. Natl. Acad. Sci. U.S.A.* **72**, 4414.

Lee, C. H., F. S. Ezra, N. S. Kondo, R. H. Sarma, and S. S. Danyluk (1976), *Biochemistry* **15**, 3627.

Mariam, Y. H., and W. D. Wilson (1979), *Biochem. Biophys. Res. Commun.* **88**, 861.

Martinson, H. G., and R. J. True (1979), *Biochemistry* **18**, 1089.

McCloskey, J. A., and S. Nishimura (1977), *Accounts Chem. Res.* **10**, 403.

Miller, J. P., Z. Hussain, and M. P. Schweizer (1976), *Nucl. Acids Res.* **3**, 1185.

Murao, K., M. Saneyoshi, F. Harada, and S. Nishimura (1970), *Biochem. Biophys. Res. Commun.* **38**, 657.

Olins, D. E., P. N. Bryan, R. E. Harrington, W. E. Hill, and A. L. Olins (1977), *Nucl. Acids Res.* **4**, 1911.

Patel, D. J. (1974), *Biochemistry* **13**, 2396.

Patel, D. J. (1976), *Biopolymers* **15**, 533.

Patel, D. J., and L. L. Canuel (1976), *Proc. Natl. Acad. Sci. U.S.A.* **73**, 674.

Patel, D. J. (1977), *Biopolymers* **16**, 1635.

Patel, D. J. (1978a), *Ann. Rev. Phys. Chem.* **29**, 337.

Patel, D. J. (1978b), *Eur. J. Biochem.* **83**, 453.

Patel, D. J. (1979a), *Biopolymers* **18**, 553.

Patel, D. J. (1979b), *Accounts Chem. Res.* **12**, 118.

Patel, D. J., and L. L. Canuel (1979), *Eur. J. Biochem.*, **96**, 267.

Patel, D. J., L. L. Canuel, and F. M. Pohl (1979), *Proc. Natl. Acad. Sci. U.S.A.*, **76**, 2508.

Quigley, G. J., and A. Rich (1976), *Science* **194**, 796.

Quigley, G. J., M. M. Teeter, and A. Rich (1978), *Proc. Natl. Acad. Sci. U.S.A.* **75**, 64.

Reddy, R. P., M. P. Schweizer, and G. B. Chheda (1979), *FEBS Letters*, **106**, 63.

Reid, B. R., S. N. Ribeiro, L. McCollum, J. Abate, and R. E. Hurd (1977), *Biochemistry* **16**, 2086.

Reid, B. R., and R. E. Hurd (1977), *Accounts Chem. Res.* **10**, 396.

Reid, B. R., L. McCollum, S. N. Ribeiro, J. Abate, and R. E. Hurd (1979), *Biochemistry*, **18**, 3996.

Rich, A., and U. L. RajBhandary (1976), *Ann. Rev. Biochem.* **45**, 805.

Robertus, J. D., J. E. Ladner, J. T. Finch, D. Rhodes, R. S. Brown, B. F. C. Clark, and A. Klug (1974), *Nature* **250**, 546.

Robillard, G. T., C. E. Tarr, F. Vosman, and H. J. C. Berendsen (1976), *Nature* **262**, 363.

Robillard, G. T., C. E. Tarr, F. Vosman, and B. R. Reid (1977), *Biochemistry* **16**, 5261.

Salemink, P. J. M., T. Swarthof, and C. W. Hilbers (1979), *Biochemistry* **18**, 3477.

Schweizer, M. P., A. D. Broom, P. O. P. Ts'o, and D. P. Hollis (1968), *J. Am. Chem. Soc.* **90**, 1042.

Schweizer, M. P., and W. D. Hamill, Jr. (1978), *Biochem. Biophys. Res. Commun.* **85**, 1367.

Schweizer, M. P., W. D. Hamill, Jr., I. Walkiw, W. J. Horton, and D. M. Grant (1980), *Nucl. Acids Res.*, in press.

Shindo, H. (1980), *Biopolymers*, **19**, 509.

Shindo, H., R. T. Simpson, and J. S. Cohen (1979), *J. Biol. Chem.*, **254**, 8125.

Shindo, H., J. D. McGhee, and J. S. Cohen (1980), *Biopolymers*, **19**, 523.

Sigler, P. R. (1975), *Ann. Rev. Biophys. Bioeng.* **4**, 477.

Simpson, R. T., and H. Shindo (1979), *Nucl. Acids Res.*, **7**, 481.

Singh, H., M. H. Herbut, C. H. Lee, and R. H. Sarma (1976), *Biopolymers* **15**, 2167.

Sobell, H. M., C. C. Tsai, S. G. Gilbert, S. C. Jain, and T. D. Sakore (1976), *Proc. Natl. Acad. Sci. U.S.A.* **73**, 3068.

Sundaralingam, M. (1969), *Biopolymers* **7**, 821.

Sussman, J. L., and S. H. Kim (1976), *Science* **192**, 853.

Tompson, J. G., F. Hayashi, J. V. Pankstelis, R. N. Loeppky, and P. F. Agris (1979), *Biochemistry* **18**, 2079.

Tritton, T. R., and I. M. Armitage (1978), *Nucl. Acids Res.* **5**, 3855.

Ts'o, P. O. P., N. S. Kondo, M. P. Schweizer, and D. P. Hollis (1969), *Biochemistry* **8**, 997.

Ts'o, P. O. P. (1974a), in *Basic Principles in Nucleic Acid Chemistry*, P. O. P. Ts'o, Ed., Vol. 1, Academic Press, NY, pp. 453–584.

Ts'o, P. O. P. (1974b), in *Basic Principles in Nucleic Acid Chemistry*, P. O. P. Ts'o, Ed., Vol. 2, Academic Press, New York, NY, pp. 305–469.

Viswamitra, M. A., O. Kennard, P. G. Jones, G. M. Sheldrick, S. Salisbury, L. Falvello, and Z. Shakked (1978), *Nature* **273**, 687.

Watts, M. T., and I. Tinoco, Jr. (1978), *Biochemistry* **17**, 2455.

Wood, D. J., R. J. Mynott, F. E. Hruska, and R. H. Sarma (1973), *FEBS Letters* **34**, 323.

Yamada, A., H. Kaneko, K. Akasaka, and H. Hatano (1978a), *FEBS Letters* **93**, 16.

Yamada, A., K. Akasaka, and H. Hatano (1978b), *Biopolymers* **17**, 749.

Subject Index